XML for Bioinformatics

Ethan Cerami

XML for Bioinformatics

 Springer

Library of Congress Cataloging-in-Publication Data

Cerami, Ethan.
  XML for bioinformatics / Ethan Cerami.
    p.   cm.
  Includes bibliographical references and index.
  ISBN 0-387-23028-9
  1. XML (Document markup language)   2. Bioinformatics.   I. Title.

QA76.76.H94C47 2005
572.8′0285′674–dc22

2004058903

ISBN 0-387-23028-9

Printed in the United States of America.

9 8 7 6 5 4 3 2 1

springeronline.com

To Lily Cerami.
Welcome to the world.

Preface

Introduction

The goal of this book is to introduce XML to a bioinformatics audience. It does so by introducing the fundamentals of XML, Document Type Definitions (DTDs), XML Namespaces, XML Schema, and XML parsing, and illustrating these concepts with specific bioinformatics case studies. The book does not assume any previous knowledge of XML and is geared toward those who want a solid introduction to fundamental XML concepts.

The book is divided into nine chapters:

- **Chapter 1: Introduction to XML for Bioinformatics.** This chapter provides an introduction to XML and describes the use of XML in biological data exchange. A bird's-eye view of our first case study, the Distributed Annotation System (DAS), is provided and we examine a sample DAS XML document. The chapter concludes with a discussion of the pros and cons of using XML in bioinformatic applications.
- **Chapter 2: Fundamentals of XML and BSML.** This chapter introduces the fundamental concepts of XML and the Bioinformatic Sequence Markup Language (BSML). We explore the origins of XML, define basic rules for XML document structure, and introduce XML Namespaces. We also explore several sample BSML documents and visualize these documents in the Rescentris Genomic Workspace™ Viewer.
- **Chapter 3: DTDs for Bioinformatics.** This chapter introduces XML Document Type Definitions (DTDs). With DTDs, you can define specific rules for XML document construction and validate XML instance documents against these rules. This chapter builds a DTD for representing protein sequences and does so in incremental stages—we therefore start out simply and add layers of complexity as new concepts are introduced. The chapter also includes an overview of XML data formats available from The National Center for Biotechnology Information (NCBI), and provides a complete description of the NCBI TinySeq DTD.
- **Chapter 4: XML Schemas for Bioinformatics.** XML Schema represents the successor to XML Document Type Definitions (DTDs). We begin by comparing the two specifications and describe some of the advantages of using XML Schema. To illustrate core concepts, we rebuild the protein sequence DTD from Chapter 3 as an XML Schema, enabling you to compare the two specifications directly. The chapter concludes with a discussion of the Proteomics Standards Initiative Molecular Interaction (PSI-MI) XML format, an XML exchange format used to encode protein–protein interactions.
- **Chapter 5: Parsing NCBI XML in Perl.** Perl remains the programming language of choice for many in bioinformatics. This chapter therefore explores several options for parsing XML in Perl,

and focuses on two standard interfaces: the Simple API for XML (SAX) and the Document Object Model (DOM). We also explore the NCBI E-Fetch service, and retrieve nucleotide sequence records in XML in real time. This chapter assumes some prior knowledge of Perl.

- **Chapter 6: The Distributed Annotation System.** This chapter provides comprehensive coverage of the Distributed Annotation System (DAS), a distributed XML protocol used to exchange genome annotation data. To put DAS in perspective, we begin by exploring the process of genome annotation, and illustrate the DAS protocol from the end-user perspective. We then describe the DAS XML protocol in detail, and examine numerous sample XML documents from the Ensembl and UCSC DAS servers. The chapter includes a reference guide to all DAS commands, and a preview of anticipated features in the next version of DAS.

- **Chapter 7: Parsing DAS Data in SAX.** Despite the popularity of Perl, Java is becoming increasingly popular in bioinformatic applications. This chapter therefore describes the mechanics of parsing XML documents using the Java Simple API for XML (SAX). SAX is the de facto event-based XML parsing standard, and is widely implemented by many XML parsers, including several open source XML parsers. Several sample DAS applications are demonstrated, including one sample application which makes use of the open source BioJava library. This chapter assumes some prior knowledge of Java.

- **Chapter 8: Parsing DAS Data in JDOM.** This chapter focuses on the fundamentals of the JDOM API, a popular alternative to the SAX API. With JDOM, Java applications can easily navigate through XML document tree structures and extract elements, attributes, and character data. JDAS, an open source DAS client library, created by the author, is explored in detail. This chapter also assumes some prior knowledge of Java.

- **Chapter 9: Web Services for Bioinformatics.** Web services represent a new paradigm for building distributed web applications, and are currently being used extensively in bioinformatics. This chapter begins by presenting two broad approaches to building web services: the Representational State Transfer (REST) approach and the SOAP approach. We explore each approach in detail, and provide complete details on the latest SOAP specification. We also explore caBio, a comprehensive web service built by the National Cancer Institute (NCI).

Companion Web Site

This book includes a companion web site, available at: *http://www.xmlbio.org*. All the sample XML documents, and example Perl/Java programs described in the book are available for download.

Acknowledgments

Many people deserve special acknowledgments for making this book happen. First, I want to thank Lincoln Stein of Cold Spring Harbor Laboratory. Lincoln's presentation at the 2002 O'Reilly Open Bioinformatics Conference and his subsequent paper in *Nature* (described in Chapter 1), inspired me to write this book in the first place. Lincoln's vision of creating a "bioinformatics nation" is a compelling one; I hope this book provides readers with the nuts and bolts information to make Lincoln's dream a reality. Lincoln also provided answers to many of my questions regarding the DAS protocol, and a complete technical review of Chapter 6. His feedback and detailed explanations were invaluable.

Second, I want to thank everyone at the Computational Biology Center (cBio) at Memorial Sloan-Kettering Cancer Center, where I work. Chris Sander has created a unique and intellectually vibrant center, where I have been able to learn and thrive, and gain hands-on experience in many of the technologies described in this book. Alex Lash provided help in understanding the database resources at NCBI, and pointed me to the NCBI E-Fetch service described in Chapter 5. Anton Enright provided detailed feedback on Chapter 5. Gary Bader provided me with much-needed background information about specific biological databases and detailed background information about the PSI-MI XML format. Gary also provided feedback on Chapter 4.

Third, I want to thank Lorrie LeJeune, Simon St. Laurent, Tracey Cranston, and Brian Gilman who helped out with an earlier incarnation of this book, before it found a new home at Springer-Verlag.

Fourth, I want to thank several additional individuals who generously agreed to review specific chapters and provided scientific and technical feedback. Peter Covitz, Director of Bioinformatics Core Infrastructure at the National Cancer Institute (NCI) provided feedback on Chapter 9, and answered many of my questions regarding the NCI caBIO bioinformatics framework. Jeff Spitzner, Chief Science Officer of Rescentris, Ltd., reviewed Chapter 2, and provided valuable feedback regarding BSML. I also want to thank Paul Farrell, my editor at Springer, for ushering the book to completion and keeping me on schedule.

Finally, I want to thank my entire family for supporting me during the whole writing process associated with this book. Thanks to Dad, who hired me at the ripe age of twelve to complete my first bioinformatics programming project (really), and instilled in me a love of scientific ideas and ideals. Thanks to Mom for buying me my first computer (a Commodore Vic 20), and always reminding me to remain balanced. Special thanks to Nelli and Carla for their support and encouragement.

Lastly, I want to thank my wife, Amy. All authors thank their wives in the acknowledgments, but Amy has the distinction of supporting me in this fourth book endeavor while also being pregnant. I do not know how she puts up with me, but she has been my rock, my soulmate, my everything. I love you.

Contents

Introduction to XML for Bioinformatics

1

Bioinformatics represents a new field of scientific inquiry, devoted to answering questions about life and using computational resources to answer those questions. A key goal of bioinformatics is to create database systems and software platforms capable of storing and analyzing large sets of biological data. To that end, hundreds of biological databases are now available and provide access to a diverse set of biological data.

Given this diverse set of biological data, the exponential growth of biological data sets, and the desire to share data for open scientific exchange, the bioinformatics community is continually exploring new options for data representation, storage, and exchange. In the past few years, many in the bioinformatics community have turned to XML to address the pressing needs associated with biological data. XML, or Extensible Markup Language, is a technical specification originally created for data representation and exchange over the Internet. XML is an open standard, officially specified by the World Wide Web Consortium (W3C), and deliberately designed to be operating system and programming language independent.

XML is extensible to many application domains and has been successfully used to represent multiple types of data, including e-commerce transactions, search engine results, scalable vector graphics, and even voice recognition and voice synthesis. Since its introduction, XML has also been successfully used to represent a growing set of biological data, including nucleotide sequences, genome annotations, protein–protein interactions, and signal transduction pathways. XML also forms the backbone of biological data exchange, enabling researchers to aggregate data from multiple heterogeneous data sources.

The goal of this book is to present the fundamentals of XML, and to demonstrate the ways in which XML is being usefully applied in the field of bioinformatics. This chapter presents the first step in this goal, and therefore focuses on three main questions:

- What exactly is XML?
- How is XML currently being used in bioinformatics?
- What are the pros and cons of using XML in bioinformatics?

To explore these issues, we examine the origins of XML, compare XML with HTML, and provide a snapshot of the XML family of specifications. We also take a bird's-eye view of our first case study in bioinformatics and explore the Distributed Annotation System (DAS).

1.1   Introduction to XML

1.1.1   XML Defined

XML is a technology specification that enables you to create highly structured documents. The ML in XML stands for *Markup Language*. A markup language is any language that takes raw text and adds annotation. For example, you could take this page, and underline some words in red and some words in green. Red might indicate bold and green might indicate italics. Along the same lines, Hypertext Markup Language (HTML) is a markup language for creating web pages. It too can be used to represent bold and italics, but it also includes many additional markup options for creating visually compelling web pages.

XML focuses on document *semantics*. This means that you can identify specific document parts and assign them specific meaning. For example, if you are representing biological sequence data, you can clearly identify which portion of the document contains sequence identifiers and cross-references to public databases, and which portion contains raw sequence data. These sections are clearly marked and organized in a hierarchical document structure. A human reader or a computer program can therefore easily traverse a complete document and extract individual pieces of data. For example, a pipeline application can extract the raw sequence data within a document and pass this information along to a BLAST sequence similarity service.

By focusing on document semantics, XML focuses on the meaning and hierarchical structure of documents and ignores the specifics of presentation and layout.* This is in sharp contrast to HTML. In its original design, HTML was created to convey very simple document structure. For example, an H1 tag indicates a first-level heading, and an H2 tag indicates a second-level heading. However, HTML has grown significantly away from document structure and now has a much greater focus on content presentation and layout. For example, HTML now supports fonts, images, tables, and colors.

One of the best ways to understand XML is to take a single set of data, encode it in HTML, and then compare it to the same data encoded in XML. Let us take a look at two very simple examples. Listing 1.1 shows a nucleotide sequence record in HTML, and Listing 1.2 shows the same nucleotide sequence record in XML.

Hopefully, you have at least a passing familiarity with HTML. If not, do not worry. There are only a few important pieces to note. First, HTML markup items are indicated with the very familiar angle brackets. For example, the `<b>` markup tag indicates the start of a bold item; likewise, the `</b>` markup tag indicates the end of a bold item. HTML documents are also formally defined with a start `<html>` tag and a corresponding end `</html>` tag. In a nutshell, markup tags are used to denote specific elements. Hence, we say that the start `<html>` tag marks the beginning of the `<html>` element. Understanding the *difference between* tags and elements is important. We shall return to the topic in Chapter 2.

If you view Listing 1.1 in a web browser, you will see something like that shown in Figure 1.1. You can now start to ask yourself some very basic questions about this document. For example, what is the accession number (or unique identifier) for this record? What organism are we dealing with? What is the raw sequence data? As a human, these answers are intuitively obvious. For example,

---

* There are some exceptions to this rule. For example, Scalable Vector Graphics (SVG) is an XML vocabulary designed precisely for presentation, layout, and even animation. XHTML is also an XML vocabulary designed precisely for presentation and layout of web pages.

Listing 1.1 A nucleotide sequence record, encoded in HTML

```
<html>
<body>
    <h1>NM_171533</h1>
    Organism: <b>Caenorhabditis  elegans</b>
    <p>
    agcacatgacatgagcagtgccccaaatgatgactgtgagatcgacaaggg
    aacaccttctaccgcttcacttttacaacgctgatgctcagtcaaccatcttcttct
    acagctgttttacagtgtacatattgtggaagctcgtgcacatcttcccaattgca
    aacatgtttattctg
    <p>
    [Full sequence has been omitted for brevity.]
</body>
</html>
```

Listing 1.2 A nucleotide sequence record, encoded in XML

```
<?xml version="1.0"  encoding="UTF-8"?>
<Sequence>
  <accession>NM_171533</accession>
  <organism>Caenorhabditis  elegans</organism>
  <sequence_data>
      agcacatgacatgagcagtgccccaaatgatgactgtgagatcgacaaggg
      aacaccttctaccgcttcacttttacaacgctgatgctcagtcaaccatcttcttct
      acagctgttttacagtgtacatattgtggaagctcgtgcacatcttcccaattgca
      aacatgtttattctg
      [Full sequence has been omitted for brevity.]
  </sequence_data>
</Sequence>
```

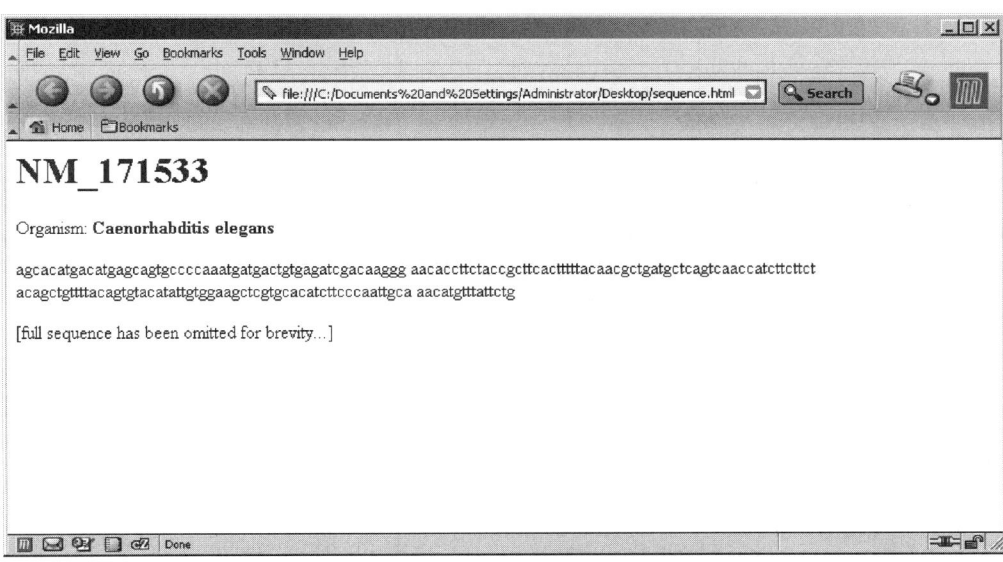

Figure 1.1 Sample HTML nucleotide sequence record, shown in the Mozilla web browser.

it's obvious that the accession number is NM_171533, and that we are dealing with a nucleotide sequence from *C. elegans*.

To a computer, however, these answers are not so obvious. For example, how do we programmatically identify the organism? There are two general approaches one could use. One option is to maintain a database of organisms (for example, one could copy the NCBI Taxonomy database), extract substrings from the HTML document, and search for matches within the database. The second option is to examine the HTML and find patterns in usage. For example, we can see that the accession number is specified within H1 tags and the organism is specified within bold tags. You could therefore write an HTML parser and map specific patterns to specific fields. This technique is frequently known as HTML "screen-scraping" and it is notoriously brittle [10]. Screen-scraping is fragile because the webmaster maintaining the HTML page may arbitrarily decide to reformat it. For example, the webmaster may decide to present organisms within italic tags instead of bold tags. If this happens, your parser breaks, and you can no longer extract the correct data.

Now, consider the XML version of the same sequence record. Listing 1.2 contains the exact same content as Listing 1.1. However, this time, instead of using HTML tags we are using XML tags. We can now revisit the same questions, e.g.: what is the accession number, what organism is this? To a human reader, the answers are still intuitively obvious. To a computer application, the answers are also now trivial. If you want to determine the accession number, simply extract the <accession> element. To determine the organism, extract the <organism> element. XML therefore makes it trivially easy for both humans and computers to identify and extract pieces of data.

HTML and XML both share a similar structure. For example, they both use the familiar angle brackets to denote markup. However, HTML is focused on document presentation, and XML is focused on document semantics. Furthermore, with HTML, you are restricted to the HTML element set such as <B> , <I> , and <H1> . With XML, you no longer have such a restriction. You can create elements for any piece of data you like. You can also organize these elements into any hierarchy of your choosing.

1.1.2   Origins of XML

XML has its roots in another markup language called Standard Generalized Markup Language (SGML). SGML has been around since the early 1970s. In fact, the predecessor to SGML, the Generalized Markup Language (GML), was first proposed at IBM as early as 1969. SGML is a markup language, which focuses on document semantics, and has been used by many organizations for document management. SGML also includes a number of companion technologies that are capable of transforming SGML into different file formats. For example, given a training manual in SGML, you can convert it to HTML, PDF, or some other format more suitable for printed, hard-copy manuals.

In 1997, the World Wide Web Consortium (W3C) set out to create a specification for XML. The goal of the W3C working group was to create a new markup language that could build on the strengths of SGML, but cut out much of its complexity. According to Tim Bray, one of the original editors of the XML 1.0 specification:

XML is an attempt to package up the important virtues and most-used features of SGML in a compact, easily-implemented package that is optimized for delivery on the WWW. [3]

In fact, one of the more curious pieces of XML trivia is that the name "Extensible Markup Language" was not the only name option under consideration. Following are the three other options that were considered:

- MAGMA: Minimal Architecture for Generalized Markup Applications
- SLIM: Structured Language for Internet Markup
- MGML: Minimal Generalized Markup Language

Each of these names conveys that XML was intended to create a minimized or "slimmed" down version of SGML. However, in the end, the name XML won with the most votes (MAGMA came in a close second, and MGML came in last) [3].

1.1.3   The XML Family of Specifications

In the early days of XML, everything you needed to know about XML appeared in just one document, the official W3C XML 1.0 specification. This document includes all the official rules for creating XML documents, and all the rules for creating Document Type Definitions (DTDs). DTDs enable you to create XML grammars which define document types or document structures. For example, you can create a DTD which requires that every `<sequence>` element must contain an `<organism>` element. It also provides a powerful mechanism for validating that documents actually include the data that they purport to hold. We cover DTDs in Chapter 3.

Since those early days, however, XML has grown considerably to now include a complete family of XML-related specifications. In fact, the list of XML-related specifications seems to grow every few months. Below is a brief summary of the most important XML specifications:

- **XML Namespaces:** The XML Namespace specification enables you to partition an XML document into distinct namespaces, and thereby prevent any possible naming conflicts. The XML Namespace specification is covered in Chapter 2.
- **XML Schema:** XML Schema represents the successor to XML Document Type Definitions (DTDs). XML Schema offers more flexibility and advanced validation options than DTDs. XML Schema is covered in Chapter 4.
- **XSLT (Extensible Style Sheet Language Transformations):** XSLT enables you to transform an XML document into a different XML format or into an HTML format. For example, you can transform an XML document into HTML and make the HTML document available via a web site.
- **XInclude (XML Inclusions):** XInclude enables you to merge multiple XML documents into one. For example, a single master XML document can import one or more XML documents or document fragments.
- **XLink (XML Linking Language):** XLink enables you to define links within XML documents. XLinks go well beyond the basic linking capabilities of HTML, and include support for multitarget, and even bidirectional links.
- **XPath (XML Path Language):** XPath enables you to locate specific elements or attributes within an XML document. For example, you can select all `sequence` elements or only the third `sequence` element. XPath is used extensively in XSLT.
- **XPointer (XML Pointer Language):** XPointer builds on XPath to enable the creation of XML specific URLs. For example, an XPointer URL is capable of pointing to a specific element within an XML document.
- **XQuery (XML Query Language):** XQuery defines a language for querying XML documents, in much the same way that SQL enables querying of relational databases.

Figure 1.2 Web services defined. A web service enables two computers to communicate using XML.

It is important to note that a specification beginning with "X" or "XML" does not imply the same wide adoption as XML itself.

1.1.4   Web Services Defined

Beyond the XML family of specifications, XML has also spawned an entirely new field of *web services*. Web services represent an important step forward in building distributed applications over the Internet. They also represent an increasingly important building block for many distributed bioinformatics applications. In this book, we cover the fundamentals of web services, and provide a case study of the caBIO web service, created by the National Cancer Institute (NCI).

In a nutshell, a web service is any service that is available over the Internet, uses a standardized XML messaging system, and is not tied to any one operating system or programming language [4]. (See Figure 1.2.) More succinctly put, one observer has defined web services as "XML in motion" [7].

In its simplest form, a web service can consist of XML documents delivered over a network protocol, such as HTTP. As we will soon see below, the Distributed Annotation System (DAS) is a prime example of one such service. At a higher level, a web service might use one of the formally defined XML protocols, such as XML-RPC or SOAP. For example, as we will see in later chapters, the XEMBL project provides a SOAP interface to the complete European Molecular Biology Laboratory (EMBL) nucleotide sequence database. When using SOAP, a web service can also be *self-describing*. In other words, if you publish a new web service, you can also publish a public interface to the service. This interface describes a list of publicly available methods, along with method parameters and return values. With a formally defined interface, new clients can more easily determine what functionality exists, and more easily connect to the service. Currently, web services can be described using the W3C Web Service Description Language (WSDL).

To appreciate the value of web services, it is useful to think of a *human-centric* web vs. an *application-centric* web. In a human-centric web, web sites are primarily designed for human consumption. Data is encoded in HTML web pages, and these pages are specifically tailored for display within standard web browsers. However, as we have already seen, applications have a hard time extracting meaningful data out of HTML documents. In an application-centric web, web sites are designed for both human consumption and application consumption. Data is encoded in HTML web pages for human users, but data is also encoded in XML for applications. In this model, there is no need for HTML screen-scraping, and applications can more easily aggregate data from a diverse set of web resources.

1.2 Using XML for Biological Data Exchange

XML is currently used to encode a wide range of biological data and has rapidly become a critical tool in bioinformatics. In fact, one recent paper on XML in bioinformatics predicted that XML will soon become "ubiquitous in the bioinformatics community" [1]. The real strength of XML is that it enables communities to create XML formats, and then use these common formats to share data. XML therefore enables individual researchers, software applications, and database systems to exchange and share biological data. In the end, this enables biologists to more easily access relevant data, aggregate data from multiple sources, and mine this data for important scientific clues.

At the 2002 O'Reilly Open Bioinformatics Conference, Lincoln Stein of the Cold Spring Harbor Laboratory delivered a keynote speech, entitled "Creating a Bioinformatics Nation." Stein's presentation and subsequent paper in *Nature* [10] describe a vision and a blueprint for creating common data formats, supporting open source software projects, and building interoperable web services for exchanging biological data. By historical analogy, Stein likened the current state of bioinformatics to the city states of Medieval Italy:

> During the Middle Ages and early Renaissance, Italy was fragmented into dozens of rival city-states controlled by such legendary families as the Estes, Viscontis and Medicis. Though picturesque, this political fragmentation was ultimately damaging to science and commerce because of the lack of standardization in everything from weights and measures to the tax code to the currency to the very dialects people spoke. [10]

In the same vein, Stein argued that bioinformatics is currently dominated by rival groups, rival data formats, and incompatible web sites, and that the lack of clear standards and interoperable software is a "significant hindrance to researchers wishing to exploit the wealth of genome data to its fullest" [10]. A recent technology feature regarding bioinformatics, published in *Nature* in 2002, echoed many of these same concerns. According to one bioinformatics expert quoted in the *Nature* feature, "To answer most interesting biological problems, you need to combine data from many data sources. However, creating seamless access to multiple data sources is extremely difficult" [5]. Echoing Stein's sentiments exactly, the researcher concludes that "The key to bioinformatics is integration, integration, integration" [5].

Academic research labs are not the only ones interested in creating interoperable bioinformatics software. The Interoperable Informatics Infrastructure Consortium (I3C) is a consortium of computer companies, biotech companies, and academic research labs devoted to supporting interoperable data and software "to accelerate discovery and solve critical problems in drug development" [8]. Several dozen organizations are currently involved, including Merck & Co., Millennium Pharmaceuticals, IBM Life Sciences, and the MIT Whitehead Center for Genome Research. Information is available online at: *http://i3c.org*.

1.2.1   Case Study: The Distributed Annotation System

By moving toward common XML data formats and open web-service protocols, the bioinformatics community can significantly lower the barriers to data integration and help build Stein's long-term vision of creating a "bioinformatics nation." To illustrate a concrete example of Stein's bioinformatics nation in action, we now turn briefly to our first case study: the Distributed Annotation System (DAS) [6]. By taking a bird's-eye view of DAS, we can gain insights into current XML usage in biological data exchange and explore the mechanics of data aggregation. We can also gain insight into the essential XML concepts and technologies that are explored throughout the remainder of this book.

DAS is an XML-based protocol that enables the distribution and sharing of genomic annotation data. Genomic annotation is the process of analyzing regions of raw sequence data, and then adding notes, observations, and predictions. For example, annotation includes the identification of exons (protein-coding portions of genes), introns (noncoding portions of genes), and the categorization of repeat-coding regions. Genomic annotation may also include the linking of sequence data to already cataloged gene sequences, making computerized predictions about the locations of novel genes, and identifying sequence similarities across species.

Despite its enormous potential, genomic annotation faces numerous technical challenges. First, annotation is highly decentralized and currently underway at hundreds of laboratories throughout the world. Second, it is not likely that one organization or institution will be able to coordinate and centralize all genomic annotations. In response to these challenges, Lincoln Stein of the Cold Spring Harbor Laboratory, along with Sean Eddy and LaDeana Hillier, both of Washington University at St. Louis, set out to build a distributed protocol for genomic annotation.

DAS is formally specified by a client/server protocol and a set of XML documents. Client applications connect to DAS servers, send queries in the form of URL parameters, and receive XML encoded data back. See Figure 1.3. For example, a client can request all genomic annotations within a specific region of human chromosome 11, or request only a subset of those annotations. This is a prime example of "XML in motion," as DAS uses XML to encode documents and then delivers those documents over Internet protocols.

Currently, clients can issue one of eight different DAS commands, and each command will trigger a different XML response from the server. For example, a client can request a list of data sources or genomes hosted by the server, retrieve annotations across a specific chromosomal region, or request raw DNA sequence data.

All DAS servers return XML data encoded in the same exact format. For example, if a client requests annotation data, the DAS server must return data encoded in the standard DASGFF, or

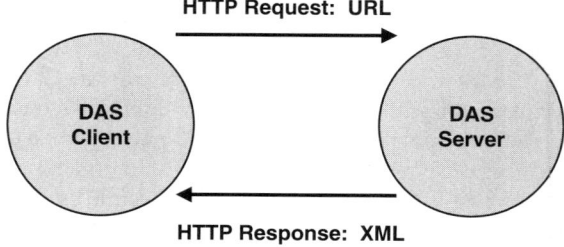

Figure 1.3 The DAS protocol. Client applications issue DAS commands in the form of URL parameters, and receive XML encoded documents back from the server.

General Feature Format. If a client requests raw sequence data, the DAS server must return data encoded in the standard DASSEQUENCE XML format.

In Chapter 6, we will discuss in detail the specific nature of DAS requests and responses. However, to get your feet wet, here is a sample DAS request:

http://servlet.sanger.ac.uk:8080/das/ensembl830/sequence?segment=1:1000, 1200

DAS requests are defined as regular web URLs. The first part of the URL indicates that we are connecting to the DAS server located at the Wellcome Trust Sanger Institute. Following this is the "/das" prefix, required by the DAS protocol. We then have three additional elements:

- The Data Source Name (DSN). Each DAS server can host multiple data sources, and clients must specify which data source they want. In the case above, we have specified "ensembl830," indicating that we want the Ensembl Human Genome assembly, version 8.30.*
- The DAS Command. In the case above, we have specified "sequence," indicating that we want to retrieve raw sequence data.
- DAS parameters. Each DAS command requires a different set of parameters. For example, to request raw sequence data, clients must specify a reference ID, such as a chromosome number, and start/stop values. In the case above, we have specified chromosome 1, base pairs 1000–1200.

In response to the DAS command defined above, the Ensembl DAS server will return the following XML document:

```
<?xml version='1.0'  standalone='no' ?>
<!DOCTYPE DASSEQUENCE SYSTEM  'dassequence.dtd' >
<DASSEQUENCE>
  <SEQUENCE id="1" version="8.30" start="1000" stop="1200"
    moltype="DNA">
taatttctcccattttgtaggttatcacttcactctgttgactttcttttgctgtgcaga
agcttttaggttgatgctattccatttgtgttttgttgcttttcttgcctgtgctttag
agtcatatcataaaatattattgcccagaccaatgtcttggagttattcccctgtttct
tctaggagttctatagtgcta
  </SEQUENCE>
</DASSEQUENCE>
```

In the next chapter, we will formally analyze all the important parts of an XML document. For now, however, you can just focus on the <SEQUENCE> element. As you can see, the <SEQUENCE> element contains the requested raw sequence data. It also includes a number of XML attributes. For example, the *id* attribute indicates the chromosome number, *version* indicates the Ensembl version number, *start* and *stop* indicate the sequence coordinates, and *moltype* indicates the type of molecule (in this case, DNA).

DAS is built to use regular web protocols, such as HTTP. You can therefore use a regular web browser to issue DAS requests. For example, to issue your very first DAS request, try typing the URL above into a web browser. A sample screenshot of the DAS response is shown in Figure 1.4.

The real power of DAS, and of XML in general, is that it enables applications to aggregate data from multiple sources. Without DAS, a user would need to manually surf to three different web sites to compare annotation data. With DAS, a user can open a single client application, and simply specify three DAS servers. Behind the scenes, the client application connects to each DAS

---

*As this book goes to press, the most recent Ensembl Human Genome assembly is version 18. For the latest version information, go to: *http://servlet.sanger.ac.uk:8080/das*.

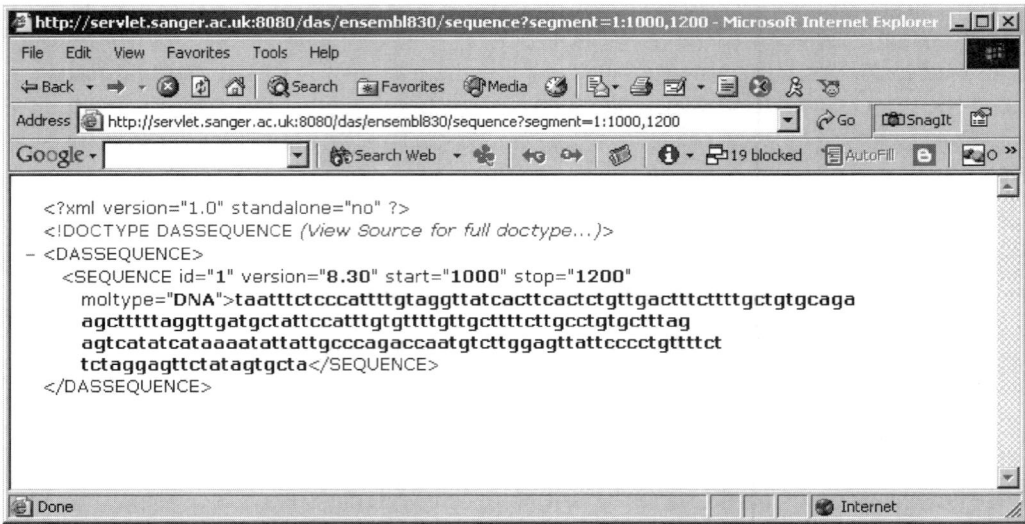

Figure 1.4 DAS is built to use regular web protocols, such as HTTP. You can therefore use a regular web browser to issue DAS requests. A sample sequence request to the DAS Ensembl server is shown.

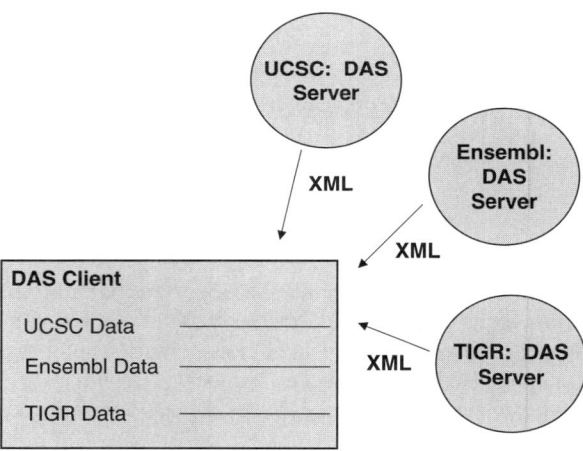

Figure 1.5 Using the DAS protocol, client applications can easily connect to multiple DAS servers and aggregate annotation data.

server, aggregates the response data, and creates a unified view of the data. Users can then see all the annotation data in one place and more easily compare the data. See Figure 1.5.

For another quick preview of DAS in action, try using the DAS viewer available at Wormbase.org, the model organism database for *Caenorhabditis elegans*. The Wormbase DAS viewer runs on the Wormbase server, but behind the scenes, it actually works as a DAS client. Here is how it works: First, the user must specify one or more DAS servers. They do this via the Wormbase web site. Second, the user specifies a specific chromosomal region of interest. The Wormbase server receives the user request and translates it into multiple DAS client commands, one for each specified DAS

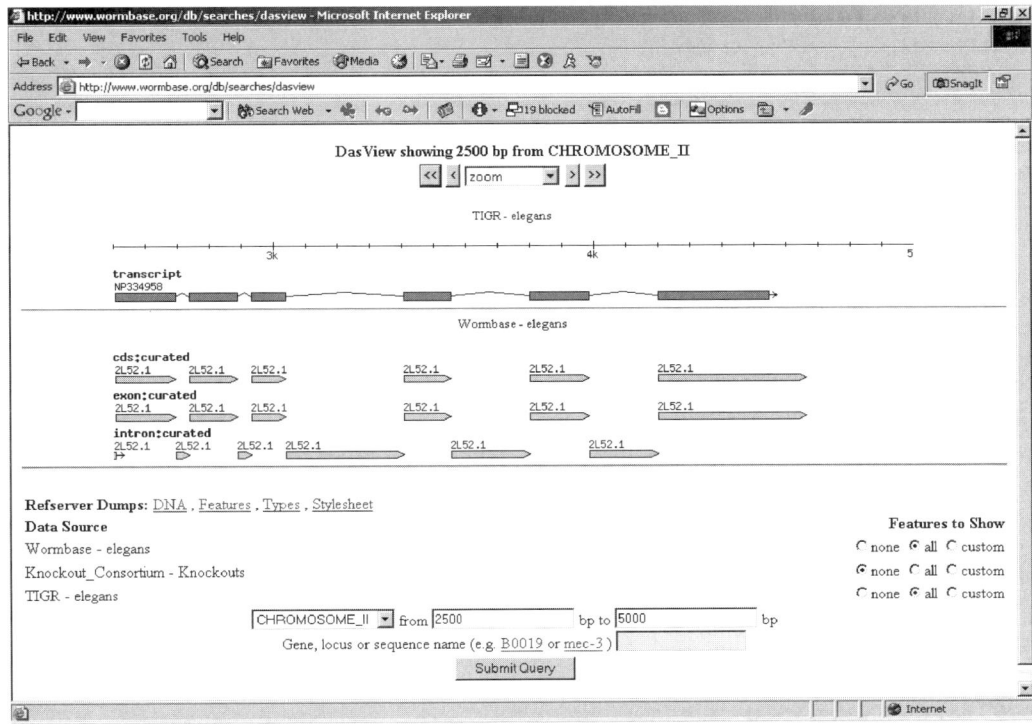

Figure 1.6 Sample screenshot of the DAS viewer available at Wormbase.org.

server. The Wormbase application then waits for responses from each server, parses the XML response data from each, and then aggregates this data into a unified view.

The unified view is then presented as a single web page to the end user. See Figure 1.6 for a sample screenshot. The bottom of our sample screenshot indicates the chosen chromosome region; in this case, we are viewing chromosome 2, 2500–5000 base pairs. Directly above this, each DAS source is represented as a separate horizontal track of data. Annotation from The Institute for Genomic Research (TIGR) is shown on the top and annotation from Wormbase.org is shown on the bottom. As you can see, the screenshot shows a coding sequence (CDS) for NP334958, and that TIGR and Wormbase agree on the location of all exons and introns.

1.2.2   XML Formats for Bioinformatics

DAS is one of the best-known examples of XML in bioinformatics. However, many other XML formats have been created. For example, the Bioinformatic Sequence Markup Language (BSML) is used to encode biological sequence and annotation data. NCBI hosts several newly released XML formats for encoding sequence data. The MicroArray Gene Expression Markup Language (MAGE-ML) is used to encode gene expression data. The UniProt XML format (formerly called SPTr-XML) is used to encode protein data. And, the Proteomics Standards Initiative Molecular Interaction XML format (PSI-MI) is used to encode protein–protein interaction networks. See Table 1.1 for a summary of current XML formats for bioinformatics.

Table 1.1 XML formats currently in use in bioinformatics. This list includes XML formats that are most widely used and most frequently cited. It is not meant to be comprehensive.

Name	Web Address
AGAVE: Architecture for Genomic Annotation, Visualization and Exchange	*http://www.animorphics.net/lifesci.html*
BioML: BIOpolymer Markup Language	*http://www.bioml.com/BIOML*
BioPAX: Biological Pathways Exchange	*http://www.biopax.org*
BSML: Bioinformatic Sequence Markup Language	*http://www.bsml.org*
CellML	*http://www.cellml.org*
DAS: Distributed Annotation System	*http://www.biodas.org*
Gene Ontology (GO) DTD	*http://www.geneontology.org/xml-dtd/go.dtd*
MAGE-ML: MicroArray Gene Expression Markup Language	*http://www.mged.org/mage*
NCBI DTDs: Numerous DTDs, including GBSeq, TinySeq, SeqSet, and NCBI Blast	*http://www.ncbi.nlm.nih.gov/dtd*
PEML: Proteomics Experiment Markup Language	*http://pedro.man.ac.uk*
PSI-MI: Proteomics Standards Initiative Molecular Interaction	*http://psidev.sourceforge.net*
SBML: The Systems Biology Markup Language	*http://www.sbw-sbml.org/sbml/docs*
UniProt XML	*http://uniprot.org*
XFF: The Extensible Feature Format	*http://www.biojava.org/thomasd/XFF*

Throughout this book, we explore many of these formats in detail. For example, in Chapter 2, we cover BSML. In Chapter 3, we cover the NCBI TinySeq XML format. In Chapter 4, we cover the PSI-MI XML format. In Chapters 6–8, we cover DAS and several options for parsing DAS data in Java.

1.3   Evaluating XML Usage in Bioinformatics

XML is a tool, and like any tool, it has specific advantages and disadvantages. This makes it very suitable for many types of bioinformatics applications, but not for all. You therefore need to evaluate each specific bioinformatics application on a case-by-case basis.

1.3.1   Advantages of XML

- **XML is extensible.** The X in XML stands for extensible. This means that you can extend XML to many areas of interest, and can create XML formats for encoding a growing set of bioinformatics data. Already, XML formats exist for representing sequence data, protein–protein interactions, and Microarray data, to name just a few. You are not beholden to any central XML committee that controls XML data formats, and can create new XML formats as needed.
- **XML provides built-in document validation.** Using Document Type Definitions (DTDs) and XML Schemas, you can create formal rules which define valid XML documents. You can then validate XML documents to make sure that they follow all the rules. This relieves you from having to write your own validation software and enables you to more easily focus on document content instead. This is particularly useful if you need to process XML documents from third parties.
- **XML is both human readable and computer readable.** XML documents are written in plain text and not in binary format. No special tools are required to view them, and you can get started with a regular text editor and a standard web browser. This lowers the barrier for getting started in XML and makes debugging XML documents much easier.

- **XML is platform and language independent.** XML was specifically designed to work on any operating system and with any programming language. It therefore works on Windows, UNIX, Linux, and Mac OS X. It also works with all your favorite programming languages, including C, C++, C#, Perl, Python, and Java. By using XML, you are not tied to any one system, and can more easily communicate with other platforms. This is particularly important in bioinformatics, as you frequently need to aggregate data from heterogenous systems and platforms.
- **XML is a public standard.** XML is an official recommendation of the World Wide Web Consortium (W3C). As a public standard, XML is not tied to any one company, and all companies that provide XML support generally find it in their best interest to fully support the W3C XML specifications. In addition to the core XML specification, the W3C also sponsors numerous other XML specifications, such as XML Schema, XLink, XQuery, and SOAP.
- **The XML tool set is large and growing.** Hundreds of XML parsers, browsers, and editors are now available. These are generally available for all the major operating systems, and all the major programming languages. Open source development around XML is also particularly strong. For example, the Apache Software Foundation hosts a number of excellent XML tools, including the Xerces XML parser, the Xalan XSLT transformation tool, and the Apache Axis SOAP toolkit. Commercial XML development is also quite strong and a number of excellent commercial XML editors are available.
- **XML works well with the Internet.** One of the original goals of XML was to create a web-friendly version of SGML. As such, XML works well with most other Internet protocols, such as HTTP. For example, you can easily serve up XML pages from a web server. You can also use XML-RPC or SOAP to create distributed applications that communicate over the Internet.
- **XML documents can be transformed.** Using XSLT, you can transform an XML document into a different format. This enables you to build conversion utilities to support multiple XML formats, and to transform XML documents into HTML documents, for presentation on the web.
- **XML is global.** The XML specification requires that all XML parsers support the global Unicode character encoding system. Unicode provides support for most of the languages on Earth, and stands in sharp contrast to ASCII, which supports only Latin or English characters. You can therefore include multiple character sets and languages within a single XML document.

1.3.2 Disadvantages of XML

The list of disadvantages is shorter, but it is important to keep the following items in mind:

- **XML is verbose.** XML documents are stored as plain text and include lots of markup tags. These markup tags add considerable overhead to the document and take up lots of memory. Very large XML documents delivered over the Internet therefore generally take up a lot of bandwidth and are hardly optimized for network delivery. As one example where this is already a problem, consider gene expression data from a microarray experiment. Just one experiment might include tens of thousands of data points and might be better off formatted in a simple tab delimited format, making it both more compact and easier to parse.
- **XML is not a cure-all for data integration.** XML certainly facilitates integration of biological data. However, there are in fact many critical data integration issues that are not addressed by XML. This includes the assignment of unique identifiers across databases, the development of ontologies for describing biological entities and phenomena, and the timely synchronization of data sources [11]. There is also a serious political dimension, as it is difficult to get database

providers to actually agree to one XML format and to convert their existing data into this new XML format.

- **XML does not guarantee unified formats.** Just because you can create common XML formats does not mean that everyone will use that format. In fact, you can frequently end up with fractured efforts and overlapping XML formats. For example, in bioinformatics there are already several options for encoding sequence data, and not all the data providers even support an overlapping set of these formats. For example, the EMBL database currently returns data in both BSML and AGAVE, but NCBI returns data in its own internally developed XML formats. If you want to retrieve XML data from both EMBL and NCBI, you have to minimally include support for at least two different XML formats.

- **XML requires a large learning curve.** XML was specifically designed to be a simpler, web-friendly version of SGML. In the early days of XML, you could therefore get by with a basic understanding of the XML specification, and perhaps one or two XML parsers. Today, however, the learning curve is much greater. To truly take advantage of XML, you need to be familiar with the core XML 1.0 specification, XML Namespaces, XML Schema, XSLT, and several XML parsing APIs. This requires a significant investment of time. Furthermore, XML specifications and technologies are constantly evolving. Staying on top of these new developments is a difficult and time-consuming task.

Hopefully, you now have a good introductory understanding of XML, can appreciate it strengths and weaknesses, and can envision ways in which XML can be usefully applied to bioinformatics. We now turn to the fundamentals of XML and explore our second case study, the Bioinformatic Sequence Markup Language (BSML).

1.4   Useful Resources

1.4.1   Articles

- F. Achard, G. Vaysseix, and E. Barillot, "XML, bioinformatics and data integration." *Bioinformatics*, 2001. **17**(2): 115–125.

  This is one of the very first papers describing the use of XML in bioinformatics. The paper provides a list of pros and cons for using XML for bioinformatics and compares XML with other data encoding systems, such as Abstract Syntax Notation One (ASN.1), CORBA, Java RMI, and object-oriented databases.

- E. Barillot and F. Achard, "XML: a lingua franca for science?" *Trends in Biotechnology*, 2000. **18**(8): 331–333.

  Provides an overview of XML and its usage in scientific data exchange.

- R.D. Dowell et al., "The distributed annotation system." *BMC Bioinformatics*, 2001. **2**(1): 7.

  Official description of the Distributed Annotation System.

- L. Stein, "Creating a bioinformatics nation." *Nature*, 2002. **417** (6885): 119–120.

  Describes the current state of bioinformatics and provides a blueprint for building interoperable software and web services. Based on Stein's keynote presentation at the 2002 O'Reilly Open Bioinformatics Conference.

- L.D. Stein, "Integrating biological databases." *Nature Reviews Genetics*, 2003. **4**(5): 337–345.

  Explores the challenges in integrating biological databases. Stein outlines the three main approaches to data integration, examines the role of biological ontologies, and assesses options for creating globally unique identifiers.

- A.C. Martin, "Can we integrate bioinformatics data on the Internet?" *Trends in Biotechnology*, 2001. **19**(9): 327–328.

  Provides an overview of the 2001 Workshop on "CORBA and XML: Towards a Bioinformatics-integrated Network Environment."

1.4.2   Web Site and Web Resources

- DAS web site: *http://www.biodas.org*

  Official home of the Distributed Annotation System. The web site includes the official DAS specification, a list of public DAS servers, links to DAS client and server software, and a Request for Comment (RFC) section with proposed features for DAS 2.

- XML.com Bioinformatics Resources: *http://www.xml.com/pub/rg/Bioinformatics*

  Provides a list of XML formats and resources for bioinformatics. The list is maintained by the editors of xml.com.

Fundamentals of XML and BSML 2

This chapter provides a detailed introduction to the fundamentals of XML. We cover all the essential concepts for understanding XML markup, creating XML elements, and working with XML Namespaces. To make the concepts concrete and focused on bioinformatics, we introduce our first case study and explore the Bioinformatic Sequence Markup Language (BSML). BSML is an open XML standard used to represent biological sequences and sequence annotation data.

The chapter begins with a bare bones BSML document used to represent raw sequence data. As we introduce this first example, we take a bird's-eye view of XML document structure in general, including start tags, end tags, elements, and attributes. We also take a quick tour of the Rescentris Genomic Workspace™, a freely available software application that visually renders BSML documents.

After our high-level introduction to XML, we turn to a detailed description of the most important XML concepts. The topics include: tag structure, comments, processing instructions, the XML prolog, options for character encoding, XML grammars, and XML Namespaces. We also explore what it means to be "well-formed" and "valid," and how to test for either property.

The chapter concludes with a more detailed overview of the BSML specification. BSML is one of the most mature XML standards in bioinformatics, and has grown to encompass a very large set of bioinformatics sequence data. We do not have space to cover BSML in its entirety, and have therefore chosen to specifically focus on core elements of the BSML specification. We also provide several more BSML examples and explore these further within Genomic Workspace™.

2.1 Getting Started with BSML

The best way to learn XML is by example. Therefore, before discussing any major concepts we will begin with a sample XML document. This initial example adheres to the Bioinformatic Sequence Markup Language (BSML) [12; 13; 25]. BSML is an open standard for representing and exchanging biological sequence data. This data can include raw sequence data, sequence features, literature references, networks of biological entities, and even graphical display widgets.

BSML is a great place to get started in XML. The first main advantage is that BSML represents one of the very first XML formats specifically created for the life sciences. Second, BSML is comprehensive in scope. Those who have ushered the BSML specification and its continuing evolution have made every effort to ensure that BSML is capable of accurately representing biological reality and all the complexity that this requires. Furthermore, the BSML web site (*http://www.bsml.org*) includes excellent documentation, including tutorial documents, a reference manual, and an FAQ. Finally, Rescentris, Ltd. makes available a free BSML viewer that enables you to visually

17

Listing 2.1 The SARS virus, encoded in BSML

```xml
<?xml version="1.0" encoding="UTF-8"?>
<!-- SARS coronavirus Urbani, complete genome. -->
<!-- Accession Number: AY278741  -->
<Bsml>
  <Definitions>
    <Sequences>
      <Sequence id="AY278741" length="29727">
      <Seq-data>
        atattaggttttacctacccaggaaaagccaaccaacctcgatctcttgtagatctgttct
        ctaaacgaactttaaaatctgtgtagctgtcgctcggctgcatgcctagtgcacctacgcagt
        ataaacaataataaattttactgtcgttgacaagaaacgagtaactcgtccctcttctgcaga
        ctgcttacggtttcgtccgtgttgcagtcgatcatcagcatacctaggtttcgtccgggtgt
        gaccgaaaggtaagatggagagccttgttcttggtgtcaacgagaaaacacacgtccaactca
        gtttgcctgtcc
        [For brevity, sequence is truncated.]
      </Seq-data>
      </Sequence>
    </Sequences>
  </Definitions>
</Bsml>
```

inspect and interact with BSML documents. This makes for much more exciting and interactive examples.

Listing 2.1 shows our first XML example, a bare bones BSML document. The document represents the raw sequence data for the coronavirus responsible for severe acute respiratory syndrome (SARS). The virus sequence is 29,727 base pairs in length, and we have taken the liberty of only displaying the first few hundred base pairs. Let us now examine Listing 2.1, and we will continue with a high-level overview of the document structure.

There is a lot going on in our first example. For now, note the following items of interest:

- Our document begins with the characters "<?xml" This is formally known as the XML prolog and is used to indicate the version of XML and the character encoding.
- The second and third lines of the document are XML comments. Comments begin with the characters "<!--" and end with the characters "-->".
- Every XML document must have a root element. In our case, Bsml is the root element, and all other elements are descendants of the root. For example, the Definitions element is a child of the root Bsml element.
- XML elements are defined with start and end tags. For example, this tag: <Seq-data> signals the start of the Seq-data element. Likewise, this tag: </Seq-data> indicates the end of the element.
- Attributes appear within start element tags, and provide additional information about that element. For example, our document includes two attributes: *id* and *length*. Within BSML, the *id* attribute is used to uniquely identify an element within a document, and the *length* attribute is used to denote the number of base pairs or residues in a sequence.

Every XML document explicitly defines a document structure or element hierarchy. The element hierarchy for our sample document is shown in Figure 2.1. As you can see, Bsml is the roots element. The root element contains a Definitions element, which in turn contains a [Sequences] element. This element then contains a [Sequence] element, which in turn contains a [Seq-data]

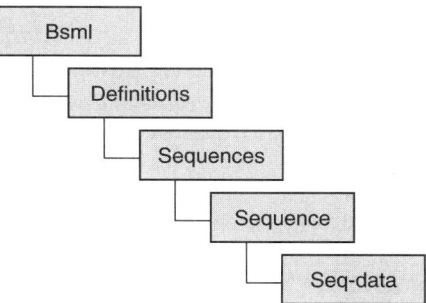

Figure 2.1 Element hierarchy of our first BSML document.

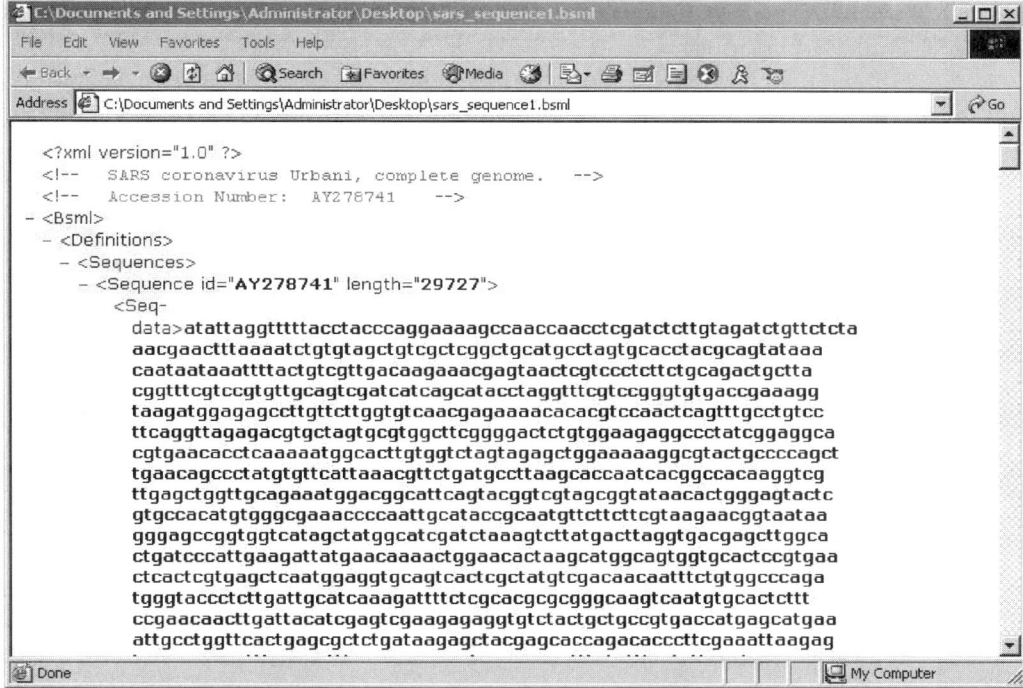

Figure 2.2 A sample screenshot of Internet Explorer. The first sample BSML document is shown. If the XML document does not reference a specific style sheet for transforming to HTML, Internet Explorer will apply a default style sheet. This default style sheet enables users to point and click their way through the element hierarchy. Clicking the + sign expands the element, revealing its direct descendents. Clicking the – sign collapses the element, hiding all its descendents.

element. Note that many BSML documents will have this same structure, and we will explore this structure in detail at the end of the chapter.

Many tools, including XML parsers and web browsers, provide complete access to the XML element hierarchy. For example, Internet Explorer provides an interactive display for browsing an XML document's structure. You can easily open and close nodes, and thereby show or hide specific branches of the element hierarchy. A sample screenshot is shown in Figure 2.2.

2.1.1   Using Genomic Workspace™

Viewing BSML documents within Internet Explorer certainly helps you understand and navigate the document structure, but it's hardly exciting. To appreciate the full power of BSML, it helps to have a BSML-aware browser. Rescentris, Ltd. provides such a browser in its Genomic Workspace™ software application. Genomic Workspace™ enables you to visually browse and interact with BSML documents. Visualization is provided by a number of specialized viewers, such as a hierarchical tree viewer, sequence viewer, sequence editor, and a multiple alignment viewer. In addition to these features, Genomic Workspace™ includes a data conversion and import utility. This enables you to import data in existing data formats, such as GenBank, Swiss-Prot, and EMBL file formats, and convert these records to BSML. This is a particularly useful feature for learning the full BSML specification.

Genomic Workspace™ is written in Java, and runs on most platforms, including Windows, Linux, and Mac OS. You can download a free copy from the Rescentris web site at: *http://www.rescentris.com.*

To explore BSML further, start Genomic Workspace™, and select File → Open, and select the example from Listing 2.1. You should now see a screen like the one shown in Figure 2.3. As you can see, the screen is divided into a number of sections. The main visual window in the center shows a snapshot of the sequence. If our sequence included annotations, such as the location of protein-coding regions, you would see these here too. However, since our example includes only

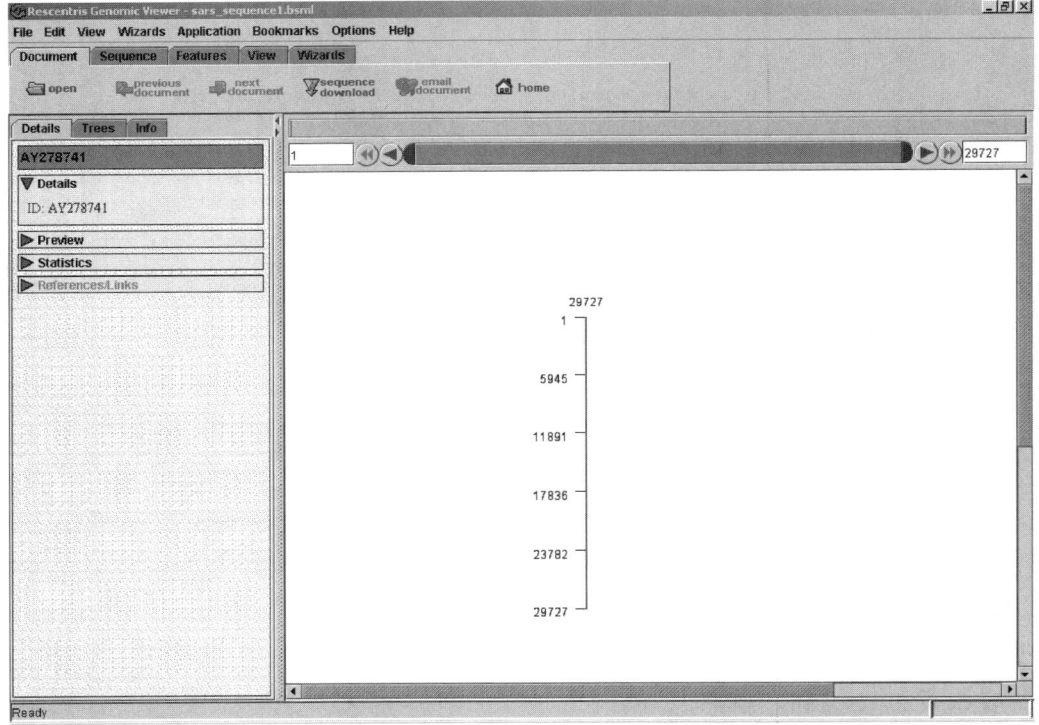

Figure 2.3 Screenshot of the Rescentris Genomic Workspace™ . First sample BSML document is shown.

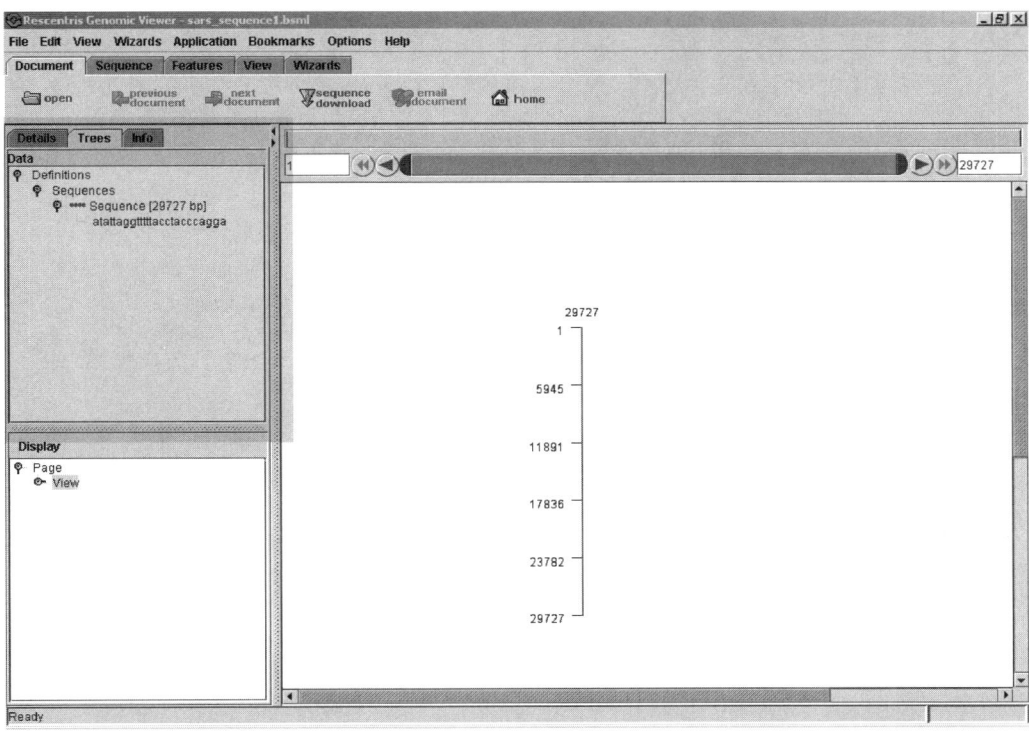

Figure 2.4 Screenshot of the Genomic Workspace™ Tree Viewer (shown highlighted). First sample BSML document is shown.

raw sequence data, we simply see a sequence widget with no annotation. The sequence widget begins at base pair 1 and ends at base pair 29,727. The navigation elements at the top of the main visual window, including the left and right arrows, enable you to zoom in and scroll through the sequence.

To the left of the main visual window, you will see three tabbed windows. By default, the "Details" tab is shown. We have not provided much information in our sample document, but you can see that the sequence ID and the sequence length are displayed. If you click on the Tree tab, you will see an interactive tree showing the complete element hierarchy. See Figure 2.4. Not surprisingly, the tree shown here matches the tree structure we saw earlier in Internet Explorer.

Next, let's explore the BSML Sequence Viewer. To access this, select View → Sequence → Sequence Viewer. Then, click the icon for "Zoom to Base Pair Level." You should now see a screen similar to that shown in Figure 2.5. As you can see, the 5′ to 3′ strand is shown, along with its complement. Below and above the strands are translation frames for amino acids. Just as in the main visual window, the sequence viewer includes navigation buttons for zooming in and scrolling through the entire sequence. Again, if this example included annotations, you would see these here. When we get to annotations at the end of the chapter, we will return to this view.

Hopefully, this gives you a taste of both BSML and the Genomic Workspace™. Once we explore the fundamentals of XML, we will return to both topics and explore them in more detail.

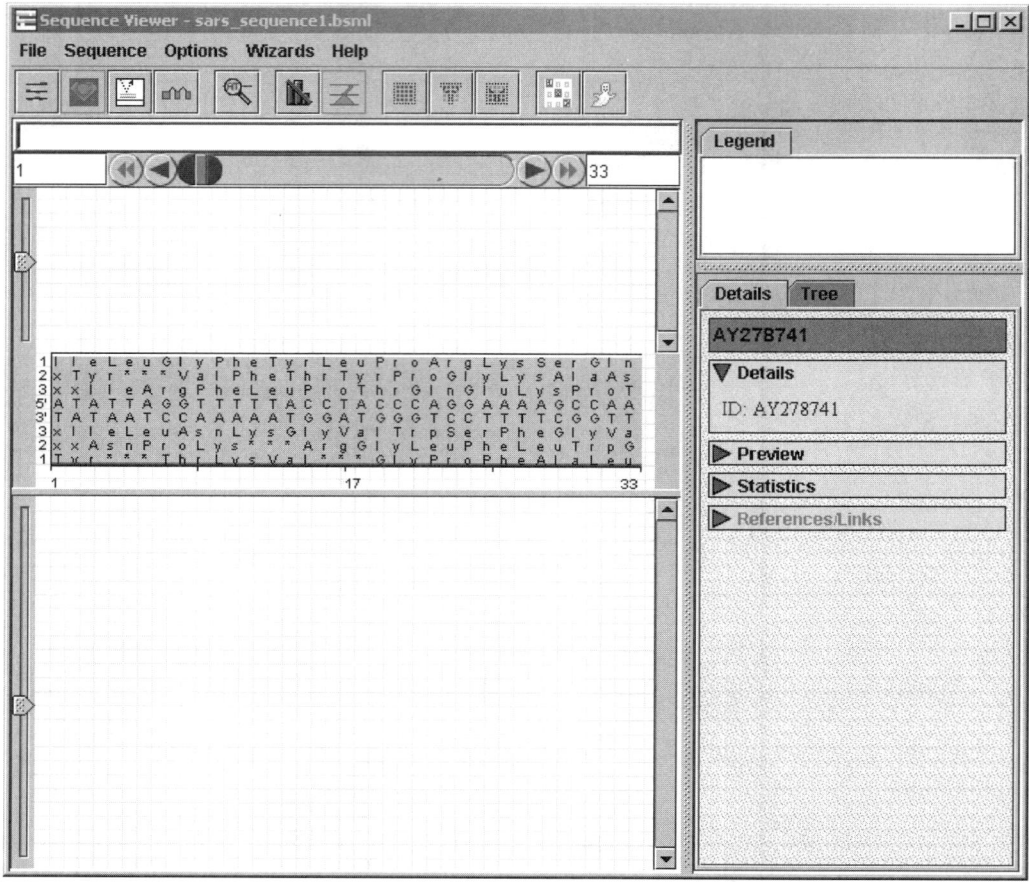

Figure 2.5 The Genomic Workspace™ sequence viewer. First sample BSML document is shown.

2.2   Fundamentals of XML

We now turn to the fundamental rules and concepts of XML. These rules apply regardless of XML application. For example, we could be dealing with e-commerce data, real estate listings, or genomics data. Some have argued that XML has grown in complexity, and that we are now inundated with too many XML specifications and XML protocols. This is certainly true, but if we stick to the core XML 1.0 specification, you may be surprised that there are only a handful of major concepts. Furthermore, the main rules for constructing XML documents are quite straightforward. We have made every effort to distill these concepts and rules into bite-size sections below.

2.2.1   Working with Elements

In its most basic form, an XML document consists of a set of elements. An element represents a discrete unit of data, such as a product listing, news headline, or biological sequence. With

XML, you can create elements for anything you want, and you are not restricted to a predefined list of elements. Furthermore, you can nest elements inside one another, and create any element hierarchy you like. For example, a product listing can include a description and a price, a news headline can include a title and a news category, and a sequence can contain references to scientific papers.

An XML element is formally defined with a start tag and a corresponding end tag. Start tags always take the form: `<ELEMENT_NAME>` , whereas end tags always take the form: `</ELEMENT_NAME>` . For example, the `Seq-data` element is defined with a start `<Seq-data>` tag and an end `</Seq-data>` tag. The complete element therefore looks like this:

```
<Seq-data>gcaggcgcagtgtgagcggcaacatggcgtccaggtc</Seq-data>
```

XML requires that every start tag must have a matching end tag. This is true even for empty XML elements. An empty element is one that does not contain any textual data or subelements, but may contain attributes. For example, the following empty element includes a cross-reference to the EMBL database:

```
<cross_reference database="EMBL" id="M29855"></cross_reference>
```

As a shortcut, you can specify empty elements with the more concise syntax: `<ELEMENT_NAME />` . For example:

```
<cross_reference database="EMBL" id="M29855"/>
```

XML has specific rules on naming XML elements. Specifically, element names must begin with a letter, an underscore character ("_"), or a colon character (":"). Names can then continue with letters, digits, hyphens, underscore, or colons. Names cannot begin with the letters "xml" or any case combination of "xml," as these are specifically reserved for use by the specification.

XML is also case sensitive. This is particularly important to remember when matching start tags with end tags. For example, the following example will result in an error:

```
<Seq-data>gcaggcgcagtgtgagcggcaacatggcgtccaggtc</SEQ-DATA>
```

In this example, `Seq-data` is not equal to `SEQ-DATA` , and the XML parser will report that the start tag is missing a matching end tag.

Every XML document must contain exactly one root element. This root element represents the entry point for traversing the entire element hierarchy. It is not legal to have more than one root element.

2.2.2   Working with Attributes

Attributes are used to provide additional information about a specific element. For example, you can specify *width* and *height* attributes for an HTML `img` element or you can add a *length* attribute to a BSML `Sequence` element. You can specify as many attributes for an element as you need, and they need not be placed in any specific order. Attributes are always placed within the start tag and never within the end tag.

XML requires that attribute values appear within quotes. You can use single quotes (') , or double quotes (") . For example, the following excerpt specifies an *id* attribute:

```
<Sequence id="AY064249">
    . . .
</Sequence>
```

2.2.3 The XML Prolog

XML documents should (but are not actually required to) begin with an XML prolog. The XML prolog includes an XML declaration and an optional reference to a Document Type Declaration (DTD). The XML declaration specifies the XML version number and optional character encoding information. The declaration must begin with the characters: `<?xml` and end with the characters `?>` .

The current version of XML is 1.1, but many individuals (and all the examples in this book) continue to use XML 1.0. XML 1.1 does not represent a significant break from XML 1.0, and primarily focuses on character encoding issues. For example, it includes revised rules on including Unicode characters and expanding the set of end-of-line characters.

As a quick example, the following XML prolog specifies XML version 1.0:

```
<?xml version="1.0"?>
```

Details on character encoding will be covered later in this section.

2.2.4 Comments

XML comments begin with the characters: `<!--` and end with the characters: `-->` . Here is an example comment:

```
<!-- SARS coronavirus Urbani, complete genome. -->
```

Comments can span multiple lines, if needed. To maintain SGML compatibility, the character sequence "- -" is not permitted within comments.

2.2.5 Processing Instructions

Processing instructions are special XML directives, used to forward information to software applications. In certain scenarios, a single XML document may be processed by one or more software applications. This XML document may include processing instructions specifically directed at these applications, possibly providing important application parameters or other hints for processing.

Processing instructions must begin with the characters `<?` , and must end with the characters `?>` . Within these tags, a processing instruction consists of two parts:

- The first part is the software target. This indicates the target of the directive, usually specifying a specific software application or a specific type of software application.
- The second part is a list of one or more processing instructions. This can be any arbitrary text, but usually takes the form of name/value pairs, called pseudo-attributes.

Processing instructions are frequently used in XSL Transformations (XSLT). With XSLT, you can transform an XML document into another XML format or to an HTML document. XML documents use processing instructions to pass application parameters to XSLT parsers. Here is an example XSLT processing instruction:

```
<?xml-stylesheet type="text/xsl" href="bsml_to_html.xsl"?>
```

In the line above, we have specified a target value of "xml-stylesheet." We have also specified two name/value pairs. The first specifies the MIME type of the transformation document, and the second specifies the name of the specific XSLT template to use. In this case, we are using the BSML to HTML XSLT style sheet.

2.2.6   Character Encoding

As stated above, the XML declaration can include optional information about character encoding. The XML specification requires that all XML parsers support Unicode. Unicode is a character-encoding standard that provides support for most languages on Earth. This is in sharp contrast to ASCII (American Standard Code for Information Interchange), which supports only English or Latin characters. Unicode is made available by the Unicode Consortium, but is also officially endorsed by the ISO (International Organization for Standardization). The terms Unicode and ISO-10646 refer to the same standard.

XML parsers are required to support two specific encodings of Unicode/ISO-10646: UTF-16 and UTF-8. UTF-16 encodes Unicode characters using 16-bit characters. For text documents which primarily consist of ASCII characters, UTF-16 can result in inefficient storage and unnecessarily large documents. For these documents, it is more efficient to use the UTF-8 encoding schema. UTF-8 uses a few tricks to more compactly store Unicode characters. Specifically, ASCII characters are stored within one byte, and other characters are stored as multibyte sequences.

Within the XML declaration, you can use the *encoding* declaration to specify the character encoding of your XML document. For example, the following XML declaration specifies XML version 1.0 and UTF-8 character encoding:

```
<?xml version="1.0" encoding="UTF-8"?>
```

Besides UTF-8 and UTF-16, you can also specify other character encodings, such as one of the ISO-8859 family of character encodings. This includes Latin 1 (ISO 8859-1), which contains characters for English and most Western European languages; Latin 2 (ISO 8859-2), which contains character for most Eastern European languages; or Cyrillic (ISO 8859-5), which contains characters for Russian and Russian-influenced languages, such as Bulgarian and Macedonian.

If you plan to create XML documents, which make extensive use of Unicode characters, it helps to use a Unicode enabled editor. For example, for Windows platforms, you might consider using the excellent UniPad editor (*http://www.unipad.org*). UniPad comes with its own set of fonts, meaning that it works out of the box without having to install separate Windows system fonts. It also provides several easy options for inputting Unicode characters, including keyboard shortcuts and virtual keyboards. A screenshot of UniPad with a sample XML document (and the Japanese virtual keyboard) is shown in Figure 2.6.

If your document consists of just ASCII characters, and you want to include an occasional non-ASCII character, you can do so with a character escape sequence. Character sequences begin with: &#[followed by a decimal value] or &#x[followed by a hexadecimal value]. Escape sequences

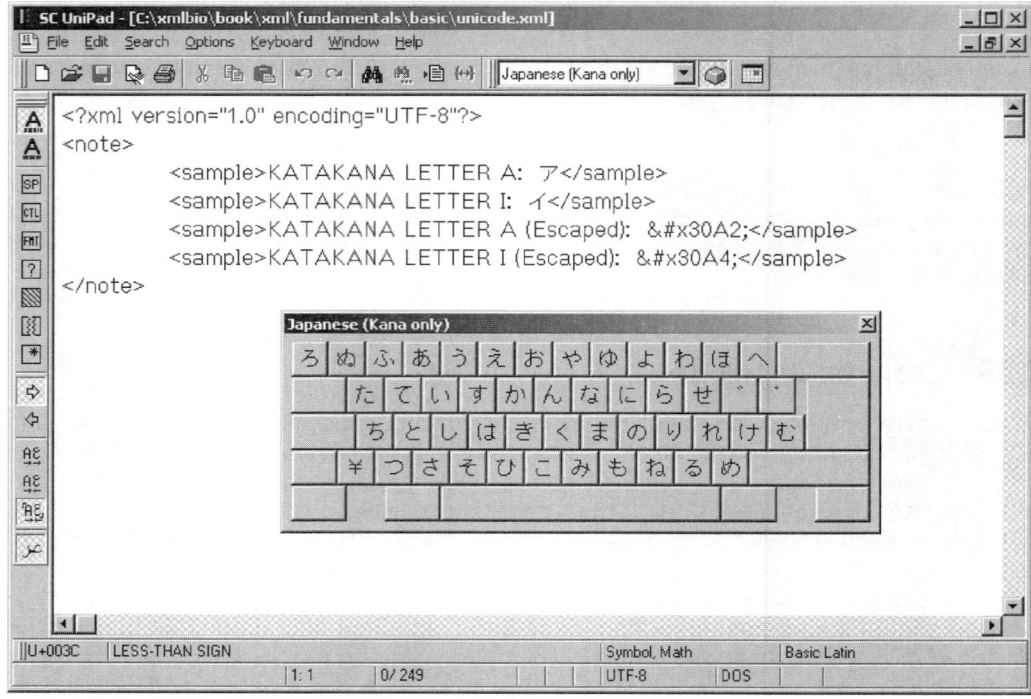

Figure 2.6 UniPad in action. UniPad provides several options for inputting Unicode characters, including virtual keyboards. The Japanese (Katana only) keyboard is shown.

must end with a semicolon (";"). For example, the following escape code references the Japanese Kana letter A:

```
<sample>KATAKANA LETTER A (Escaped): &#x30A2;</sample>
```

Certain characters in XML have special importance, because they are used to denote XML markup. For example, the less than sign (<) is used as the first character for XML tags. If you want to use one of these reserved characters within element text, you must use its corresponding character escape sequence. There are only five reserved characters in XML, and each of these has a corresponding character escape sequence. These are defined as follows:

- &   ampersand sign (&)
- <   less than sign (<)
- >   greater than sign (>)
- '   apostrophe (')
- &quote; quote (")

2.2.7   CDATA Sections

Occasionally, you may want to escape an entire section of text. Text that is stored within a CDATA section is preserved exactly as it is. Reserved characters, such as the less than sign (<), which would normally be interpreted as markup characters, are no longer interpreted as such. This can be useful if you want to include sample XML or HTML markup examples within your XML document. CDATA sections must begin with the characters `<![CDATA[`, and must end with the characters `]]>`.

Here is a sample XML document with a CDATA section:

```
<note>
    <section>
    <![CDATA[
        In XML, start tags always take the  form: <ELEMENT_NAME>.
    ]]>
    </section>
</note>
```

Without the CDATA section, this XML document would result in an error; specifically, an XML parser would complain that the `<ELEMENT_NAME>` element was missing a corresponding end tag. However, because this text is actually contained within a CDATA section, the parser knows to ignore all the markup characters and preserve the text as it is.

2.2.8   Creating Well-Formed XML Documents

XML has very strict requirements on what constitutes a legal XML document. If an XML document meets these specific requirements, it is said to be *well-formed*.

To be well-formed, an XML document must meet the following requirements:

- Every start tag must have a corresponding end tag. The only exception to this rule is the empty element tag syntax, e.g., `<ELEMENT_NAME/>`.
- Elements must be properly nested. In other words, a subelement must have its start and end tags defined within the scope of the parent element. For example, this example is nested properly and therefore well-formed:

```
<Sequence>
    <Seq-data>atggcgtccaggtctaagcggcgtgccgtg</Seq-data>
</Sequence>
```

However, this example is not property nested, and therefore not well-formed:

```
<Sequence>
    <Seq-data>atggcgtccaggtctaagcggcgtgccgtg
</Sequence>
</Seq-data>
```

- All attribute values must appear within quotes.
- Every XML document must have exactly one root element.
- Reserved characters, such as the less than sign, are always treated as markup. If they appear on their own, they must be specified with character escape sequences, or placed within a CDATA section.

There are lots of XML editors and command line tools, which can test your XML documents for well-formedness. We explore some of these tools in the next chapter. However, you may have a convenient XML tool already loaded on your machine. In fact, if you have one of the more current

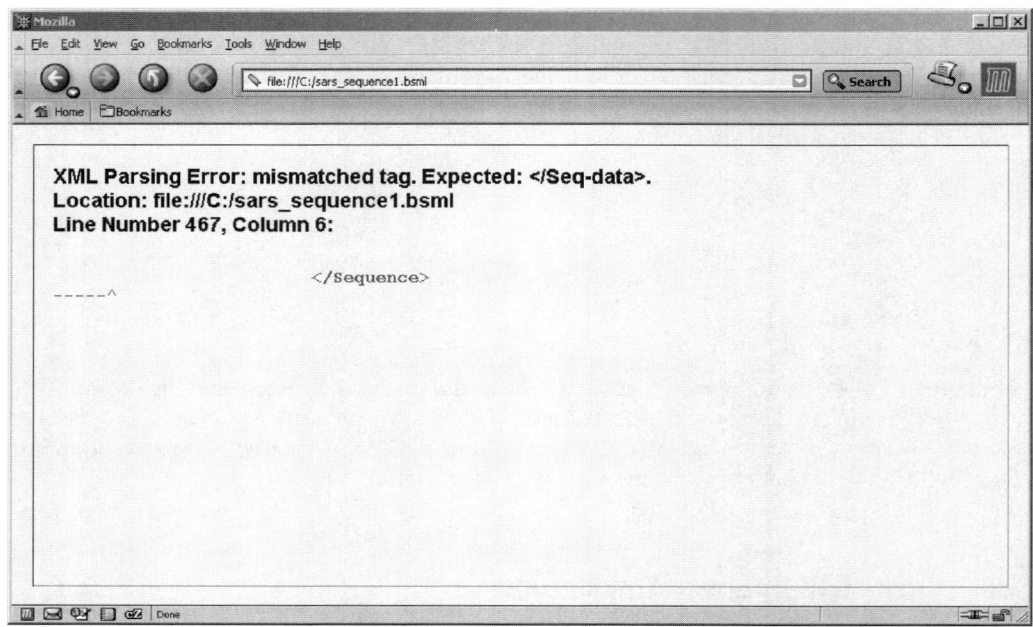

Figure 2.7 Mozilla Web Browser in action. Upon loading an XML document, the built-in parser will automatically check the document for well-formedness and immediately report any errors.

web browsers, such as Internet Explorer 6.0 or Mozilla 1.4 or later, both of these now include lots of built-in XML features and a built-in XML parser. When these browsers load an XML document, the internal XML parser will automatically check for well-formedness and report any errors. For example, Figure 2.7 shows a screenshot of the Mozilla browser. We have just opened a sample XML document, and it immediately reports a missing end tag:

```
XML Parsing Error: mismatched tag. Expected: </Seq-data>.
```

It also reports the exact location of the error. If no errors are encountered, Mozilla uses a default style sheet to render the XML document with a simple tree view. See Figure 2.8.

2.2.9 Creating Valid XML Documents

An XML grammar defines rules for creating XML documents. For example, the BSML specification is actually a grammar, and this grammar spells out specific rules for creating BSML documents. For example, it defines that a Sequences element can contain one or more Sequence elements, and that the Sequence element contains a *length* attribute. If we know that a document adheres to a specific grammar, we already know what type of data we are dealing with, and we can accurately predict the extract structure of the document. We can therefore build applications that are specifically designed to consume specific types of XML documents. Furthermore, if multiple documents adhere to the same grammar, we can process all these documents using the same software application. We can even create new software applications, which aggregate data from multiple disparate sources.

Figure 2.8 Mozilla Web Browser in action (continued). If Mozilla does not find any errors in well-formedness, it will use a default style sheet to display your XML document. In other words, if you see this view, you can be certain that your XML document is in fact well-formed.

There are two main types of XML grammars: Document Type Definitions (DTDs) and XML Schemas. DTDs have been around since the very beginning of XML and are formally specified within the W3C XML 1.0 specification. XML Schemas are a newer specification and provide considerably more features than DTDs. For example, XML Schema supports data typing and enables you to specify that certain elements can only contain integer or float values. With XML Schema, you can also specify regular expression patterns and require that certain elements match those patterns.

We discuss both DTDs and XML Schemas in detail in the next two chapters. For now, we only want to cover one essential point: document *validity*. A document that adheres to all the critical rules in XML is said to be *well-formed*. A document that adheres to all the rules of a specific grammar is said to be *valid*. This is a critical distinction, and one that we will explore many times in the next few chapters.

For now, a very simple example should make the distinctions very clear. First, consider this sample document:

```
<?xml version="1.0" encoding="UTF-8"?>
<!DOCTYPE Bsml PUBLIC "-//Labbook, Inc. BSML DTD//EN"
"http://www.labbook.com/dtd/bsml3_1.dtd">
<Bsml>
    <Definitions>
        <Sequences>
            <Sequence id="AY064249" length="1245" molecule="rna">
                <Seq-data>gcaggcgcagtgtgagcggcaacatggcg....
```

```
      </Sequence>
    </Sequences>
  </Definitions>
</Bsml>
```

See any problems here? The end `</Seq-data>` tag is missing. The document is therefore not well-formed. To fix this problem, we simply add the end tag, like this:

```
<Seq-data>gcaggcgcagtgtgagcggcaacatggcg....</Seq-data>
```

Now, consider this sample document:

```
<?xml version="1.0" encoding="UTF-8"?>
<!DOCTYPE Bsml PUBLIC "-//Labbook, Inc. BSML DTD//EN"
"http://www.labbook.com/dtd/bsml3_1.dtd">
<Bsml>
    <Definitions>
        <Sequences>
            <Dna id="AY064249" length="1245">
                <Seq-data>gcaggcgcagtgtgagcggcaacatggcg....</Seq-data>
            </Dna>
        </Sequences>
    </Definitions>
</Bsml>
```

See any problems here? All the start tags have matching end tags, everything is nested properly, and all attribute values appear in quotes. It is therefore well-formed. However, you can now see that the `Sequences` element contains a `Dna` element. This may seem just fine, but the BSML grammar does not actually specify a `Dna` element. Therefore, this document does not follow all the rules of the BSML grammar and is considered invalid.

How do we actually know that BSML does not specify a `Dna` element? This is the topic that we explore in great detail in the next two chapters. For now, understand that there is a fundamental difference between well-formedness and validity. To recap, we define these two terms below:

- Well-formed: a document is said to be well-formed if it follows all the main rules defined by the XML specification. For example, every start tag must have a matching end tag, elements must be properly nested, and all attribute values must appear within quotes.
- Valid: a document is said to be valid if it follows all the rules of the referenced XML grammar.

A document can be well-formed, but invalid. This means that the document follows all the main XML rules, but fails to follow the rules of the XML grammar.

2.2.10   Working with XML Parsers

An XML parser (or XML processor) is responsible for parsing an XML document and making its contents available to a calling application. Specific responsibilities include: retrieving XML documents from a local file system or from a network connection, checking to make sure that the document is well-formed, and making the contents of the document available via a standard Application Programming InterFace (API). If you have lots of time to spare, you could, of course, write your own XML parser. However, this may not make the best use of your time! A much more convenient

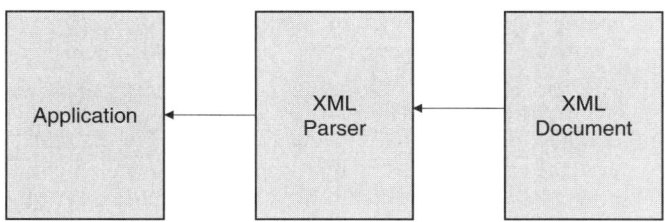

Figure 2.9 A typical XML application consists of three distinct layers.

option is to find an existing XML parser, and plug this into your application. XML parsers are freely available for dozens of programming languages, including C, C++, C#, Java, Perl, and Python.

A typical XML application consists of three distinct layers, see Figure 2.9. Working from right to left, the first layer is an XML document or a set of XML documents. These documents contain useful information, which you want to extract; for example, the documents may contain useful BSML data that you want to analyze further. The second layer is the XML parser. The parser consumes XML documents and makes the content available to the third layer, which is your software application. The XML parser takes care of all XML specific details and enables your application to more easily focus on content and programming logic.

XML parsers are broadly divided into two types:

- validating parser: this parser is capable of validating a document against an XML grammar, such as a DTD or an XML Schema.
- nonvalidating parser: this parser is not capable of validating a document against an XML schema.

As a general rule of thumb, nonvalidating parsers tend to be faster and take up less memory. However, validating parsers tend to be more useful, as you can use them to validate documents, and you don't need to include any validation code within your software application.

We will explore XML parsers in great detail in later sections of this book.

2.3   Fundamentals of XML Namespaces

XML Namespaces were not defined in the original W3C XML 1.0 specification. However, the namespace specification was finalized soon after, and namespaces are now considered a crucial element in the XML family of protocols. They are also a critical building block for other XML specifications, including XML Schemas, XSL Transformations, SOAP, and the Web Service Description Language (WSDL). In this section, we explore why XML Namespaces are important and then describe the mechanics of declaring and using namespaces.

2.3.1   Why We Need XML Namespaces

XML Namespaces are designed to address two very specific issues. First, namespaces prevent name conflicts. If your XML document references a single DTD or XML Schema, this is never an issue. However, if your document references two or more XML grammars, you have the potential for name conflicts. For example, two DTDs might define a `Sequence` element. XML Namespaces lets you attach a namespace to each `Sequence` element, and therefore uniquely identify each element.

Second, namespaces enable you to mark certain elements for processing by a specific software application. From the software module perspective, an XML document consists of actionable elements and nonactionable elements. By filtering for elements from a specific namespace, the software module can determine which elements are actionable and take the appropriate action.

In practice, name conflicts do not actually occur that often. For example, if your document references two grammars, the chance that they both define the same element is small. This is not to minimize name conflicts. It is just to point out that the second scenario of the software module perspective is more common. Hence, let's dig a little deeper into this scenario.

First, consider the following XML document:

```
<?xml version="1.0" encoding="UTF-8"?>
<stylesheet version="1.0">
    <template match="/">
        <html>
            <body>
            <h1>BSML Sequence Data:</h1>
            <value-of select="Bsml/Definitions/Sequences/Sequence"/>
            </body>
        </html>
    </template>
</stylesheet>
```

This is an example XSLT document. The document consists of two sets of elements. The first set consists of XSLT specific instructions. For example, `stylesheet`, `template`, and `value-of` are all XSLT instructions. The second set consists of HTML elements. For example, `html`, `body`, and `h1` are all HTML elements. An XSLT application will consume this document and apply the XSLT transformations. In this specific example, the style sheet is responsible for transforming BSML documents into HTML.

From the XSLT software module perspective, it needs an easy way to identify which elements are XSLT and which are not. In other words, it needs an easy way to determine which elements are actionable and which are nonactionable. As the document exists right now, the elements are not clearly partitioned.

Now, consider this XML document:

```
<?xml version="1.0" encoding="UTF-8"?>
<xsl:stylesheet version="1.0" xmlns:xsl="http://www.w3.org/1999/
  XSL/Transform">
    <xsl:template match="/">
        <html>
            <body>
            <h1>BSML Sequence Data:</h1>
                    <xsl:value-of select="Bsml/Definitions/Sequences/
                      Sequence"/>
            </body>
        </html>
    </xsl:template>
</xsl:stylesheet>
```

This document now contains an XML namespace declaration for XSLT. Furthermore, all XSLT elements now have an *xsl* prefix. For example, the `template` element is now defined as `xsl:template`. From the software module perspective, it is now a trivial task to determine which elements are XSLT instructions and which are not. It can therefore more easily carry out the XSL transformation.

The "Namespaces in XML" specification [14; 15] is currently available as an official W3C Recommendation. The complete specification is available online at: *http://www.w3.org/TR/REC-xml-names/*.

2.3.2   Declaring and Using XML Namespaces

Now that you understand the rationale for namespaces, let's look into the mechanics of declaring and using XML namespaces.

To use an XML namespace, you must first declare it. XML namespace declarations can occur within any XML element, but in practice, most developers place them at the top of their document usually within the root XML element. A namespace declaration is scoped to the element wherein the declaration occurs and all its subelements.

An XML declaration is a special XML attribute consisting of three parts. The first part is the reserved prefix *xmlns*. The second part is a namespace prefix of your choosing, and the third part is a Uniform Resource Identifier (URI). For example, the following element declares a namespace for XSLT:

```
<xsl:stylesheet version="1.0" xmlns:xsl="http://www.w3.org/1999/XSL/
   Transform">
```

The namespace prefix serves as a shortcut to the namespace declaration. You can use whatever namespace prefix you like. However, there are a few common conventions. For instance, the XSLT prefix is usually specified as *xsl* and the XML Schema prefix is usually specified as *xs* or *xsd*.

The URI value serves as a unique identifier and enables you or a software module to unambiguously partition elements into discrete namespaces. Values are most often represented as absolute URLs, e.g., *http://www.w3.org/1999/XSL/Transform*. If you are creating your own namespace, you should have control over the referenced host or URL. Otherwise, you may not be able to ensure absolute uniqueness.

It is important to note that the URI value does not necessarily point to anything meaningful. For example, if you copy and paste a namespace URI value into a web browser, you may or may not find a meaningful resource there. Therefore, the URI value serves as a unique identifier and nothing more.

Having declared a namespace, you later reference the namespace via a *Qualified Name*. A Qualified Name consists of two parts: a namespace prefix and a local element name. The two parts are delimited with a colon character. For example, the following start tag now includes a Qualified Name:

```
<xsl:template match="/">
```

In plain English, this start tag now references the `template` element in the *xsl* namespace. We already know that every start tag must have a matching tag. In this case, the end tag must also include a Qualified Name:

```
</xsl:template>
```

The complete XSLT example above should now make a lot more sense. It is repeated below:

```
<?xml version="1.0" encoding="UTF-8"?>
<xsl:stylesheet version="1.0" xmlns:xsl="http://www.w3.org/1999/XSL/
  Transform">
    <xsl:template match="/">
        <html>
          <body>
          <h1>BSML Sequence Data:</h1>
          <xsl:value-of select="Bsml/Definitions/Sequences/Sequence"/>
          </body>
        </html>
    </xsl:template>
</xsl:stylesheet>
```

We now know that the root element contains a namespace declaration for XSLT. We also know that all XSLT elements are specified with Qualified Names. All other elements, e.g., html , body, and h1, are not namespace qualified and therefore, do not exist within any namespace. Also note that the *select* attribute in the xsl:value-of element does not exist in any namespace either. To place an attribute within a namespace, you must explicitly specify it with a Qualified Name. For example:

```
<xsl:value-of xsl:select="Bsml/Definitions/Sequences/Sequence"/>
```

Now, both the element and the attribute share the same namespace.

2.3.3   Declaring a Default Namespace

The XML Namespaces specification supports default namespaces. A default namespace applies to the element where the declaration occurs and all its subelements. All unqualified elements within this scope are assumed to be part of the default namespace.

Default namespaces are specified with the special *xmlns* attribute—this is a special case of the namespace declaration defined above, except that there is no namespace prefix. For example, the following declares a default namespace for the XHTML specification:

```
<html xmlns="http://www.w3.org/1999/xhtml">
```

All unqualified subelements will therefore belong to the XHTML namespace. Here is a slightly longer example:

```
<?xml version="1.0" encoding="UTF-8"?>
<xsl:stylesheet version="1.0" xmlns:xsl="http://www.w3.org/1999/XSL/
  Transform" xmlns="http://www.w3.org/1999/xhtml">
    <xsl:template match="/">
        <html>
            <body>
            <h1>BSML Sequence Data:</h1>
            <xsl:value-of select="Bsml/Definitions/Sequences/
              Sequence"/>
            </body>
        </html>
    </xsl:template>
</xsl:stylesheet>
```

The root `stylesheet` element now contains two namespace declarations. The first is for XSLT; the second is a default namespace for XHTML. All elements beginning with the *xsl* prefix are explicitly defined to exist within the XSLT namespace. All unqualified elements, e.g., `html`, `body`, and `h1`, now exist within the default XHTML namespace.

Note that default namespaces do not apply to attributes. If you want an attribute to exist within a specific namespace, you must always specify it with a Qualified Name.

2.4   Fundamentals of BSML

As stated at the beginning of the chapter, BSML is an open standard for representing and exchanging bioinformatics sequence data. Since its inception, BSML has grown to accommodate a wide range of bioinformatics data. This now includes:

- raw sequence data, including the ability to reference sequence data stored in other external data files. Sequences can also be represented at several levels, including at the individual sequence record level, chromosome level, and whole genome level.
- sequence annotation, enabling you to attach positional and nonpositional sequence features, such as coding regions, promoter sequences, Single Nucleotide Polymorphisms (SNPs), etc.
- scientific literature references, including the ability to reference full journal citations, along with Pub Med identifiers.
- networks of biological entities, enabling you to encode metabolic and signaling pathways.
- multiple sets of tabular data, enabling you to encode gene expression or Microarray data.
- display widgets, used to store visual representations of sequences. This includes the ability to store image captions, draw sequence features, and reference external GIFs and JPEGs.
- resource information, enabling you to store information about individual investigators, research organizations, and copyright availability.

BSML was originally created by Visual Genomics and was first funded in 1997 by the National Human Genome Research Institute (NHGRI). The goal of the initial NHGRI grant was to develop a standard for representing sequence data in XML, and to release the standard to the public domain. Joseph H. Spitzner, Ph.D. was the primary author of BSML at Visual Genomics. Spitzner continued work on BSML while working at LabBook, Inc., and now works at Rescentris, Ltd. BSML is currently available as a Document Type Definition (DTD), but the data model is at least partially based on preexisting data formats, including the GenBank ASN.1 file format.

The main BSML web site at: *http://www.bsml.org* includes an FAQ, an introductory tutorial, and a complete reference guide.

As this book goes to press, BSML is currently available as version 3.1. Since its original release, a number of organizations have announced support for BSML, including Bristol-Meyers Squibb, IBM, Accelrys, Inc., and the European Bioinformatics Institute (EBI). A number of other organizations have also released BSML conversion programs. For example, the Cold Spring Harbor Laboratory has released a utility for converting GenBank ASN.1 sequence data to BSML. The EBI has also released a utility for converting European Molecular Biology Laboratory (EMBL) documents to BSML.

BSML is not the only XML format for representing sequence data. In fact, there are several alternatives to BSML, including the NCBI DTDs, the Architecture for Genomic Annotation, Visualization and Exchange (AGAVE), Genome Annotation Markup Elements (GAME), and Biopolymer Markup Language (BioML).

As stated in the introduction, the BSML specification is quite large and we do not have the space to explore the specification in full. Instead, we will focus on the core elements, and on the representation of sequences and sequence features.

2.4.1   BSML File Formats

The BSML specification recommends three file extensions for use with BSML. These are defined in Table 2.1.

2.4.2   BSML Document Structure

Every BSML document shares properties and a similar structure. The first property is that every BSML document must begin with an XML prolog and must include a reference to the BSML DTD. Every BSML document will therefore begin like this:

```
<?xml version="1.0"?>
<!DOCTYPE Bsml PUBLIC "-//Labbook,  Inc. BSML DTD//EN"
"http://www.rescentris.com/dtd/bsml3_1.dtd">
```

In the next chapter, we will discuss the exact mechanics of referencing DTDs. For now, note that we are referencing the BSML 3.1 DTD, available on the Rescentris.com web site.

Second, every BSML document must begin with a root `Bsml` element. Following the root element, BSML is divided into three main sections.

- **Definitions**: this section stores biological sequences and sequence annotations. The section can also include tables of associated data and network graphs.
- **Research**: this section stores information about experimental research, such as experimental conditions, program queries, or search parameters. For example, you can store query parameters for a specific BLAST search.
- **Display**: this section stores display widgets and references external image files. This section is primarily used by software applications that are capable of visually rendering sequence data. For example, the Rescentris Genomic Workspace™ application uses this section to store visual representations of sequences and their features.

Table 2.1 BSML file extensions

File Extension	Description
*.bsml	A regular BSML file
.bsmz	This is a gzipped archive of one or more BSML documents, along with related resources, such as external data files and images
.bso	This is a BSML overlay file. An overlay file consists of BSML fragments, which can be overlaid onto an existing BSML document. For example, two researchers can swap overlay files for a base sequence record, and more easily exchange and compare sequence annotations

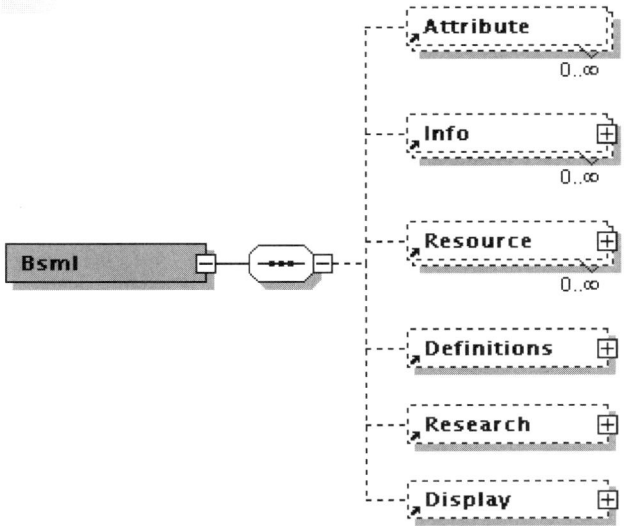

Figure 2.10 A bird's-eye view of the BSML DTD. The first level of elements is shown. (Document was created with XML Spy®.)

For a bird's-eye view of the BSML document structure, refer to Figure 2.10. As you can see, the root of the document structure is specified with a `Bsml` element. Under this, you have three common elements: `Attribute`, `Info`, and `Resource`. The `Attribute` element is used to store arbitrary name/value pairs which do not fit into any other elements. For example:

```
<Attribute name="definition" content="SARS coronavirus Urbani,
    complete genome."/>
```

Most BSML elements can contain 0 or more `Attribute` elements, providing a catch-all category for any data that doesn't fit the existing BSML data model. Furthermore, the BSML `Info` element can be used to store sets of `Attribute` elements.

The BSML `Resource` element is used to store metadata about the BSML document. In general, metadata elements are used to store "data about data." For example, a web page can contain metadata that describes the author, description, keywords, and the date last updated. The BSML metadata elements are based on the Dublin Core Metadata Initiative [19; 22; 24], one of the most popular metadata standards, used primarily to describe web pages and XML documents. Dublin Core is a minimal specification consisting of just 15 basic elements, such as Title, Description, Creator, Rights, and Date. By using `Resource` elements, you can therefore add important metadata to your BSML documents and record authorship, organizational affiliation, and copyright availability. Furthermore, since most BSML elements contain a `Resource` element, you can even add metadata to specific portions of the document. For example, you can record who made a specific sequence annotation and when they made it.

For more information about the Dublin Core Metadata Initiative, see: *http://www.dublincore.org*. The web site includes a number of relevant documents, including a guide to "Using Dublin Core" and a list of "Guidelines for implementing Dublin Core in XML."

Table 2.2 Text/binary formats for storing raw sequence data. For use with the BSML seq-data-import element

Format Name	Description
IUPACna	One-letter IUPAC codes for nucleic acids (see Appendix A for IUPAC codes)
IUPACaa	One-letter IUPAC codes for amino acids (see Appendix A for IUPAC codes)
NCBI2na	NCBI compact binary representation for nucleic acids. Each nucleic acid is represented with just 2 bits. Does not allow for ambiguity or gaps
NCBI4na	NCBI moderately compact binary representation for nucleic acids. Each nucleic acid is represented with 4 bits. Does include provisions for ambiguity and gaps

2.4.3 Representing Sequences

Sequence elements represent the heart of any BSML document. As we have already seen, these elements are used to store raw sequence data and can also be used to store annotations about the raw sequence. For example, we can add positional and nonpositional sequence features. Within this section, we consider the mechanics of representing raw sequences. In the next section, we move on to discuss sequence features.

The first important detail to note is that all sequence data must appear with the BSML Definitions section. This section contains a Sequences element, which can contain any number of Sequence-import or Sequence elements. Sequence-import elements are used to reference sequence data stored within other BSML files. For example, consider the following document fragment:

```
<Definitions>
   <Sequences>
        <Sequence-import source="sars_sequence1.bsml" id="AY278741"/>
   </Sequences>
</Definitions>
```

This document references sequence id=AY27841 in the sars_sequence1.bsml file.

In contrast to the Sequence-import element, the Sequence element is used to define sequence data within a BSML document. However, even in this case, the actual raw sequence data can be stored within the BSML document itself or within an external text or binary file. To represent raw sequence data, use the Seq-data element. To import data from an external text or binary file, use the Seq-data-import element. When importing data, you must specify a *source* attribute specifying the location of the file, and a *format* attribute specifying the text/binary format (see Table 2.2 for details). For example, consider the following document excerpt:

```
<Definitions>
   <Sequences>
        <Sequence id="AY278741" length="29727">
            <Seq-data-import format="IUPACaa" source="
             sars_sequence.txt"/>
        </Sequence>
   </Sequences>
</Definitions>
```

Table 2.3 Main attributes of the `Sequence` element

Attribute Name	Description
comment	Usually used to indicate a displayable description of the sequence record. See also the *title* attribute
db-source	Used to identify a public database, such as GenBank, EMBL, or the DNA Database of Japan (DDBJ). See also the *ic-acckey*
ic-acckey	An accession number used to uniquely identify a sequence record within the international consortium of nucleotide sequence databases. The consortium consists of GenBank, the EMBL Nucleotide Sequence Database, and the DNA Database of Japan (DDBJ). This attribute is usually used in conjunction with the *db-source* attribute.
length	Indicates the length of the sequence
local-acckey	An accession number used to uniquely identify a sequence record within a local or private database
molecule	Indicates the type of molecule represented. Options include: "dna," "rna," "aa" (amino acid), "na" (nucleic acid), "other-mol," and "mol-not-set." If you do not specify a molecule attribute, it defaults to "dna"
title	A displayable name for the sequence record. See also the *comment* attribute
topology	Specifies the topology of the sequence. Usually indicated with the values "linear" or "circular"

In this case, we are defining a new sequence for the same SARS virus as Listing 2.1, but specifying that the actual sequence data is stored in an external text file.

When using the `Seq-data` element, you must stick to IUPAC codes for nucleic acids and amino acids (see Appendix A). However, the data can include white space characters and numbers. For example, the following document excerpt is considered valid:

```
<Definitions>
    <Sequences>
            <Sequence id="AY278741" length="29727">
                    <Seq-data>
    1 atattaggtt tttacctacc caggaaaagc caaccaacct cgatctcttg tagatctgtt
   61 ctctaaacga actttaaaat ctgtgtagct gtcgctcggc tgcatgccta gtgcacctac
  121 gcagtataaa caataataaa ttttactgtc gttgacaaga aacgagtaac tcgtccctct
  181 tctgcagact gcttacggtt tcgtccgtgt tgcagtcgat catcagcata cctaggtttc
                [For brevity, sequence is truncated.]
                    </Seq-data>
            </Sequence>
    </Sequences>
    </Definitions>
</Bsml>
```

Each `Sequence` element can include a number of attributes. The main attributes are defined in Table 2.3. A more complete example of the SARS virus, along with more fully detailed attributes, is also provided in Listing 2.2.

2.4.4 Representing Sequence Features

In addition to raw sequence data, BSML can also represent sequence features. A sequence feature is any piece of annotation that provides additional details regarding a specific location or range of sequence data. When we get to Chapter 6, we will spend more time formally defining sequence annotation, and discuss in detail the Distributed Annotation System (DAS). For now, it is simplest to think of sequence annotation as any piece of data that provides additional details regarding a raw sequence record. For example, we can take a raw sequence record, and identify important parts, such as promoter regions, protein-coding regions, and 5′ and 3′ untranslated regions. We can also annotate sequence records with important references to scientific articles. Sequence features

Listing 2.2 The SARS virus, Take 2. This example is identical to Listing 2.1, except that we have now added additional attributes.

```
<?xml version="1.0"?>
<!DOCTYPE Bsml PUBLIC "-//Labbook, Inc. BSML DTD//EN"
"http://www.rescentris.com/dtd/bsml3_1.dtd">
<Bsml>
  <Definitions>
    <Sequences>
      <Sequence id="AY278741" title="AY278741" molecule="rna"
           length="29727" db-source="GenBank" ic-acckey="AY278741"
           topology="linear" strand="ss" representation="raw">
        <Attribute name="definition" content="SARS coronavirus
           Urbani, complete genome."/>
        <Attribute name="submission-date" content="21-APR-2003"/>
        <Attribute name="version" content="AY278741.1 GI:30027617"/>
        <Attribute name="source" content="SARS coronavirus Urbani"/>
        <Seq-data>
        atattaggtttttacctacccaggaaaagccaaccaacctcgatctcttgtagatctgtt
        ctctaaacgaactttaaaatctgtgtagctgtcgctcggctgcatgcctagtgcacctac
        gcagtataaacaataataaattttactgtcgttgacaagaaacgagtaactcgtccctct
        tctgcagactgcttacggtttcgtccgtgttgcagtcgatcatcagcatacctaggtttc
        gtccgggtgtgaccgaaaggtaagatggagagccttgttcttggtgtcaacgagaaaaca
        cacgtccaactcagtttgcctgtcc
        [For brevity, sequence is truncated.]
        </Seq-data>
      </Sequence>
    </Sequences>
  </Definitions>
</Bsml>
```

are an important element in other file formats as well. For example, the GenBank Flat File Format includes extensive support for sequence features and includes a recommended list of feature types.

In BSML, each sequence can contain any number of features. Features are formally nested within a `Feature-tables` element and individual features are defined within a `Feature` element. Two types of features are supported: positional and nonpositional. Positional features are tied to specific sequence locations and can be used to represent a host of sequence annotations, including protein-coding regions, locations of predicted genes, single nucleotide polymorphisms (SNPs), etc. Nonpositional features are not tied to any specific region of sequence, but are instead associated with the sequence record as a whole. For example, you can attach literature references that are associated with the entire sequence record.

Nonpositional features are slightly less complex than positional features. Let's take a look at an example, shown in Listing 2.3. This new example adds a single nonpositional feature detailing the direct submission to GenBank. More specifically, it lists the primary contributors of the work and their affiliation with the Centers for Disease Control and Prevention. As you can see, the `Reference` element contains a list of authors, a title, and the complete journal reference. For references to published material, you can include cross-reference identifiers to MEDLINE and PubMed.

Listing 2.3 The SARS virus, Take 3. The record now includes a single nonpositional feature, describing the direct submission to GenBank.

```
<?xml version="1.0"?>
<!DOCTYPE Bsml PUBLIC "-//Labbook, Inc . BSML DTD//EN"
"http://www.rescentris.com/dtd/bsml3_1.dtd">
<Bsml>
  <Definitions>
    <Sequences>
      <Sequence id="AY278741" title="AY278741" molecule="rna"
        length="29727" db-source="GenBank" ic-acckey="AY278741"
        topology="linear" strand="ss" representation="raw">
       <Attribute name="definition" content="SARS coronavirus
        Urbani, complete genome."/>
       <Attribute name="submission-date" content="21-APR-2003"/>
       <Attribute name="version" content="AY278741.1 GI:30027617"/>
       <Attribute name="source" content="SARS coronavirus Urbani"/>
       <Feature-tables id="AY278741.FTS1">
         <Feature-table id="AY278741.FTS1.FTB1" title="Genbank
           References" class="GB_REFERENCES">
          <Reference id="REF1" title="Direct Submission">
            <RefAuthors>Bellini,W.J., Campagnoli,R.P.,
               Icenogle,J.P., Monroe,S.S., Nix,W.A., Oberste,M.S.,
               Pallansch,M.A. and Rota,P.A.
            </RefAuthors>
            <RefTitle>Direct Submission</RefTitle>
            <RefJournal>Submitted (17-APR-2003) Division of Viral
            and Rickettsial Diseases, Centers for Disease Control
            and Prevention, 1600 Clifton RD, NE, Atlanta, GA
            30333, USA</RefJournal>
          </Reference>
         </Feature-table>
        </Feature-tables>
        <Seq-data>
        atattaggtttttacctacccaggaaaagccaaccaacctcgatctcttgtagatctgttctc
        taaacgaactttaaaatctgtgtagctgtcgctcggctgcatgcctagtgcacctacgcagta
        taaacaataataaattttactgtcgttgacaagaaacgagtaactcgtccctcttctgcagac
        tgcttacggtttcgtccgtgttgcagtcgatcatcagcatacctaggtttcgtccgggtgtga
        ccgaaaggtaagatggagagccttgttcttggtgtcaacgagaaaacacacgtccaactcagt
        ttgcctgtcc
        [For brevity, sequence is truncated.]
        </Seq-data>
      </Sequence>
    </Sequences>
  </Definitions>
</Bsml>
```

Positional features are just slightly more complicated. Each feature can contain any number of *Qualifier* and location elements. A *Qualifier* element describes a name/value attribute that describes the feature. A location element describes the location of the feature. Two types of locations can be specified: *Site-loc* and *Interval-loc*. A *Site-loc* identifies a single point within a raw sequence; an

Listing 2.4 SARS virus, Take 4. The record now includes a single positional feature.

```
<?xml version="1.0"?>
<!DOCTYPE Bsml PUBLIC "-//Labbook, Inc. BSML DTD//EN"
"http://www.rescentris.com/dtd/bsml3_1.dtd">
<Bsml>
  <Definitions>
    <Sequences>
      <Sequence id="AY278741" title="AY278741" molecule="rna"
          length="29727" db-source="GenBank" ic-acckey="AY278741"
          topology="linear" strand="ss" representation="raw">
        <Attribute name="definition" content="SARS coronavirus
         Urbani, complete genome."/>
        <Attribute name="submission-date" content="21-APR-2003"/>
        <Attribute name="version" content="AY278741.1 GI:30027617"/>
        <Attribute name="source" content="SARS coronavirus Urbani"/>
        <Feature-tables  id="AY278741.FTS1">
        <Feature-table id="AY278741.FTS1.FTB2" title="Genbank
            Features" class="GB_FEATURES">
        <Feature id="AY278741.FTS1.FTB2.FTR9" title="envelope
            protein" class="CDS" comment="envelope protein"
            display-auto="1">
        <Interval-loc startpos="26117" endpos="26347"/>
        <Qualifier value-type="note" value="envelope protein"/>
        <Qualifier value-type="codon_start" value="1"/>
        <Qualifier value-type="product" value="E protein"/>
        <Qualifier value-type="protein_id" value="AAP13443.1"/>
        <Qualifier value-type="db_xref" value="GI:30027622"/>
        <Qualifier value-type="translation" value="MYSFVSEETGTLIVNSVL
            LFLAFVVFLLVTLAILTALRLCAYCCNIVNVSLVKPTVYVYSRVKNLNSSEGV
            PDLLV"/>
          </Feature>
        </Feature-table>
      </Feature-tables>
    <Seq-data>
    atattaggttttaacctacccaggaaaagccaaccaacctcgatctcttgtagatctgttctcta
    aacgaactttaaaatctgtgtagctgtcgctcggctgcatgcctagtgcacctacgcagtataaa
    caataataaattttactgtcgttgacaagaaacgagtaactcgtccctcttctgcagactgctta
    cggtttcgtccgtgttgcagtcgatcatcagcatacctaggtttcgtccgggtgtgaccgaaagg
    taagatggagagccttgttcttggtgtcaacgagaaaacacacgtccaactcagtttgcctgtcc
    [For brevity, sequence is truncated.]
    </Seq-data>
    </Sequence>
    </Sequences>
  </Definitions>
</Bsml>
```

Interval-loc identifies a specific interval or range of raw sequence data. Again, a specific example should clarify the most important points. Take a look at Listing 2.4.

If you download the full SARS virus genome record from GenBank, you will see that it includes dozens of features. However, to keep the example more manageable, we have chosen to just include one positional feature in Listing 2.4. As you can see, this feature identifies a single coding sequence

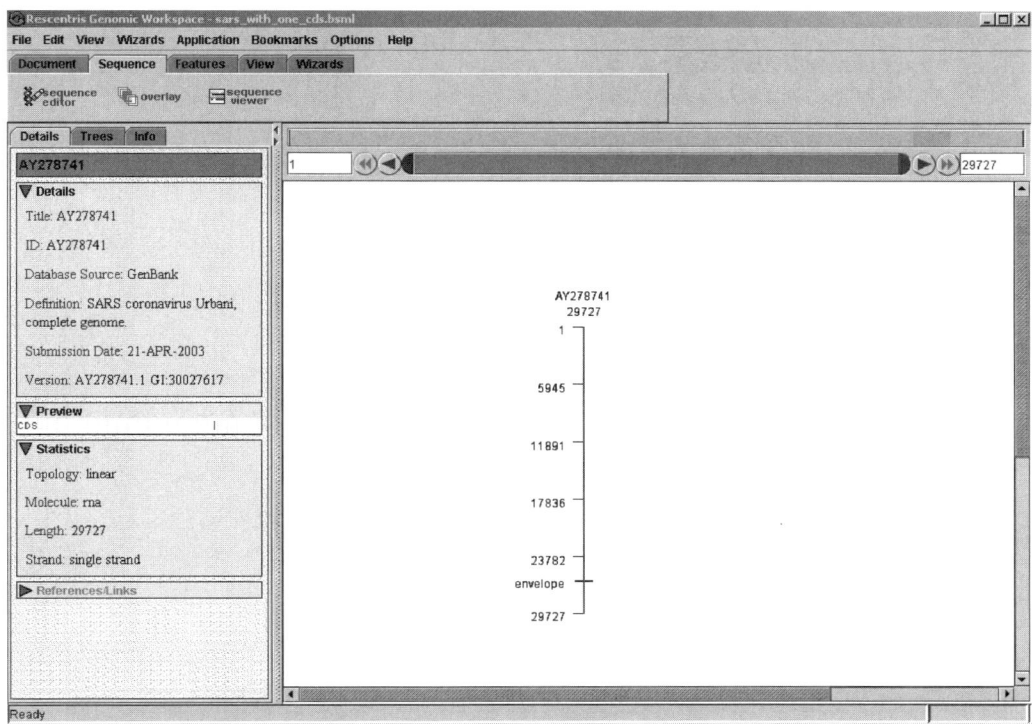

Figure 2.11 Sample screenshot of the Rescentris Genomic Workspace™ application. We have just loaded the SARS example from Listing 2.4. Note that our envelope protein is now included in the main sequence window (it is denoted with a single line between the markers 23,782 and 29,727).

region, identifying the SARS virus envelope protein. The coding region spans a specific interval of sequence data and we therefore use the `Interval-loc` element:

```
<Interval-loc startpos="26117"  endpos="26347"/>
```

As stated above, each feature can include any number of `Qualifier` elements. In this case, we use `Qualifier` elements to denote important attributes. For example, we identify the protein ID, a cross-reference to the protein GI number in GenBank, and the amino acid sequence of the translated region.

The Rescentris Genomic Workspace™ application will automatically draw all sequence features for you. For example, Figure 2.11 shows a screenshot of our revised SARS example. As you can see, our single feature is overlaid onto the main sequence widget in the center of the screen. Of course, this is one of the simplest possible feature examples. If you import a fully annotated sequence with multiple features, Genomic Workspace™ will draw all these features for you as well. You can then interactively select specific features and drill down to an increased level of detail. For example, Figure 2.12 shows a screenshot of one of the sample BSML files that comes bundled with the viewer. All features are displayed around the perimeter of the main circular sequence widget. If you select one of the features in the main window, detailed feature information is immediately displayed in the "Details" panel on the left.

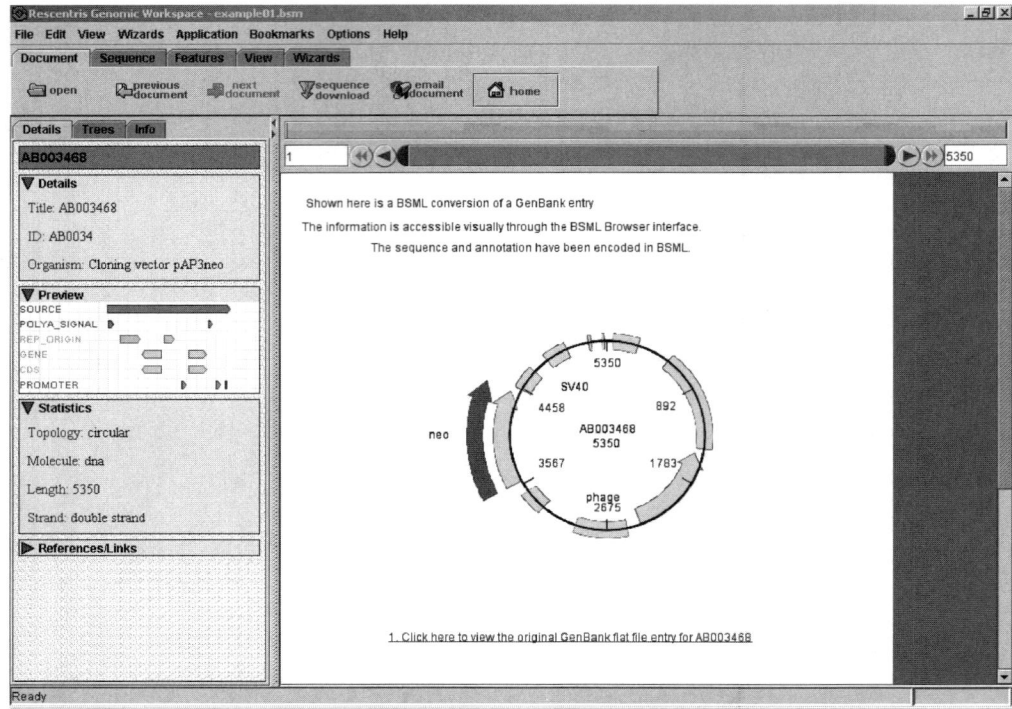

Figure 2.12 Sample screenshot of the Rescentris Genomic Workspace™ application. A fully annotated BSML sequence is shown.

You may have noticed that the `Feature` element contains a *display-auto* attribute. When set to "1," this provides a hint to the BSML rendering software that you want to automatically display the feature with a separate graphical widget. For example, Genomic Workspace™ uses this information to automatically render and visualize all BSML files.

In BSML 3.1, you can explicitly denote that some features span multiple regions. For example, you can specify all the exons for a protein-coding sequence. To do so, you must specify a *join* attribute, and set it to "1." Following this, the first `Interval-loc` specifies the complete range of the sequence, and each subsequent `Interval-loc` element specifies a specific subrange of data. For example, the following excerpt describes a protein-coding sequence with three exons:

```
<Feature-table>
    <Feature id="sample_protein" class="CDS" display-auto="1"
        join="1">
      <Interval-loc startpos="100" endpos="400"/>
      <Interval-loc startpos="120" endpos="150"/>
      <Interval-loc startpos="190" endpos="210"/>
      <Interval-loc startpos="300" endpos="400"/>
    </Feature>
</Feature-table>
```

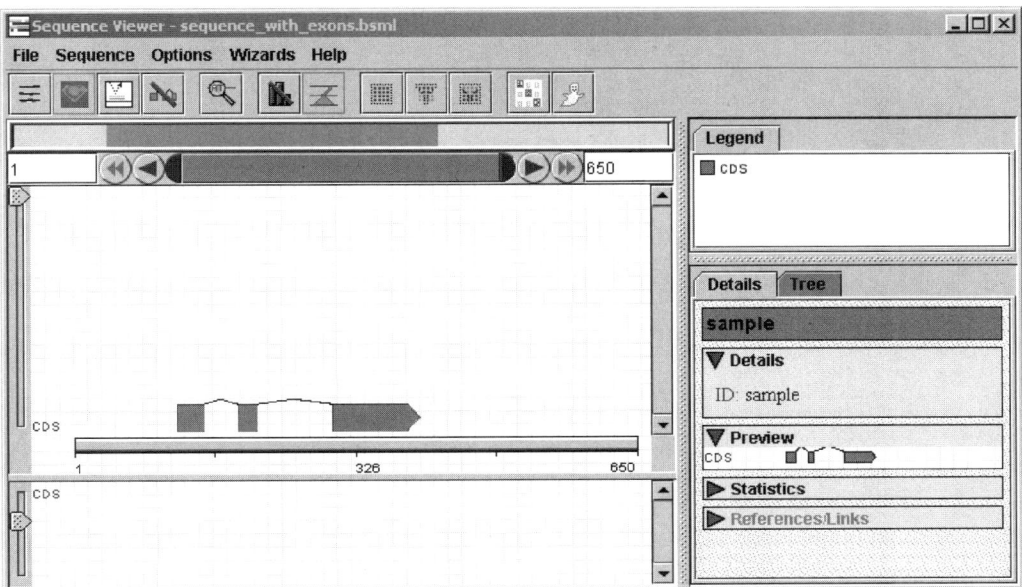

Figure 2.13 Sample screenshot of the Genomic Workspace™ Sequence Viewer. A sample protein-coding sequence with three exons is shown.

Within Genomic Workspace™ Sequence Viewer you can then choose to explicitly draw all individual exons. A sample screenshot is shown in Figure 2.13.

> Genomic Workspace™ includes a number of utilities for importing existing data and converting it to BSML. It also includes functionality for searching and importing data directly from public databases, search as GenBank, Ensembl, Swiss-Prot, and EMBL. To get started, select Wizards → Import and follow the onscreen instructions.

2.4.5   Retrieving Live BSML Data via XEMBL

Before ending our discussion of BSML, we will take a quick tour of the XEMBL service, from the European Bioinformatics Institute. XEMBL provides complete access to the EMBL Nucleotide Sequence Database. This database is produced in collaboration with GenBank and the DNA Database of Japan, and currently provides access to millions of nucleotide sequence records. It also provides access to completed genomes, including the human genome, the fruit fly, and *C. elegans*.

XEMBL [26] is a recently released interface that provides easy XML access to the complete EMBL database. Access is provided via two main methods. The first is a URL interface whereby users specify parameters within a URL and XEMBL returns a complete XML document. The second

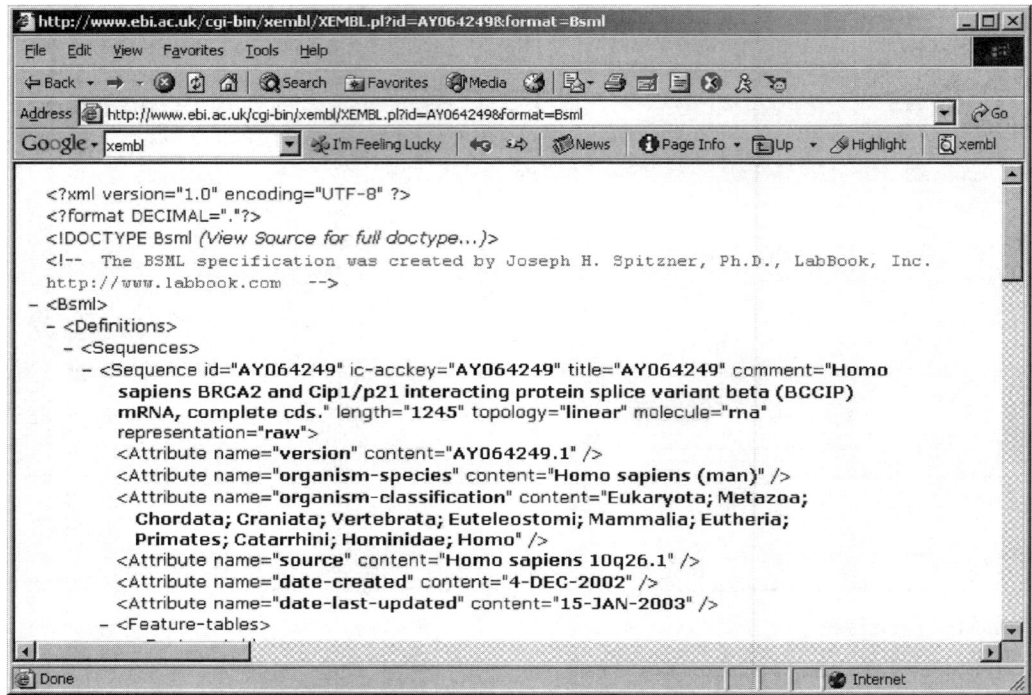

is a formal web services interface that uses the SOAP protocol and the Web Services Description Language (WSDL). (For details on SOAP, refer to Chapter 9.)

The XEMBL services expect two main parameters: an accession ID and a format. The ID specifies a unique international accession code; for example, SC49845 specifies the AXL2 gene in *Saccharomyces cerevisiae*. The format indicates the XML format of the returned document. Two format options are currently supported: BSML and AGAVE (Architecture for Genomic Annotation, Visualization and Exchange). To retrieve data in BSML format, you must specify format=Bsml; to retrieve data in AGAVE format, you must specify format=sciobj.

The XEMBL home page is available at: *http://www.ebi.ac.uk/xembl*.

You don't need any special tools or toolkits to access the XEMBL URL interface. All you need is a web browser. Simply start your browser, enter the main XEMBL URL and append the ID and format parameters. For example, the URL *http://www.ebi.ac.uk/cgi-bin/xembl/XEMBL.pl?id=AY064249&format=Bsml* retrieves the complete AY064249 record in BSML format. A sample screenshot of the XEMBL response is shown in Figure 2.14.

BSML is currently available as version 3.1. However, as this book goes to press, XEMBL is currently using BSML 2.2.

2.5   Useful Resources

Articles and Tutorials

- *Bioinformatic Sequence Markup Language—BSML 3.1 Tutorials.* LabBook, Inc.

  Provides a general introduction to the Bioinformatic Sequence Markup Language (BSML). Tutorial is available in MS Word and PDF formats at: *http://www.bsml.org/Resources/default.asp.*

- Cibulskis Kristian, "An Introduction to BSML." *XML J.* 4(3).

  Provides a concise introduction to BSML. Article is available online at: *http://www.sys-con.com/xml/archivesa.cfm?volume=04&Issue=03.*

- L. Wang, J. J. Riethoven, and A. Robinson, "XEMBL: distributing EMBL data in XML format." *Bioinformatics* 2002; 18(8): 1147–1148.

  Provides a short description of the XEMBL web service. Includes a discussion of the supported XML formats, and the CORBA back-end. Article can be downloaded from the *Bioinformatics* web site at: *http://bioinformatics.oupjournals.org.*

Web Sites and Web Services

- BSML web site: *http://www.bsml.org*

  Official home of the Bioinformatic Sequence Markup Language. The web site includes a short BSML Overview, an FAQ, a BSML Tutorial, and the official BSML Reference Manual.

- Rescentris, Ltd.: *http://www.rescentris.com*

  Official home of the Rescentris Genomic Workspace™ BSML Viewer.

- Unicode web site: *http://www.unicode.org*

  Official home of the Unicode specification. The site includes an introduction to Unicode, FAQ, and Glossary. Code charts in PDF format are available at: *http://www.unicode.org/charts.*

- XEMBL web service: *http://www.ebi.ac.uk/xembl*

  XEMBL provides XML access to the complete European Molecular Biology Laboratory (EMBL) Nucleotide Sequence Database. BSML and AGAVE formats are currently supported.

- The XML FAQ: *http://www.ucc.ie:8080/cocoon/xmlfaq*

  This is an excellent resource for those new to XML. It includes answers to dozens of questions, including: "What is XML?" "Where can I get an XML Browser?" "Does XML Replace HTML?"

XML Specifications

- XML 1.0 Specification: *http://www.w3.org/TR/REC-xml*

When in doubt, head to the official specification. The specification can be difficult to digest on the first reading. For excellent commentary and jargon-free side notes, check out Tim Bray's Annotated XML Specification at: *http://www.xml.com/axml/testaxml.htm.*

- Namespaces in XML: *http://www.w3.org/TR/REC-xml-names*

  The namespace specification is only 12 pages long and highly readable. The introductory section, "Motivation and Summary," is well worth the read.

DTDs for Bioinformatics **3**

Document Type Definitions (DTDs) describe XML document structures. This chapter provides a comprehensive overview of reading and creating DTDs, along with specific applications to bioinformatics. We begin with a bird's-eye view of a simple DTD for exchanging protein data, and explore several options for validating XML documents. We then dive into the specific details of DTD declarations, including the mechanics of declaring elements and attributes. We also explore the use of XML entities as a tool for creating reusable text and for building modular DTDs. Entities exist in several flavors, and this chapter describes when and how to use each one.

The chapter concludes with a case study of the NCBI TinySeq DTD. We begin with an overview of NCBI support for XML in general, and conclude with an in-depth discussion of TinySeq, a concise DTD used to represent biological sequence data.

3.1 Introduction to DTDs

Document Type Definitions (DTDs) contain specific rules, which constrain the content of XML documents. These rules are extremely specific—for example, a PROTEIN element must contain an ORGANISM element, and the ORGANISM element must include a *taxonomy_id* attribute. Within the world of XML, a set of these constraint rules is formally known as a *grammar*. Language grammars specify rules for constructing sentences. XML grammars specify rules for constructing documents. Grammars can be written in DTDs (explored in this chapter) or in XML Schemas (explored in the next chapter).

DTDs (and grammars in general) are important for several reasons. First, DTDs define specific constraints on XML documents and provide an easy method to verify that all the constraints are followed. A document which purports to follow a specific DTD is formally known as an XML *instance document*. An instance document which follows all the grammar rules is said to be *valid*. All the validity checking can be performed by a validating XML parser, freeing you from the tedious task of writing your own validation code.

The second reason DTDs are important is that they can easily be shared within communities or industries. Ideally, interested parties gather together, share opinions, and hammer out a common grammar. Ideally, they also create a stable process for soliciting further opinion and evolving the specification to support new functionality. With a common grammar, people, research labs, and software applications can more easily exchange and distribute data.

A number of DTDs have been developed for bioinformatics and computational biology. A list of the best-known DTDs for bioinformatics is presented in Table 3.1. Note that a number of the newest bioinformatics formats now use XML Schemas. A list of these is provided in the next chapter.

Table 3.1 DTDs for bioinformatics

Name	Web Address

AGAVE: Architecture for Genomic Annotation, Visualization and Exchange [27; 28]
    Note: AGAVE 2.3 is written as an XML DTD. The latest version, AGAVE 3.0, is written as an
    XML Schema
	http://www.animorphics.net/lifesci.html

BioML: BIOpolymer Markup Language [32]	*http://www.bioml.com/BIOML*
BSML: Bioinformatic Sequence Markup Language [12; 25]	*http://www.bsml.org*
CellML [31]	*http://www.cellml.org*
DAS: Distributed Annotation System [6]	*http://www.biodas.org*
Gene Ontology (GO) DTD [33]	*http://www.geneontology.org/xml-dtd/go.dtd*
MAGE-ML: MicroArray Gene Expression Markup Language [39]	*http://www.mged.org/mage*
NCBI DTDs: Numerous DTDs, including GBSeq, TinySeq, SeqSet, and NCBI Blast [36]	*http://www.ncbi.nlm.nih.gov/dtd*

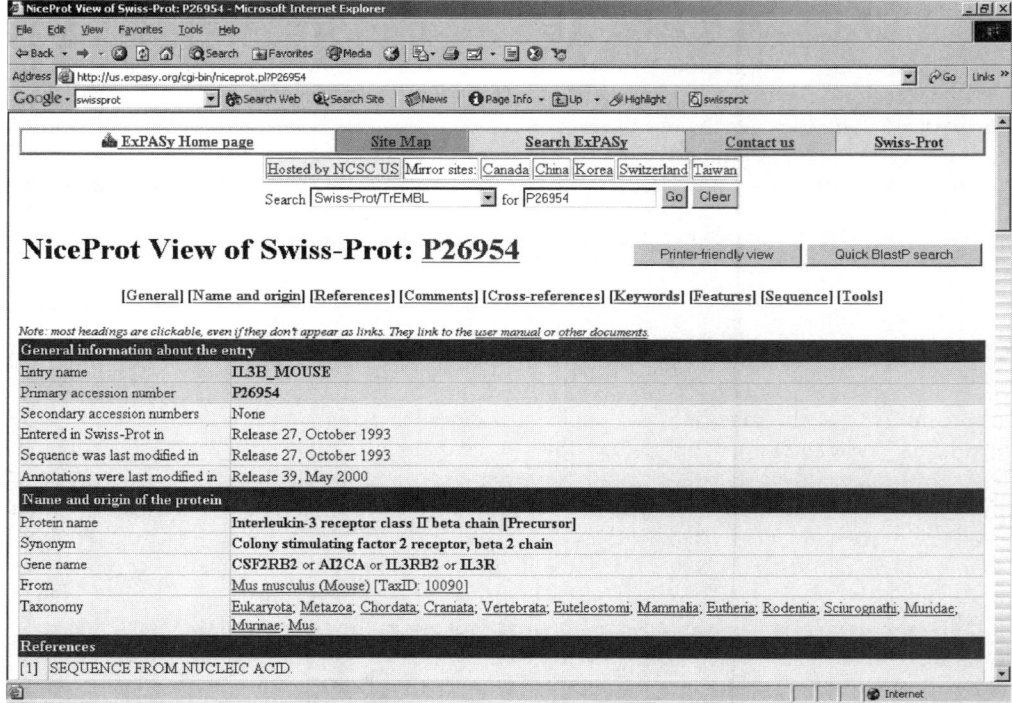

Figure 3.1 A sample protein record from the Swiss-Prot database.

3.1.1   A Bird's-Eye View: Protein DTD

The best way to learn DTDs is by example. We therefore begin with a simple DTD for representing protein data. Don't worry if all the details don't make sense just yet—the premise of the first example is to present a bird's-eye view of a sample DTD. All of the details are explored in the sections that follow.

Our task is to build a very simple DTD for exchanging protein data. For example, the screenshot in Figure 3.1 shows a sample protein record from Swiss-Prot. Swiss-Prot is a database of curated

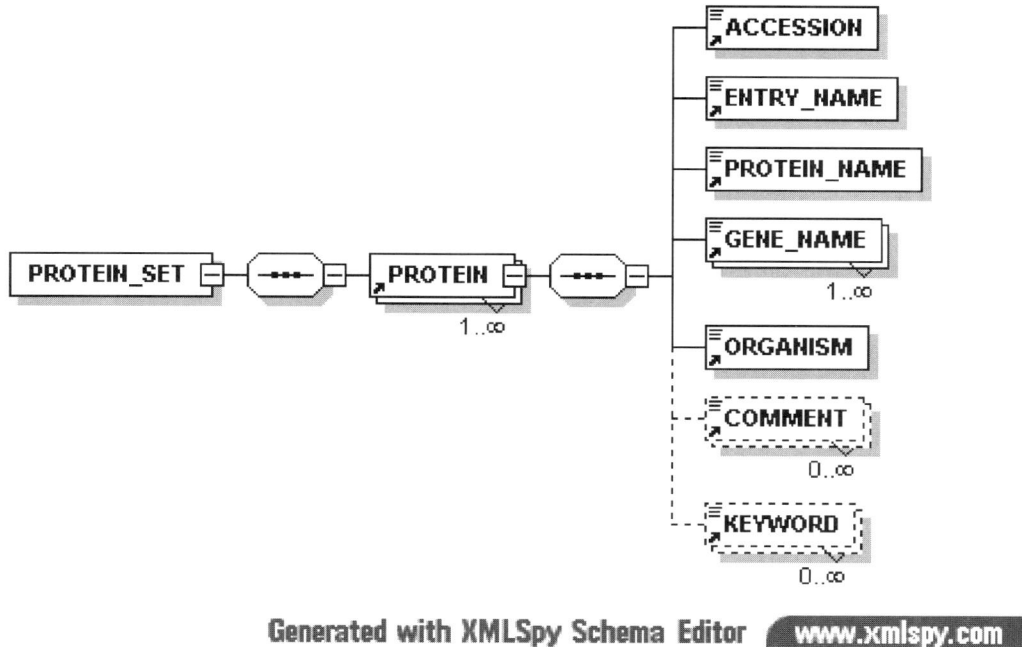

Figure 3.2 Schematic of the proposed protein DTD. Required elements are outlined in solid rectangular lines. Optional elements are outlined in dashed rectangular lines. (Diagram was created with XMLSpy®).

protein sequences, maintained by the Swiss Institute of Bioinformatics (SIB) and the European Bioinformatics Institute (EBI).

To simplify things, we will take a subset of the Swiss-Prot fields, and draw a schematic of our envisioned DTD (see Figure 3.2). As you create new DTDs, it helps to have a basic visual schematic of your DTD. It also helps to have a textual description, written in plain English. For example, our schematic diagram has a few rules: the root of the document is specified with a PROTEIN_SET element. The root element contains one or more PROTEIN elements. The PROTEIN element in turn contains a number of subelements, such as accession number, protein name, organism, and comment.

The protein DTD described here is purely for instructive purposes. However, Swiss-Prot does have it own grammar, called UniProt XML (formerly called SPTr-XML). UniProt XML is written in XML Schema and is capable of representing the full set of Swiss-Prot/TrEMBL data. Information is available at: *http://www.uniprot.org*.

The complete DTD is presented in Listing 3.1. Again, don't be too concerned with the details. Here's a bird's-eye view of what to look for now:

- DTDs are written in their own special format, and do not actually use the familiar XML format of start and end tags.
- DTDs can contain comments. Make sure to include concise, readable documentation within any DTD.

Listing 3.1 Sample DTD for representing protein data: protein.dtd

```
<!-- Sample DTD for representing protein data-->
<!-- A PROTEIN_SET can have one or more PROTEIN elements -->
<!ELEMENT PROTEIN_SET (PROTEIN+)>
<!-- Main PROTEIN Element -->
<!ELEMENT PROTEIN (ACCESSION, ENTRY_NAME, PROTEIN_NAME, GENE_NAME+,
ORGANISM, COMMENT*, KEYWORD*)>
<!-- Sub Elements containing PCDATA -->
<!ELEMENT ACCESSION (#PCDATA)>
<!ELEMENT ENTRY_NAME (#PCDATA)>
<!ELEMENT PROTEIN_NAME (#PCDATA)>
<!ELEMENT GENE_NAME (#PCDATA)>
<!ELEMENT COMMENT (#PCDATA)>
<!ELEMENT KEYWORD (#PCDATA)>
<!-- ORGANISM for referencing NCBI Taxonomy ID -->
<!ELEMENT ORGANISM (#PCDATA)>
<!ATTLIST ORGANISM
    taxonomyid NMTOKEN #REQUIRED
>
```

- Elements are declared with the prefix <!ELEMENT. For example, the third line declares the PROTEIN_SET element. Each element is declared with a specific content model—this defines a set of valid content that can exist within the specified element. For example, the PROTEIN_SET element can only contain PROTEIN elements.
- Attributes are declared with the prefix <!ATTLIST. For example, the final three lines specify that all ORGANISM elements must include a *taxonomy_id* attribute.
- A number of elements are defined to contain #PCDATA. The full details of this are explored below. However, the basic idea is that these elements can only contain textual data, but may not contain any other subelements.

A sample instance document that adheres to the protein DTD is presented in Listing 3.2. Note that the XML prolog now includes two lines. The first line is the XML declaration and is used to specify the XML version and the character encoding. The second line is known as the *Document Type Declaration*, and is used to specify the location of the document's DTD. In Listing 3.2, the instance document points to the protein DTD.

3.1.2   Validating XML Documents

A document that adheres to all the rules of its specified DTD is valid. To validate a document, you need to run it through a validating XML parser. Dozens of validating XML parsers exist, but you are probably best off if you pick an XML editor with a built-in validation feature. For example, Figure 3.3 shows a screenshot of the <oXygen/> XML editor. To validate a document, select menu XML → Validate XML. Figure 3.3 shows the validation results of a sample invalid protein document. As shown in the screenshot, the <oXygen/> validator has discovered a single validation error: "Attribute 'taxonomy_id' is required and must be specified for element type 'ORGANISM'."

If you haven't already chosen a good XML editor, take some time now to select one. A full and growing list of XML editors (both open source and commercial) is provided

Listing 3.2 Sample instance document adhering to the protein DTD.

```xml
<?xml version="1.0" encoding="UTF-8"?>
<!DOCTYPE PROTEIN_SET SYSTEM "protein.dtd">
<PROTEIN_SET>
    <PROTEIN>
        <ACCESSION>P26954</ACCESSION>
        <ENTRY_NAME>IL3B_MOUSE</ENTRY_NAME>
        <PROTEIN_NAME>Interleukin-3 receptor class II beta chain
          [Precursor]
        </PROTEIN_NAME>
        <GENE_NAME>CSF2RB2</GENE_NAME>
        <GENE_NAME>AI2CA</GENE_NAME>
        <GENE_NAME>IL3RB2</GENE_NAME>
        <GENE_NAME>IL3R</GENE_NAME>
        <ORGANISM taxonomy_id="10090">Mus musculus</ORGANISM>
        <COMMENT>FUNCTION: IN MOUSE THERE ARE TWO CLASSES OF
          HIGH-AFFINITY IL-3 RECEPTORS. ONE CONTAINS THIS
          IL-3-SPECIFIC BETA CHAIN AND THE OTHER CONTAINS THE BETA
          CHAIN ALSO SHARED BY HIGH-AFFINITY IL-5 AND GM-CSF
          RECEPTORS.</COMMENT>
        <COMMENT>SUBUNIT: Heterodimer of an alpha and a beta
          chain.</COMMENT>
        <COMMENT>SUBCELLULAR LOCATION: Type I membrane
          protein.</COMMENT>
        <COMMENT>SIMILARITY: BELONGS TO THE CYTOKINE FAMILY OF
          RECEPTORS.
        </COMMENT>
        <KEYWORD>Receptor</KEYWORD>
        <KEYWORD>Glycoprotein</KEYWORD>
        <KEYWORD>Signal</KEYWORD>
    </PROTEIN>
</PROTEIN_SET>
```

at: *http://www.xmlsoftware.com/editors.html.* Whichever editor you choose, make sure that it has a simple, easy to use XML validation feature.

As you evaluate XML editors, note that some editors will automatically scan a DTD file, and use this information to provide automatic code suggestions. For example, XMLSpy® will read DTD and XML Schema files, and provide automatic code suggestions and completion. A sample screenshot of XMLSpy's editing capabilities is shown in Figure 3.4. In this example, the user is creating a new sample protein document. After typing "<P", XMLSpy® automatically suggests the PROTEIN element. After completing the PROTEIN start tag, XMLSpy® automatically creates a skeleton of all the required subelements. A screenshot of the auto-generated code is provided in Figure 3.5.

Information about the <oXygen/> XML editor is available at: *http://www.oxygenxml.com.* Information about XMLSpy® is available at: *http://www.xmlspy.com.*

If you don't want to bother with installing a full-blown XML editor, a number of command line and web service based validators are available. For example, ElCel Technology offers a free command line XML validator at: *http://www.elcel.com/products/xmlvalid.html.* To use it, simply specify the

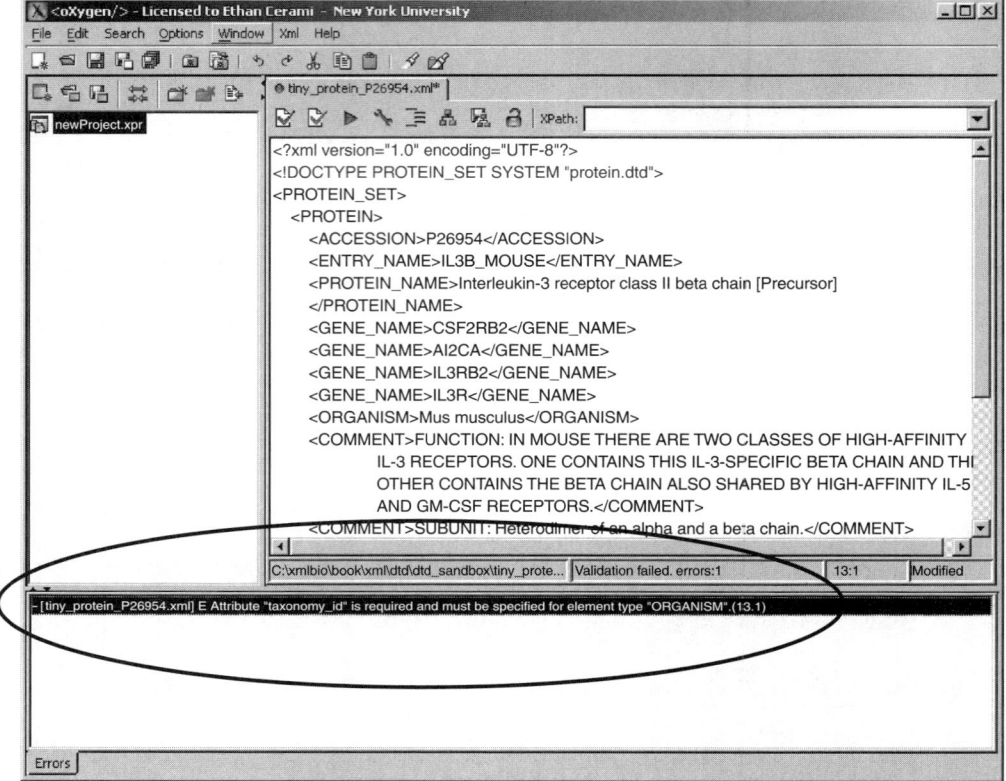

Figure 3.3 Sample screenshot of the <oXygen/> XML editor. Results of XML validation for a sample protein document are shown.

xmlvalid executable, followed by the file name. For example:

```
xmlvalid protein_P26954_2.xml
protein_P26954_2.xml [13:19] : Error: required attribute 'taxonomy_id'
has not been supplied for element 'ORGANISM'
```

Brown University also hosts an easy-to-use XML validation service at: *http://www.stg. brown.edu/service/xmlvalid.* You can specify the URL of an XML document, upload a local XML file, or simply cut and paste the document into the HTML form. Upon submission, the Brown service will automatically parse the XML file and report any validation errors.

If you are creating a new DTD from scratch, a good first step is to create one or more sample instance documents. You can then send these sample documents through an auto-DTD generator. For example, XMLSpy® can examine several sample documents, and then produce a DTD or an XML Schema. You are not likely to get perfect results, but the generated DTD can serve as a good first draft and will save you a lot of time. For a web-based DTD generator, check out *http://www.pault.com/pault/dtdgenerator.* Simply specify a local XML file and hit the "Generate DTD!" button.

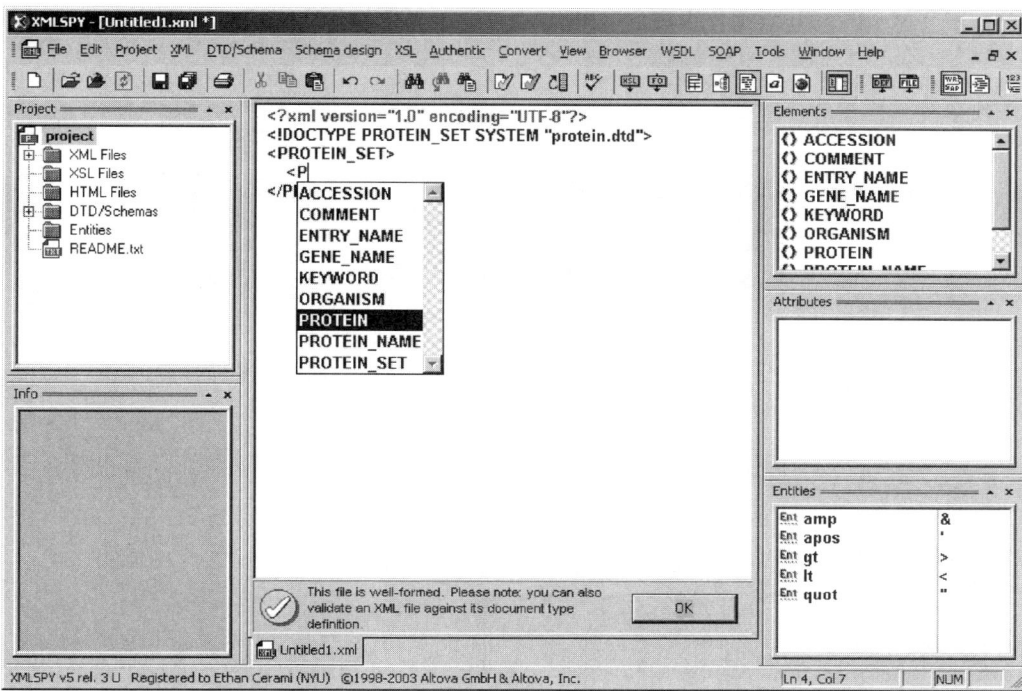

Figure 3.4 Sample screenshot of XMLSpy. XMLSpy will automatically read the protein DTD and use this to provide coding suggestions to the user.

3.2   Document Type Declarations

XML instance documents reference DTDs via a Document Type Declaration. The declaration is part of the XML prolog and must be specified before the root XML element. Instance documents can include internal DTDs, reference external DTDs, or both. For example, the following document includes an internal DTD:

```
<?xml version="1.0" encoding="UTF-8"?>
<!DOCTYPE dna [
     <!ELEMENT dna (#PCDATA)>
]>
<dna>catctcgcacttccaactgc</dna>
```

The <!DOCTYPE prefix indicates the Document Type Declaration, and "dna" specifies the name of the root element. All the actual DTD rules are specified between the opening and closing square brackets. Following the Document Type Declaration, the document continues with instance data—in this case, beginning with the root dna element. By definition, internal DTDs are tied to specific instance documents and cannot be shared among multiple documents. Internal DTDs are therefore most useful during the initial stages of DTD development, where you may want to keep DTD rules together with a sample instance document.

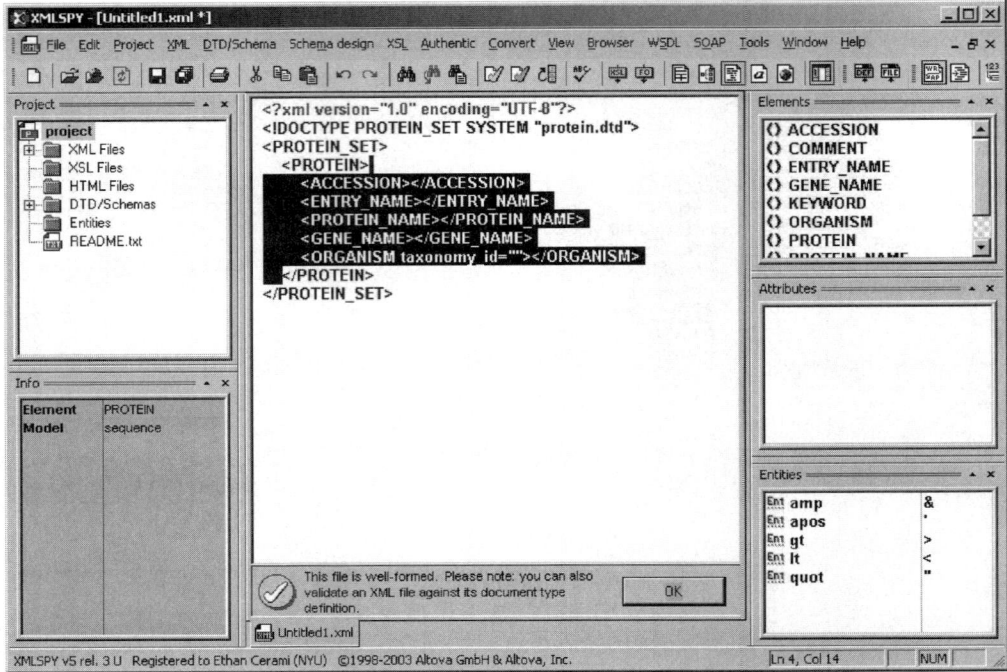

Figure 3.5 Sample screenshot of XMLSpy® (continued). After completing the <PROTEIN> start tag, XMLSpy® automatically creates a skeleton of all the required subelements.

After the initial stage of DTD development, you nearly always want to separate your DTD into a separate file where it can be used by multiple documents. For example, you can separate the DTD above into a separate file, named dna.dtd:

```
<!-- External DTD -->
<!ELEMENT dna (#PCDATA)>
```

You can then create an external reference within your instance document:

```
<?xml version="1.0" encoding="UTF-8"?>
<!DOCTYPE dna SYSTEM "dna.dtd">
<dna>catctcgcacttccaactgc</dna>
```

Note that the Document Type Declaration now includes a SYSTEM keyword, followed by a Uniform Resource Identifier (URI). In this case, we specify a relative file location, but you could just as easily specify an absolute URL to a specific web location.

The SYSTEM keyword generally designates DTDs that are used locally within a specific application or organization. To reference a DTD that is publicly available, use the PUBLIC keyword. For example, the following document references the NCBI TinySeq DTD (discussed later in the chapter):

```
<?xml version="1.0"?>
<!DOCTYPE TSeq PUBLIC "-//NCBI//NCBI TSeq/EN"
"http://www.ncbi.nlm.nih.gov/dtd/NCBI_TSeq.dtd">
```

```
<TSeq>
  <TSeq_seqtype value="nucleotide"/>
  <!-- Content Continues... -->
</TSeq>
```

When referencing a public DTD, you must specify the PUBLIC keyword, followed by a public identifier, followed by a URI. Although not required by the XML specification, most public identifiers are specified as Formal Public Identifiers (FPIs), as defined by the International Organization for Standardization (ISO). FPIs have a peculiar syntax, where the / and // characters serve as token delimiters. The first token is specified with a + or −. + indicates that the organization is formally registered with ISO (as the example above shows, NCBI is not registered with ISO). The second token indicates the owner of the DTD. The third token indicates the name of the DTD, and the fourth token indicates the natural language of the DTD—in this case, English.

In theory, an XML parser could extract the public identifier and look up the DTD in a DTD catalog. However, the XML specification requires that public identifiers must also include a URI. In practice, therefore, most XML parsers simply ignore the public identifier and retrieve the DTD from the URI.

3.3 Declaring Elements

When reading a DTD, or creating a new DTD from scratch, the first step is to understand element declarations. Element declarations provide the backbone to a document's content model, giving it structure and hierarchy. This backbone defines a valid list of XML elements, and a specific set of content rules for each of those elements. For example, some elements can have text, while others can only contain specific child elements. Furthermore, the DTD can specify the specific order of child elements and the occurrence with which they may appear. For example, some elements are required and must be specified once and only once. Other elements are optional and may appear zero or more times.

The general rules for declaring elements are actually quite straightforward, and this section includes details on all the various options. We begin with an overview of element content. Every time you declare a new element, it can be defined with one of five options:

- EMPTY: the element cannot contain any text or any child elements.
- ANY: the element can contain any text or any child element(s).
- #PCDATA: the element can contain regular text.
- Child Elements: the element can contain a specific set of child elements.
- MIXED: the element can contain text data interspersed with child elements.

Information on each option is provided in the sections below.

3.3.1 EMPTY

The EMPTY keyword is used to indicate that the declared element cannot contain any text or any child elements. It is, by definition, the simplest possible option for element declaration. EMPTY element declarations are most frequently used to define elements which do not have content, but may have attributes. For example, a DB_REFERENCE element could be used to reference external biological databases. It may have no content, but it might have two required attributes. Its element declaration would therefore be defined as:

```
<!ELEMENT DB_REFERENCE EMPTY>
```

Information on attribute declaration is provided later in the chapter.

3.3.2 ANY

The ANY keyword is used to indicate that the declared element can contain any text or any defined child element. It is the least restrictive option for element declaration and should therefore be strenuously avoided. The use of ANY can sometimes be useful during the early stages of DTD development, where it can be used to indicate a placeholder for an element which needs further refinement[*]. However, ANY is usually much too open-ended and you are not likely to find its use in many public DTD standards.

Below is a simple example of the ANY keyword, used to create a subset of XHTML:

```
<!ELEMENT BODY ANY>
<!ELEMENT H1 (#PCDATA)>
<!ELEMENT H2 (#PCDATA)>
<!ELEMENT B (#PCDATA)>
<!ELEMENT I (#PCDATA)>
```

The BODY tag uses the ANY keyword and can therefore contain any of the defined elements, including H1, H2, B, and I. Note, however, that ANY restricts content to *elements defined within the DTD*. The following is therefore invalid, because it references an undeclared H3 element:

```
<BODY>
    <H1>XML for Bioinformatics</H1>
    <H3>Springer Verlag</H3>
</BODY>
```

3.3.3 #PCDATA

The #PCDATA keyword is used to indicate that the declared element can contain text. Formally, PCDATA stands for *Parsed Character Data*, indicating that the text data will be parsed by the XML processor. For example, the XML parser will analyze PCDATA text and replace all entities with their defined text substitution strings (information on entities is provided later in the chapter).

Most DTDs have lots of #PCDATA elements and we will see many examples throughout the chapter. For example, the declaration below defines a SEQUENCE element:

```
<!ELEMENT SEQUENCE (#PCDATA)>
```

---

[*] Tim Bray, one of the original editors of the XML specification, actually voted against including the ANY keyword in the XML 1.0 specification. Bray originally argued that ANY was too open-ended and he "couldn't see any excuse" for including it in the specification [16]. Bray now concedes that ANY is a useful tool for building placeholder elements. Complete details are available in the Annotated XML Reference [16], available online at: *http://www.xml.com*.

Below is a sample document fragment:

```
<SEQUENCE>
MINIRKTHPLMKILNDAFIDLPTPSNISSWWNFGSLLGLCLIMQILTGLFLAMHYTPDTS
TAFSSVAHICRDVNYGWFIRYLHANGASMFFICLYAHIGRGLYYGSYMFQETWNIGVLLL
LTVMATAFVGYVLPWGQMSFWGATVITNLLSAIPYIGTTLVEWIWGGFSVDKATLTRFFA
</SEQUENCE>
```

One caveat to using #PCDATA: the declaration states that the element *may* contain text data, but a validating parser will not actually enforce that text is actually provided. For example, consider the following fragment:

```
<SEQUENCE></SEQUENCE>
```

This element contains no textual data, but is considered valid. The XML parser will return the text content as an empty string and your application will need the correct logic to act appropriately.

3.3.4   Child Elements

Elements can be declared to contain other elements, thereby creating hierarchical content models. For example, consider the following declaration:

```
<!ELEMENT PROTEIN (ACCESSION, NAME, DESCRIPTION)>
```

This defines a protein element, which must contain three subelements: an accession number, name, and description. Each subelement is separated by a comma, and instance documents must follow the exact same order of elements. For example, the following instance document is valid:

```
<PROTEIN>
   <ACCESSION>Q9TDL5</ACCESSION>
   <NAME>CYB_CEPHA</NAME>
   <DESCRIPTION>Cytochrome b</DESCRIPTION>
</PROTEIN>
```

However, this document is invalid:

```
<PROTEIN>
   <ACCESSION>Q9TDL5</ACCESSION>
   <DESCRIPTION>Cytochrome b</DESCRIPTION>
   <NAME>CYB_CEPHA</NAME>
</PROTEIN>
```

The exact validity message will vary by parser. For example, in XMLSpy®, the error message states: "Mandatory element 'NAME' expected in place of 'DESCRIPTION'."

You can also specify either/or options via the vertical bar operator. For example, consider the following declaration:

```
<!ELEMENT PROTEIN ( (ACCESSION | NAME), DESCRIPTION)>
```

This protein element can have either an accession number or a name (but not both), followed by a description.

Occurrence Operators

When declaring child elements, each element can be appended with an *occurrence operator*. This operator determines the number of times the element may appear and is based on regular expression syntax. The operators are defined as follows:

Operator	Description
No Operator	Indicates that exactly one instance of the element is required
?	Indicates that zero or one instance of the element may appear
*	Indicates that zero or more instances of the element may appear
+	Indicates that one or more instances of the element may appear

To make these concepts more concrete, let's return to the protein DTD in Listing 3.1. For now, we focus on the first six lines:

```
<!-- Sample DTD for representing protein data-->
<!-- A PROTEIN_SET can have one or more PROTEIN elements -->
<!ELEMENT PROTEIN_SET (PROTEIN+)>
<!-- Main PROTEIN Element -->
<!ELEMENT PROTEIN (ACCESSION, ENTRY_NAME, PROTEIN_NAME, GENE_NAME+,
ORGANISM, COMMENT*, KEYWORD*)>
```

The PROTEIN_SET element uses the + occurrence operator, indicating that it can have one or more PROTEIN elements. In turn, the PROTEIN element specifies its own content model. In plain English, the element declaration specifies that a PROTEIN must have exactly one of each, ACCESSION , ENTRY_NAME , and PROTEIN_NAME . Following this, the PROTEIN can have *one or more* GENE_NAME elements and a required ORGANISM element. Finally, a PROTEIN can have *zero or more* COMMENT elements, followed by *zero or more* KEYWORD elements.

3.3.5   Mixed Content

The final option for element declaration is *mixed content*. This indicates that an element can contain text data interspersed with specific child elements. Mixed content declarations require a special syntax, defined as follows:

```
<!ELEMENT ELEMENT_NAME (#PCDATA | CHILD1 | CHILD2, etc.)* >
```

When using mixed content, you are not permitted to determine the sequence of child elements or to specify any occurrence operators.

As a simple example, consider the following DTD that defines a subset of XHTML:

```
<!ELEMENT BODY (#PCDATA | H1 | H2 | B | I)* >
<!ELEMENT H1 (#PCDATA)>
<!ELEMENT H2 (#PCDATA)>
<!ELEMENT B (#PCDATA)>
<!ELEMENT I (#PCDATA)>
```

The BODY element is defined to have mixed content and the following instance document is therefore valid:

```
<?xml version="1.0" encoding="UTF-8"?>
<!DOCTYPE BODY SYSTEM "xhtml_mixed_content.dtd">
```

```
<BODY>
This text is <B>Bold</B>.
This text is <I>Italics</I>
</BODY>
```

There is in fact no way to specify a root element within a DTD. For example, a document that adheres to the protein DTD in Listing 3.1 could specify a root PROTEIN_SET element, a root PROTEIN element, or even a root COMMENT element. Any of these options would be considered valid. The choice of the root element is entirely determined by the instance document within the Document Type Declaration. For example, the following declaration specifies PROTEIN_SET as the root element:

```
<!DOCTYPE PROTEIN_SET SYSTEM "protein.dtd">
```

In contrast, the following declaration specifies PROTEIN as the root element:

```
<!DOCTYPE PROTEIN SYSTEM "protein.dtd">
```

3.4   Declaring Attributes

Once you have defined your elements, the next step is to define attributes for those elements. The syntax for declaring attributes is specified in Figure 3.6. The basic syntax requires that you first specify the element name, followed by the attribute name. Following this, you specify an *attribute type*. Attribute types are primarily used to restrict the range of value for attributes. For example, you can restrict your attribute values to a specific set by using an enumerated list, or you can specify that the attribute must specify a unique identification value. In total, the XML specification defines ten different attribute types and each of these is defined in the section below. Following the attribute type, you specify an *attribute behavior*. Generally, this enables you to specify if the attribute is required or optional, and if it has a default value.

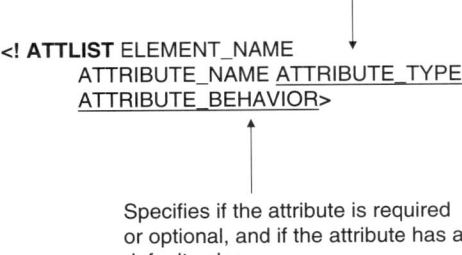

Figure 3.6 DTD syntax for defining XML attributes.

Tim Bray, one of the original technical editors for the XML 1.0 specification, provides an excellent "Annotated XML" [3] reference on the web. It is available online at: *http://www.xml.com.* Throughout the specification, you can click on annotation icons and immediately view historical commentary, technical explanations, and examples. If you happen to view the annotations for attribute types, Tim Bray has written a witty note on "Attribute Declarations and Proust." Bray has taught in numerous one-day seminars on XML and most of these seminars explore attribute types right after lunch:

> "The consequence of this is that when the class comes back after lunch, to listen to the discussion of attribute types and their declarations, most of them go to sleep. The fact of the matter is that there are a lot of attribute types (I voted against a few of them), there are lots of relevant details, and it is pretty tedious" [3].

If you find all this tedious too, or find yourself falling asleep in the next section, focus on the first three subsections: CDATA, Enumeration Lists, and IDs. You are most likely to use these attributes and will only rarely use the others.

3.4.1   Attribute Types

XML 1.0 defines a total of 10 different attribute types. Each of these types is detailed below.

CDATA

This is the most general and commonly used attribute type. It indicates that the attribute value may be set to any arbitrary text string. For example, the following rule defines a CDATA *species* attribute for a PROTEIN element:

```
<!ATTLIST PROTEIN species CDATA #REQUIRED>
```

The following PROTEIN element is therefore valid:

```
<PROTEIN species="Homo Sapiens"/>
```

It is perfectly legal to use entities within attribute values. For example, consider the following DTD declarations:

```
<!ENTITY ecoli "Escherichia coli">
<!ATTLIST PROTEIN species CDATA #REQUIRED>
```

You can then define the following XML document:

```
<PROTEIN species="&ecoli;"/>
```

Your XML parser will automatically replace the entity with its defined value and return the following element to you:

```
<PROTEIN species="Escherichia coli" />
```

Enumeration List

Attributes can be restricted to a specific set of values by using the enumeration construct. Valid values must be placed within parentheses and separated by a vertical bar. For example, the following defines an enumerated *source* attribute, which is restricted to a list of four possible values:

```
<!ATTLIST SEQUENCE source (WormBase | FlyBase | Ensembl | UCSC)
    #REQUIRED>
```

The following SEQUENCE element is therefore valid:

```
<SEQUENCE source="WormBase"/>
```

Enumeration values are restricted to XML *name tokens*. Within the official XML 1.0 specification, name tokens are restricted to specific characters, including letters, digits, periods, dashes, underscores, and colons. However, name tokens may *not* contain white space characters. Therefore, although this looks tempting, the following attribute rule is actually illegal:

```
<!ATTLIST PROTEIN species
      (Mus musculus | Homo sapiens) #REQUIRED>
```

This rule is illegal because the enumerated values contain whitespace characters. One option to fix this is to replace whitespace characters with underscores. For example, this rule is now legal:

```
<!ATTLIST PROTEIN species
      (Mus_musculus | Homo_sapiens) #REQUIRED>
```

ID, IDREF, and IDREFs

The third attribute type is ID, which requires that the attribute value contain a unique identifier. By using an ID attribute, each XML element can be assigned a unique identifier. You can then later reference those elements with IDREF attributes. This enables you to create a web of internal links within a single document.

To make these concepts concrete, consider a simple DTD for defining protein–protein interactions. The DTD consists of PROTEIN elements and INTERACTION elements. Each protein is assigned a unique identifier and each interaction contains INTERACTOR elements, which reference those identifiers. Here is the complete DTD:

```
<!-- Sample DTD for representing protein-protein interactions -->
<!ELEMENT SUBMISSION (PROTEIN+, INTERACTION+) >

<!-- Proteins must have a unique ID, and a text description -->
<!ELEMENT PROTEIN EMPTY>
<!ATTLIST PROTEIN id ID #REQUIRED>
<!ATTLIST PROTEIN description CDATA #REQUIRED>

<!-- Interactions use IDREF attributes to reference proteins -->
<!ELEMENT INTERACTION (INTERACTOR+)>
<!ELEMENT INTERACTOR EMPTY>
<!ATTLIST INTERACTOR reference IDREF #REQUIRED>
```

Note that the PROTEIN *id* attribute is defined as the unique identifier. Also note that the INTERACTOR *reference* attribute is defined to reference unique identifiers.

Below is a sample XML file, which adheres to the interaction DTD:

```
<?xml version="1.0" encoding="UTF-8"?>
<!DOCTYPE SUBMISSION SYSTEM "interaction.dtd">
<SUBMISSION>
    <PROTEIN id="lat" description="Linker for Activation of T-cells"/>
    <PROTEIN id="itk" description="IL-2 Inducible Tyrosine Kinase"/>
    <PROTEIN id="grap" description="Grb2-like adaptor protein"/>
    <PROTEIN id="grb2" description="Growth factor Receptor Bound
      protein 2"/>
    <INTERACTION>
        <INTERACTOR reference="lat"/>
        <INTERACTOR reference="itk"/>
    </INTERACTION>
    <INTERACTION>
        <INTERACTOR reference="lat"/>
        <INTERACTOR reference="grap"/>
    </INTERACTION>
    <INTERACTION>
        <INTERACTOR reference="lat"/>
        <INTERACTOR reference="grb2"/>
    </INTERACTION>
</SUBMISSION>
```

This document declares four proteins and three interactions. For example, you can see that the LAT protein interacts with ITK, GRAP, and GRB2. By using IDs and IDREFs, your XML validator will ensure referential integrity. For example, if a INTERACTOR element references a "gads" protein, the validator will immediately indicate that the referenced ID does not exist within the document.

NMTOKEN

The NMTOKEN attribute type restricts values to XML *name tokens.* Within the official XML 1.0 specification, name tokens are restricted to specific characters, including letters, digits, periods, dashes, underscores, and colons. However, name tokens may *not* contain white space characters. The main difference between a CDATA type and an NMTOKEN type is that CDATA values may include whitespace characters, whereas NMTOKEN values cannot. This may not seem like much of a difference, but it does provide a very rudimentary filter for attribute values. For example, an NMTOKEN might be useful for describing specific numerical attributes, which by their very nature never contain whitespace. Examples might include: accession number, sequence length, or version number.

NMTOKENS

The NMTOKENS attribute type restricts values to a list of XML name tokens. Each token must be separated by one or more whitespace characters.

ENTITY

The ENTITY attribute type provides a mechanism for referencing *unparsed external entities* defined elsewhere in the DTD. Although it is not commonly used, it provides a mechanism for referencing external data, such as images, sounds, or videos. For details on parsed vs. unparsed entities, refer to the section on "External Entities," later in this chapter.

ENTITIES

The ENTITIES attribute type provides a mechanism for referencing multiple unparsed external entities. Each entity reference must be separated by one or more whitespace characters. For details on parsed vs. unparsed entities, refer to the section on "External Entities," later in this chapter.

NOTATION

The NOTATION attribute type provides a mechanism for referencing notations defined elsewhere in the DTD. This attribute type is very rarely used.

3.4.2   Attribute Behaviors

The XML 1.0 specification defines four kinds of attribute behaviors. Each of these is defined below.

#IMPLIED

The attribute is optional.

#REQUIRED

The attribute is required and must be specified.

Default Value

The attribute has a defined default value. If no value is specified, the default value is automatically used. For example, the following attribute declaration specifies *Homo Sapiens* as the default species:

```
<!ATTLIST PROTEIN species CDATA "Homo Sapiens">
```

Therefore, if you have the following XML element:

```
<PROTEIN/>
```

and run it through an XML parser, the end result will look like this:

```
<PROTEIN species="Homo Sapiens" />
```

You can also specify default values for enumeration lists. For example, the following attribute defaults to UCSC:

```
<!ATTLIST SEQUENCE source (WormBase | FlyBase | Ensembl | UCSC) "UCSC">
```

#FIXED

The attribute is hard coded to a specific value. This may seem odd, but could be useful in some situations. For example, the following declaration provides a hard coded version number:

```
<!ATTLIST SEQUENCE version NMTOKEN #FIXED "1.0">
```

If your document attempts to override the fixed value, it is considered a validation error.

3.5   Working with Entities

For many simple DTDs, you can get by with just element and attribute declarations. For added convenience and power, however, it is useful to delve into the details of XML entities. An XML entity is any label that references another piece of data. For example, in the simplest case, the entity `&author;` may reference the text "James Watson." Upon parsing your XML document, the XML parser will automatically undertake a global search and replace operation and replace all `&author;` references with the text "James Watson." Simple enough. However, entities are actually much more powerful than this, and are capable of referencing data that is outside the XML document, and even referencing non-XML data, such as images and videos. They are also a key tool in building modular DTDs.

Very broadly, there are two types of entities: *general entities* and *parameter entities*. General entities are used *within XML documents* and provide a simple mechanism for referencing common data. Parameter entities are used *within DTDs*, and provide a general mechanism for creating modular DTDs that are easier to maintain over the long haul. We examine each type of entity in the sections below. We also examine conditional DTD sections, particularly when used in combination with parameter entities.

3.5.1   General Entities

General entities encode simple rules for text replacement. The XML standard includes five built-in entities:

- &   ampersand sign (&)
- <   less than sign (<)
- >   greater than sign (>)
- '   apostrophe (')
- &quote;   quote (")

The standard includes entities for each of these special characters, because each one conveys special meaning to the XML parser. For example, the less than sign (<) indicates the start of an XML tag. Therefore, if you use the less than sign within regular element text, the parser will interpret this as the start of a tag, and will probably report a well-formedness error. To prevent the parser from processing special text, you have two options. The fist is to include your text within a CDATA section. For example:

```
<![CDATA[
Error: You must specify a <PROTEIN> element.
]]>
```

The second option is to use the built-in entities. For example:

```
Error: You must specify a &lt;PROTEIN&gt; element.
```

Of course, you are not just limited to the five built-in entities and are free to define your own entities within your DTD. The general syntax for declaring general entities is as follows:

```
<!ENTITY entity_name "[Entity Value Goes Here]">
```

For example, the lines below define entities for common organisms:

```
<!ENTITY mouse "Mus musculus (house mouse)">
<!ENTITY human "Homo sapiens (human)">
<!ENTITY fly "Drosophila melanogaster (fruit fly)">
```

To reference a general entity, the reference must begin with an ampersand character (&), followed by the entity name, and end with a semicolon (;). More concisely, the general syntax for referencing a general entity is as follows:

```
&entity_name;
```

For example, the following element references the mouse entity defined above:

```
<ORGANISM taxonomy_id="10090">&mouse;</ORGANISM>
```

After XML parsing, the element will look like this:

```
<ORGANISM taxonomy_id ="10090">Mus musculus (house mouse)</ORGANISM>
```

External Entities

General entities are not just confined to referencing simple strings of text. They can, in fact, reference data outside of an XML document and even non-XML data. External entities are further subdivided into two additional categories: *parsed external entities* and *unparsed external entities*. A parsed external entity is actually parsed by the XML processor and must contain either XML or regular text. An unparsed external entity is not parsed by the XML processor and usually contains non-XML data, such as images or audio files.

The syntax for defining a parsed external entity is very similar to a regular entity, except that you must specify the SYSTEM keyword, followed by an absolute or relative path to the data. The general syntax is as follows:

```
<!ENTITY entity_name SYSTEM "[Path to External Data]">
```

To make this clear, let's consider a complete example. First, consider that we have a file NM_000854.txt that contains DNA sequence data only. For example, the first three lines of the file look like this:

```
ctccataagg cacaaacttt cagagacagc agagcacaca agcttctagg acaagagcca
ggaagaaacc accggaagga accatctcac tgtgtgtaaa catgacttcc aagctggccg
tggctctctt ggcagccttc ctgatttctg cagctctgtg tgaaggtgca gttttgccaa
```

Our goal is to include the same sequence data within multiple XML documents. To do so, we need to create an external entity. For example, consider the following simple, but complete DTD:

```
<!ELEMENT DNA (SEQUENCE)>
<!ELEMENT SEQUENCE (#PCDATA)>
<!-- External Parsed Entities -->
<!ENTITY NM_000584 SYSTEM "NM_000584.txt">
<!ENTITY NM_000584_REMOTE SYSTEM
    "http://www.xmlbio.org/NM_000584.txt">
```

The first entity declaration assumes the sequence data is located within the same directory as the DTD, but you could just as easily specify an absolute URL (as seen in the second entity declaration).

To include the sequence data within your XML document, simply include an entity reference. For example:

```
<?xml version="1.0" encoding="UTF-8"?>
<!DOCTYPE DNA SYSTEM "seq_ext_entity.dtd">
<DNA>
    <SEQUENCE>
    &NM_000584_REMOTE;
    </SEQUENCE>
</DNA>
```

Upon processing the document, the XML parser will read the DTD, discover the entity declarations, download the contents of the external data files, and paste everything together. After processing the document will therefore look like this:

```
<?xml version="1.0" encoding="UTF-8" ?>
<!DOCTYPE DNA SYSTEM "seq_ext_entity.dtd">
<DNA>
  <SEQUENCE>ctccataagg cacaaacttt cagagacagc agagcacaca agcttctagg
    acaagagcca ggaagaaacc accggaagga accatctcac tgtgtgtaaa catgacttcc
    aagctggccg tggctctctt ggcagccttc ctgatttctg
    For brevity, complete contents are omitted...
  </SEQUENCE>
</DNA>
```

Of course, the example above will only work if the DTD you are using defines the external entity for you. Frequently, this is not the case, and you may want to create external entities for a specific subset of XML documents. To do so, your instance document can extend an external DTD with an internal DTD subset. For example, consider the following instance document:

```
<?xml version="1.0" encoding="UTF-8"?>
<!DOCTYPE DNA SYSTEM "seq_ext_entity.dtd" [
  <!ENTITY NM_000576_REMOTE SYSTEM
  "http://www.xmlbio.org/NM_000576.txt">
]>
<DNA>
 <SEQUENCE>
 &NM_000576_REMOTE;
 </SEQUENCE>
</DNA>
```

> This document references the previous DTD, but also includes an internal DTD sub-set. This internal DTD includes its own entity declaration, which references a second sequence document.

By using external parsed entities, you can create modular, reusable components of text or XML. The advantage of this approach is that you generally only need to make changes in one place. For example, to update the sequence content, you just update the sequence file, instead of dozens of instance documents.

In theory, *unparsed* external entities work in a similar manner, but can be used to reference non-XML data, such as images and audio. In practice, however, very few applications actually use them, and we shall spare you the complex details here. If you are looking for an in-depth discussion of unparsed external entities, there are several excellent references, including: *XML in a Nutshell* [34] and *XML: A Primer* [40].

3.5.2   Parameter Entities

Parameter entities provide text substitution *within DTDs*. This is especially useful for two specific situations. In the first situation, you may have two or more elements, which share common properties. For example, consider the following element declarations:

```
<!ELEMENT PROTEIN (ACCESSION, NAME, DESCRIPTION, ORGANISM, SEQUENCE)>
<!ELEMENT SMALL_MOLECULE (ACCESSION, NAME, DESCRIPTION,
  SMILES_STRUCTURE)>
```

Proteins and small molecules share three common properties: accession number, name, and description. If you are used to thinking in terms of object-oriented programming, you might want to extract these properties to a superclass, and then create subclasses for proteins and small molecules. However, DTDs are not object oriented and the best you can do is extract common properties into a parameter entity.

To declare a parameter entity, use the following syntax:

```
<!ENTITY % entity_name "[Entity Value Goes Here]">
```

Note that the percent sign (%) is required, and there must be a space between the percent sign and the entity name. To reference a parameter entity, use this syntax:

```
%entity_name;
```

In this case, note that the percent sign (%) is again required, but there is no space between the percent sign and the entity name.

Using parameter entities, we can rewrite the element declarations for proteins and small molecules like this:

```
<!ENTITY % general_info "ACCESSION, NAME, DESCRIPTION" >
<!ELEMENT PROTEIN        (%general_info;, ORGANISM, SEQUENCE)>
<!ELEMENT SMALL_MOLECULE (%general_info;, SMILES_STRUCTURE)>
```

In the text above, we have extracted the common elements into a parameter entity, named general_info. This entity is later referenced in the element declarations for PROTEIN and SMALL_MOLECULE .

There are several advantages to using parameter entities in a situation like this. First, the DTD becomes more concise and easier to maintain. For example, if you want to add a new common element to proteins and small molecules, you only need to add it to the general_info entity. Second, it helps you to think about your content model in a more structured manner. When you are first developing a DTD, you will probably start with a series of individual elements. As you add more elements, you may notice certain commonalities and work to extract those commonalities into parameter entities.

The second situation when parameter entities are useful is when you want to break a large monolithic DTD into a series of smaller, modular DTDs. To do so, you create *external* parameter entities. For example, we might want to break our protein DTD into two files: one file which contains the protein grammar rules and another file that contains common organisms. We can tie these two DTDs together via a master DTD document that uses external parameter entities. For example, we could create a protein_root.dtd file:

```
<!-- Example of Using External Parameter Entities -->
<!-- Include Main Protein DTD via Parameter Entity-->
<!ENTITY % protein_set SYSTEM "protein.dtd">
%protein_set;
<!-- Include Common Organisms via Parameter Entity-->
<!ENTITY % organisms SYSTEM "organisms.dtd">
%organisms;
```

As you can see, the root DTD declares external parameter entities for two external files. It then includes these files by referencing the entities directly. When the XML parser validates documents against the main DTD, it automatically retrieves the external entities, and pastes them together to create one single DTD in memory. Breaking your DTD into submodules like this makes it possible to reuse specific components. Further examples of this are provided in the NCBI case study at the end of the chapter.

3.5.3 Entity Summary

Many new XML developers tend to find the complete suite of entity options a bit confusing at first. To help you along, we have created an entity overview diagram, presented in Figure 3.7. The overview diagram includes a complete hierarchy of entity options along with a few simple rules of thumb.

3.5.4 Conditional DTD Sections

XML 1.0 provides support for conditional DTD sections. This enables you to include or ignore specific sections of a DTD. For example, to include the ORGANISM *taxonomy_id* attribute, use the following markup:

```
<![INCLUDE[
    <!ATTLIST ORGANISM
        taxonomy_id NMTOKEN #REQUIRED
    >
]]>
```

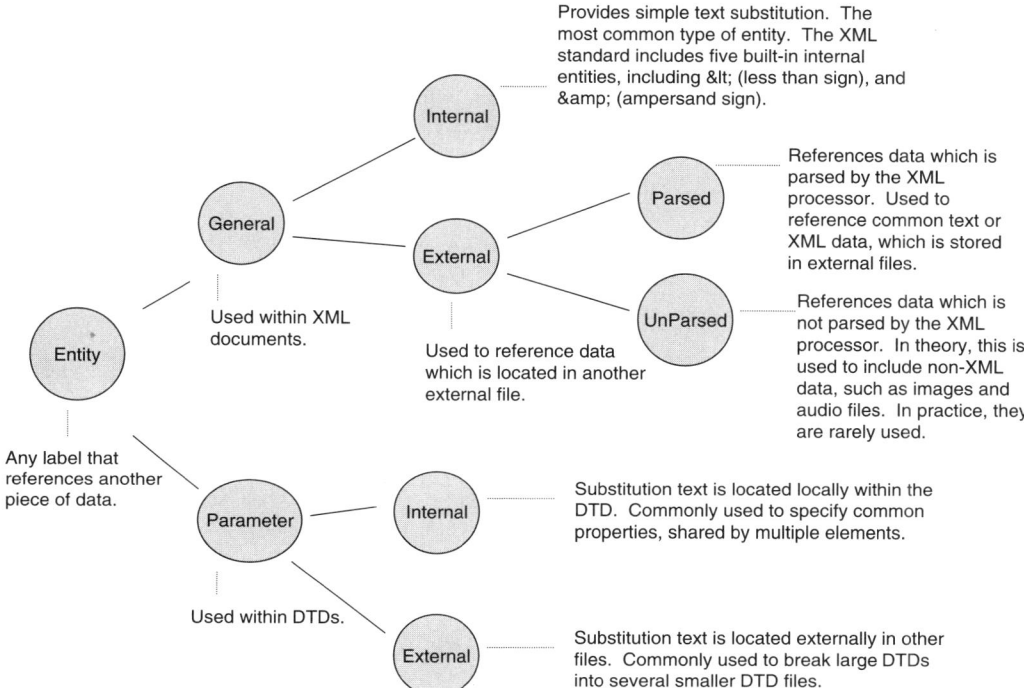

Figure 3.7 Entity overview diagram. The diagram includes the complete hierarchy of entity options along with a few rules of thumb.

To ignore this section, use the IGNORE directive:

```
<![IGNORE[
    <!ATTLIST ORGANISM
        taxonomy_id NMTOKEN #REQUIRED
    >
]]>
```

When using the IGNORE directive, the *taxonomy_id* attribute is completely ignored, and any instance document that includes a *taxonomy_id* attribute will be considered invalid.

By themselves, the IGNORE/INCLUDE directives are not very useful. However, when combined with parameter entities, you can create more powerful conditional sections. For example, consider the following:

```
<!ENTITY % specifyNcbiTaxonomyId "IGNORE">
<![%specifyNcbiTaxonomyId; [
    <!ATTLIST ORGANISM
        taxonomy_id NMTOKEN #REQUIRED
    >
]]>
```

In this example, the `specifyNcbiTaxonomyId` entity can be specified as "IGNORE" or "IN-CLUDE." Depending on the setting, the *taxonomy_id* attribute can be turned on or off. This is particularly useful for developing a new DTD, as you can test out only certain sections at a time. It can also be particularly useful when you are extending an existing DTD, and you want to conditionally activate specific sections. For example, assume that the example above is defined in protein_conditional.dtd, and that `specifyNcbITaxonomyId` is set to "IGNORE." You can create a new DTD, protein_with_tax_id.dtd, and now activate the conditional section, like this:

```
<!ENTITY % specifyNcbiTaxonomyId "INCLUDE">
<!ENTITY % protein_conditional SYSTEM "protein_conditional.dtd">
%protein_conditional;
```

If an entity with the same name is declared more than once, the *first* entity declaration has precedence. Hence, in the example above, our new `specifyNcbiTaxonomyId` entity will override the default value. The *taxonomy_id* attribute declaration section is therefore dynamically included.

3.6 Case Study: NCBI TinySeq

Now that we have covered the basics of DTDs, we turn to a sample DTD for storing and exchanging biological sequence data. A number of XML formats actually exist for representing biological sequences, including Bioinformatic Sequence Markup Language (BSML) [12; 13; 25] and Architecture for Genomic Annotation, Visualization and Exchange (AGAVE) [27; 28]. For this section, however, we will focus on TinySeq [37], a very simple DTD made available by NCBI. The TinySeq DTD is itself quite concise and yet it uses most of the concepts discussed within this chapter. Furthermore, NCBI makes its sequence data available in TinySeq format, and it is therefore very easy to find real, live TinySeq examples. For all of these reasons, TinySeq is an ideal candidate for an in-depth case study. We begin with a brief overview of the NCBI data model and then continue with the details of the TinySeq DTD.

3.6.1 NCBI and XML

The National Center for Biotechnology Information (NCBI) is the central location for numerous bioinformatics databases, including Entrez, GenBank, LocusLink, RefSeq, and PubMed. Due to its central role in storing and exchanging sequence data, NCBI set out over 12 years ago to create a comprehensive data model for representing biological sequences. The data model includes specific details regarding sequence identifiers, sets of sequences, scientific literature citations, cross database references, organism information, and sequence features.[*]

Specific instances of sequence data are required to adhere to the NCBI data model, and this data is stored internally within NCBI in ASN.1 format. ASN.1, or Abstract Syntax Notation 1, is

---

[*] For a complete description of the NCBI Data Model, see Chapter 2 of *Bioinformatics: A Practical Guide to the Analysis of Genes and Proteins*, edited by Andreas D. Baxevanis and B.F. Francis Oullette (John Wiley: New York, 2001) [3].

a standard for defining abstract data types and predates XML. ASN.1 includes support for specific predefined data types, such as integers, and Booleans, and makes it possible to create compound data types, such as structures and lists [35]. NCBI uses ASN.1 to store data internally, but also includes utility tools for converting ASN.1 data to more easily readable formats, such as the familiar GenBank and FASTA flat file formats.

Recently, NCBI introduced a new set of conversion tools for converting to XML [36]. The current NCBI plan is to offer two levels of XML support. The first level is a complete translation from ASN.1 directly to XML. This level, simply known as NCBI XML, provides a complete mapping of the entire NCBI data model. However, the NCBI DTD is quite lengthy, and according to the official NCBI README file, "not for the faint of heart" [36].

The second level of XML support provided by NCBI is a set of smaller DTDs, targeted for specific audiences. Within this set, NCBI has released GBSeq [38] and TinySeq [37]. GBSeq is an attempt to provide the same information as that provided in GenBank flat files. In contrast, TinySeq is a minimal set of elements for describing a set of biological sequences. It essentially conveys the same information as that found in a FASTA file, plus a few additional fields.

Complete details regarding NCBI support for XML are available online at: *http://www.ncbi.nlm.nih.gov/IEB/ToolBox/XML/ncbixml.txt*. This document also includes information regarding the NCBI `asn2xml` tool for converting ASN.1 data to XML.

Fortunately for end users, NCBI stores its sequence data in ASN.1, but provides easy access to the same data in multiple formats. For example, if you search NCBI for a specific sequence record, you can immediately select a data format from the "display" pull-down menu. For example, as you can see in Figure 3.8, NCBI currently offers 10 different display options, including ASN.1, FASTA, and GenBank flat file format. It also offers three options for XML output: TinySeq XML, GBSeq XML, and "XML." The final XML option is the full ASN.1-to-XML conversion.

3.6.2   The TinySeq DTD

With that quick overview of NCBI and XML, we now turn to the TinySeq DTD. The TinySeq DTD actually consists of three separate files (see Figure 3.9). The first file, NCBI_TSeq.dtd, is the root DTD file and it references two other modules. The first module, NCBI_Entity.mod, contains definitions for basic data types, such as INTEGER, REAL, and BOOLEAN (more on this below). The second module, NCBI_TSeq.mod, contains the bulk of rules for defining biological sequences.

By convention, most other NCBI DTDs are structured in a similar manner—the root DTD is usually specified with a .dtd file extension and this root document references module DTDs, specified with a .mod file extension. By breaking their DTDs into modules, NCBI can reuse modules across several applications. For example, several root documents reference the same NCBI_Entity.mod file.

The full contents of the three TinySeq DTD files are provided in Listings 3.3–3.5. You might want to take a first quick look at each file, before delving into the detailed discussion below. If you want to download the files, and try them out locally, they are all available online at: *http://www.ncbi.nlm.nih.gov/dtd*.

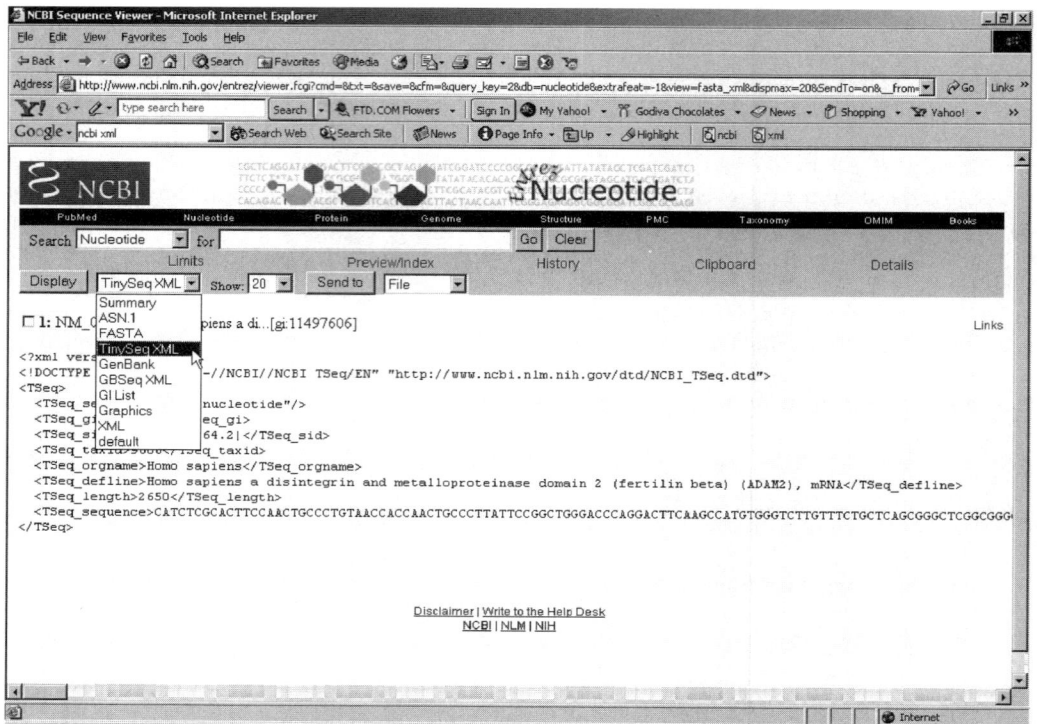

Figure 3.8 A sample screenshot from the NCBI web site. The TinySeq XML data format is highlighted.

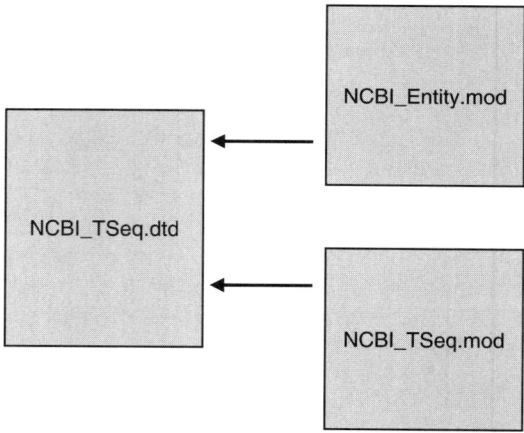

Figure 3.9 A graphical representation of the three files which make up the complete TinySeq DTD. The root DTD, NCBI_TSeq.dtd includes references to two other modular DTD components. The modular components are included via external parameter entities.

Listing 3.3 NCBI_TSeq.dtd

```
<!-- NCBITSeq.dtd
  This file is built from a series of basic modules.
  The actual ELEMENT and ENTITY declarations are in the modules.
  This file is used to put them together.
-->

<!ENTITY % NCBI_Entity_module PUBLIC "-//NCBI//NCBI Entity Module//EN"
"/dtd/NCBI_Entity.mod">
%NCBI_Entity_module;

<!ENTITY % NCBI_TSeq_module PUBLIC "-//NCBI//NCBI TSeq Module//EN"
"/dtd/NCBI_TSeq.mod">
%NCBI_TSeq_module;
```

Listing 3.4 NCBI_Entity.mod

```
<!-- ======================= -->
<!-- NCBI DTD                -->
<!-- NCBI ASN.1 mapped to XML -->
<!-- ======================= -->

<!-- Entities used to give specificity to #PCDATA -->
<!ENTITY % INTEGER '#PCDATA'>
<!ENTITY % ENUM 'EMPTY'>
<!ENTITY % BOOLEAN 'EMPTY'>
<!ENTITY % NULL 'EMPTY'>
<!ENTITY % REAL '#PCDATA'>
<!ENTITY % OCTETS '#PCDATA'>
<!-- =========================================== -->
```

Listing 3.5 NCBI_TSeq.mod

```
<!-- =========================================== -->
<!-- This section mapped from ASN.1 module NCBI-TSeq -->


<!-- =========================================== -->
<!-- Definition of TSeq -->
<!--
*******************************************************************

  ASN.1 for a tiny Bioseq in XML
    basically a structured FASTA file with a few extras
    in this case we drop all modularity of components
      All ids are Optional - simpler structure, less checking
      Components of organism are hard coded - can't easily
      add or change sequence is just string whether DNA or protein
  by James Ostell, 2000


*******************************************************************
-->
```

Listing 3.5 *(cont.)*

```
<!ELEMENT TSeq (
                TSeq_seqtype ,
                TSeq_gi? ,
                TSeq_accver? ,
                TSeq_sid? ,
                TSeq_local? ,
                TSeq_taxid? ,
                TSeq_orgname? ,
                TSeq_defline ,
                TSeq_length ,
                TSeq_sequence )>

<!ELEMENT TSeq_seqtype %ENUM; >
<!ATTLIST TSeq_seqtype value (
                nucleotide |
                protein ) #REQUIRED >
<!ELEMENT TSeq_gi ( %INTEGER; )>
<!ELEMENT TSeq_accver ( #PCDATA )>
<!ELEMENT TSeq_sid ( #PCDATA )>
<!ELEMENT TSeq_local ( #PCDATA )>
<!ELEMENT TSeq_taxid ( %INTEGER; )>
<!ELEMENT TSeq_orgname ( #PCDATA )>
<!ELEMENT TSeq_defline ( #PCDATA )>
<!ELEMENT TSeq_length ( %INTEGER; )>
<!ELEMENT TSeq_sequence ( #PCDATA )>

<!-- Definition of TSeqSet -->

<!ELEMENT TSeqSet ( TSeq+ )>
```

NCBI_TSeq.dtd

The root NCBI_TSeq.dtd file declares two external parameter entities. As discussed in the entity section above, parameter entities are used within DTDs, and external parameter entities can be used to merge multiple DTD modules into a single DTD. For example, the first entity definition references the NCBI_Entity.mod file:

```
<!ENTITY % NCBI_Entity_module PUBLIC "-//NCBI//NCBI Entity Module//EN"
"/dtd/NCBI_Entity.mod">
```

Note that the declaration includes a PUBLIC identifier and a file location. Most XML parsers will ignore the PUBLIC identifier and jump directly to the file location. In this instance, the file location is specified with a relative path. If your XML parser has downloaded the root DTD file directly from NCBI, the file location is relative to the NCBI web site, and your parser will therefore automatically reconnect to NCBI to download the module files. The advantage of using relative file locations is that you can easily copy the NCBI DTDs locally to your file system and the entity references still work.

Having defined the NCBI_Entity_module, the root DTD file includes the complete file contents by referencing the parameter entity directly:

```
%NCBI_Entity_module;
```

The same process is repeated for the NCBI_TSeq.mod file. By using external parameter entities, the root DTD basically copies and pastes the two module files together and creates a single, larger "virtual" DTD.

NCBI_Entity.mod

At first glace, the NCBI_Entity.mod file may seem counterintuitive. The file consists of six parameter entities, each of which defines a specific data type. For example, the following declarations define INTEGERs and REALs:

```
<!ENTITY % INTEGER '#PCDATA'>
<!ENTITY % REAL '#PCDATA'>
```

The DTD specification has no built-in support for data types—everything is basically a string. It, therefore has no facility for verifying that a specific element or attribute can only contain integer or real values. However, as we will soon see in the next chapter, the XML Schemas specification does support data typing, and therefore provides significantly more validation features than DTDs.

While DTDs do not support data typing, data typing is central to ASN.1 and to the NCBI data model. NCBI therefore created the NCBI_Entity.mod file to provide hints about mapping ASN.1 data types to XML. These hints are purely for the benefit of the human reader. For example, certain elements in TinySeq are defined as INTEGERs, and human users are advised to provide only integer values for these elements. However, in the actual entity declaration, INTEGERs are simply defined as #PCDATA, meaning that the element can contain any string value. The XML parser therefore cannot enforce that integer values are actually specified. If and when NCBI migrates to XML Schemas, these data types will become more meaningful and enforceable by XML parsers.

NCBI_TSeq.mod

The third and final DTD, NCBI_Tseq.mod, contains the actual rules for specifying one or more biological sequences. A graphical representation of these rules is presented in Figure 3.10 and a sample instance document is provided in Listing 3.6.

The root of the TinySeq DTD is specified with a TSeqSet element declaration. The root TSeqSet element can contain one or more TSeq elements. Alternatively, an instance document can specify a root TSeq element and forgo the use of the TSeqSet element altogether (see Listing 3.6 for an example). The TSeq element in turn contains a number of subelements:

- TSeq_seqtype (required): specifies the type of sequence represented. The *seqtype* attribute uses an enumeration list to restrict sequence types to "nucleotide" or "protein."
- TSeq_gi (optional): specifies the NCBI GenInfo (GI) number. Every sequence stored in NCBI receives a unique GI number. If the sequence data is updated (even by one base pair), a new record is created and a new GI number is assigned. However, the accession number (defined below) remains constant.
- TSeq_accver (optional): specifies the NCBI accession number, followed by a version number. Accession numbers represent the most stable NCBI identifiers and are used by most NCBI end-users. As indicated above, if the sequence data is updated, a new GI number is assigned; however, the accession number remains constant, and the version number is incremented by one.

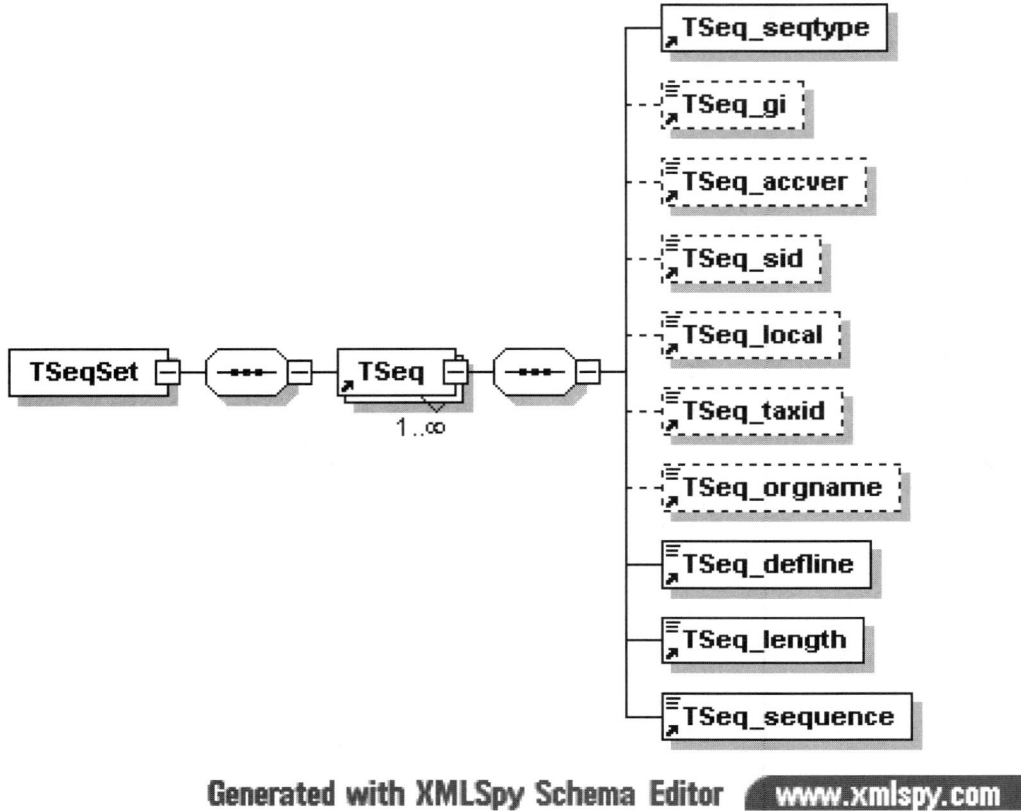

Figure 3.10 A graphical representation of the TinySeq DTD. Required elements are outlined in solid rectangular lines. Optional elements are outlined in dashed rectangular lines. (Diagram was created with XMLSpy®.)

- TSeq_sid (optional): specifies other meaningful sequence identifiers associated with this sequence record. Sequence identifiers are specified using the "vertical bar format," commonly found in other NCBI flat file formats. For example, in the instance document provided in Listing 3.6, the TSeq_sid element is specified as:

```
<TSeq_sid>ref|NM_001464.2|</TSeq_sid>
```

The first part of the string before the vertical bar specifies the database name. In this case, the record references the NCBI RefSeq database of curated genes. The value after the vertical bar indicates the accession number, followed by the version number.

- TSeq_taxid (optional): specifies the taxonomy ID of the source organism. Taxonomy IDs are available at the NCBI Taxonomy database at: *http://www.ncbi.nlm.nih.gov/Taxonomy.* For example, the instance document in Listing 3.6 specifies 9606 as the taxonomy ID for *Homo.*
- TSeq_orgname (optional): specifies the species name of the source organism.
- TSeq_defline (required): specifies a short description of the sequence record. This roughly corresponds to the first line of the FASTA flat file format.

Listing 3.6 Sample instance document adhering to the TinySeq DTD.

```
<?xml version="1.0"?>
<!DOCTYPE TSeq PUBLIC "-//NCBI//NCBI TSeq/EN"
"http://www.ncbi.nlm.nih.gov/dtd/NCBI_TSeq.dtd">
<TSeq>
    <TSeq_seqtype value="nucleotide"/>
    <TSeq_gi>11497606</TSeq_gi>
    <TSeq_sid>ref|NM_001464.2|</TSeq_sid>
    <TSeq_taxid>9606</TSeq_taxid>
    <TSeq_orgname>Homo sapiens</TSeq_orgname>
    <TSeq_defline>Homo sapiens a disintegrin and metalloproteinase
domain 2 (fertilin beta) (ADAM2), mRNA</TSeq_defline>
    <TSeq_length>2650</TSeq_length>
    <TSeq_sequence>CATCTCGCACTTCCAACTGCCCTGTAACCACCAACTGCCCTTATTCCGGCTG
GGACCCAGGACTTCAAGCCATGTGGGTCTTGTTTCTGCTCAGCGGGCTCGGCGGGCTGCGGATGGACAGT
AATTTTGATAGTTTACCTGTGCAAATTACAGTTCCGGAGAAAATACGGTCAATAATAAAGGAAGGAATTG
AATCGCAGGCATCCTACAAAATTGTAATTGAAGGGAAACCATATACTGTGAATTTAATGCAAAAAAACTT
TTTACCCCATAATTTTAGAGTTTACAGTTATAGTGGCACAG
```
For brevity, full sequence has been omitted.
```
</TSeq_sequence>
</TSeq>
```

- `TSeq_length` (required): specifies the length of the sequence data.
- `TSeq_sequence` (required): contains the actual sequence data.

You may be curious to check out the other NCBI DTDs, including GBSeq and the full ASN.1 to XML translation. All of the NCBI DTDs are available in one central directory at: *http://www.ncbi.nih.gov/dtd/*. Each of these DTDs is considerably longer than the TinySeq DTD, but they each use the same basic DTD concepts discussed in this chapter. You are therefore well equipped to dive right in.

XML Schemas for Bioinformatics 4

XML Schema represents the successor to Document Type Definitions (DTDs), offering more features, flexibility, and complexity. This chapter provides an overview of the XML Schema specification, with specific applications to bioinformatics. The XML Schema specification is one of the largest ever produced by the World Wide Web Consortium (W3C), and several entire books have been devoted to the subject. Therefore, rather than covering the complete specification, the chapter focuses on the most important and essential Schema concepts, and illustrates those concepts with numerous examples. When possible, we provide references to other sources for more in-depth information.

The chapter begins with an overview of the main features of XML Schemas, especially as contrasted with DTDs. We also provide an overview of the best-known bioinformatics standards that currently use XML Schema. Following the introduction, we examine our first XML Schema— a relatively simple schema used to represent protein data. Although relatively concise (especially when compared to full-blown bioinformatics schemas), the protein schema provides us with a point of departure for discussing several critical schema issues. This includes the differences between simple and complex types and global versus local element declarations. Each of these topics along with several intermediate topics is explored in detail throughout the remainder of the chapter.

The chapter concludes with a case study discussion of the Proteomics Standards Initiative Molecular Interaction (PSI-MI) XML Schema. The PSI-MI format is a recent initiative of the Human Proteome Organization (HUPO), and is used to represent and exchange protein–protein interactions. We explore the main goals of the PSI-MI format and provide an overview of its main schema components.

4.1 Introduction to XML Schemas

The XML Schema specification [42; 43; 52] is an official recommendation of the World Wide Web Consortium (W3C). The W3C specifically created XML Schemas to address several deficiencies with XML1.0 Document Type Definitions (DTDs). Both specifications provide rules for building valid instance documents. For example, both specifications can dictate that a `protein` element must have an `accession` element and may have 0 or more `keyword` elements.

Beyond this overlapping functionality, Schemas provide several features beyond those defined by DTDs. These features include the following:

- With DTDs, elements and attributes are all treated as strings. In contrast, XML Schema provides built-in data types, such as integers, floats, and dates. This is an extremely powerful feature, as it enables you to provide extra validation rules for instance documents. It also enables tools to automatically map between XML documents and primitive data types provided in programming languages and databases.

- DTDs are written in their own peculiar syntax, which is largely inherited from SGML. In contrast, XML Schemas use regular XML syntax. As a result, any tool that can read or process an XML document, can read and process an XML Schema.
- Schemas support several object-oriented practices, which are simply not provided by DTDs. For example, Schemas have a notion of named types, which are akin to classes; and element declarations, which are akin to object instances. Schemas also support basic inheritance concepts, which enable you to derive new types from existing base types.
- Schemas provide several additional validation rules, which are not supported by DTDs. For example, values can be restricted to a minimum or maximum length of characters, or can be restricted to match a specific regular expression pattern.
- Schemas provide full support for XML Namespaces. This is critically important, as it enables a single instance document to reference two or more schemas, each of which is defined in a separate namespace.

4.1.1   XML Schemas for Bioinformatics

With all these features, XML Schemas provide considerably more power than DTDs. For this reason, many of the newest file formats for bioinformatics are now being created as XML Schemas, and many of the older DTDs are being upgraded to schemas. A list of the best-known XML Schemas for bioinformatics is presented in Table 4.1. A case study of the PSI-MI format is provided at the end of the chapter.

4.2   Essential Concepts: Representing Protein Data

We are now ready to jump into our first example XML Schema. Listing 4.1 provides an XML Schema for representing basic protein data (a visual representation of the Schema is also provided in Figure 4.1). It is similar to the protein DTD explored in the previous chapter, but it also includes a few additional features.

There is actually a lot going on in our first sample schema. Don't be too concerned if it doesn't all make sense just yet. We will dissect its major parts below and explore some of the more advanced concepts in later sections.

Table 4.1 XML Schemas for bioinformatics

Name	Description

AGAVE: Architecture for Genomic Annotation, Visualization and Exchange [28]
    Note: AGAVE 2.3 is written as an XML DTD. The latest version, AGAVE 3.0, is written as an
    XML Schema.
	http://www.animorphics.net/lifesci.html

CML: Chemical Markup Language [49]	*http://www.xml-cml.org/*
PEML: Proteomics Experiment Markup Language	*http://pedro.man.ac.uk*
PSI-MI: Proteomics Standards Initiative Molecular Interaction [46]	*http://psidev.sourceforge.net/*
SBML: The Systems Biology Markup Language [44; 47]	*http://www.sbw-sbml.org/sbml/docs/*
UniProt XML (formerly SPTr-XML)	*http://www.pir.uniprot.org/*
XFF: The Extensible Feature Format	*http://www.biojava.org/thomasd/XFF/*

Listing 4.1 protein.xsd

```xml
<?xml version="1.0" encoding="UTF-8"?>
<xs:schema xmlns:xs="http://www.w3.org/2001/XMLSchema">
    <xs:annotation>
        <xs:documentation>
            Sample XML Schema for representing Protein data.
        </xs:documentation>
    </xs:annotation>
    <xs:element name="protein_set">
        <xs:annotation>
            <xs:documentation>
                A protein set can have one or more protein elements.
            </xs:documentation>
        </xs:annotation>
        <xs:complexType>
            <xs:complexContent>
                <xs:restriction base="xs:anyType">
                    <xs:sequence>
                        <xs:element ref="protein"
                            maxOccurs="unbounded"/>
                    </xs:sequence>
                </xs:restriction>
            </xs:complexContent>
        </xs:complexType>
    </xs:element>
    <xs:element name="protein">
        <xs:annotation>
            <xs:documentation>Main Protein Element</xs:documentation>
        </xs:annotation>
        <xs:complexType>
            <xs:complexContent>
                <xs:restriction base="xs:anyType">
                    <xs:sequence>
                        <xs:element name="accession"
                            type="xs:string"/>
                        <xs:element name="entry_name"
                            type="xs:string"/>
                        <xs:element name="protein_name"
                            type="xs:string"/>
                        <xs:element name="gene_name" type="xs:string"
                            maxOccurs="unbounded"/>
                        <xs:element ref="organism"/>
                        <xs:element ref="cross_reference"
                            minOccurs="0" maxOccurs="unbounded"/>
                        <xs:element name="comment" type="xs:string"
                            minOccurs="0" maxOccurs="unbounded"/>
                        <xs:element name="keyword" type="xs:string"
                            minOccurs="0" maxOccurs="unbounded"/>
                    </xs:sequence>
                </xs:restriction>
            </xs:complexContent>
```

Listing 4.1 *(cont.)*

```
        </xs:complexType>
    </xs:element>
    <xs:element name="organism">
        <xs:annotation>
            <xs:documentation>
                Organism for referencing NCBI Taxonomy ID
            </xs:documentation>
        </xs:annotation>
        <xs:complexType>
            <xs:simpleContent>
                <xs:extension base="xs:string">
                    <xs:attribute name="taxonomy_id" type="xs:integer"
                        use="required"/>
                </xs:extension>
            </xs:simpleContent>
        </xs:complexType>
    </xs:element>
    <xs:element name="cross_reference">
        <xs:annotation>
            <xs:documentation>Cross reference to other database.
                </xs:documentation>
        </xs:annotation>
        <xs:complexType>
            <xs:complexContent>
                <xs:restriction base="xs:anyType">
                    <xs:attribute name="database" type="xs:string"
                        use="required"/>
                    <xs:attribute name="id" type="xs:string"
                        use="required"/>
                </xs:restriction>
            </xs:complexContent>
        </xs:complexType>
    </xs:element>
</xs:schema>
```

As noted in the previous chapter, the protein DTD and schema are designed to illustrate basic concepts, and are not meant to provide a complete description of protein data. Readers looking for a comprehensive format should check out UniProt XML (formerly SPTr-XML), available at: *http://www.uniprot.org*.

4.2.1   The <schema> element

Like every other XML document, an XML Schema document begins with an XML prolog and a root XML element. For XML Schemas, the root element is the schema element. For example, Listing 4.1 has the following root element:

```
<xs:schema xmlns:xs="http://www.w3.org/2001/XMLSchema">
```

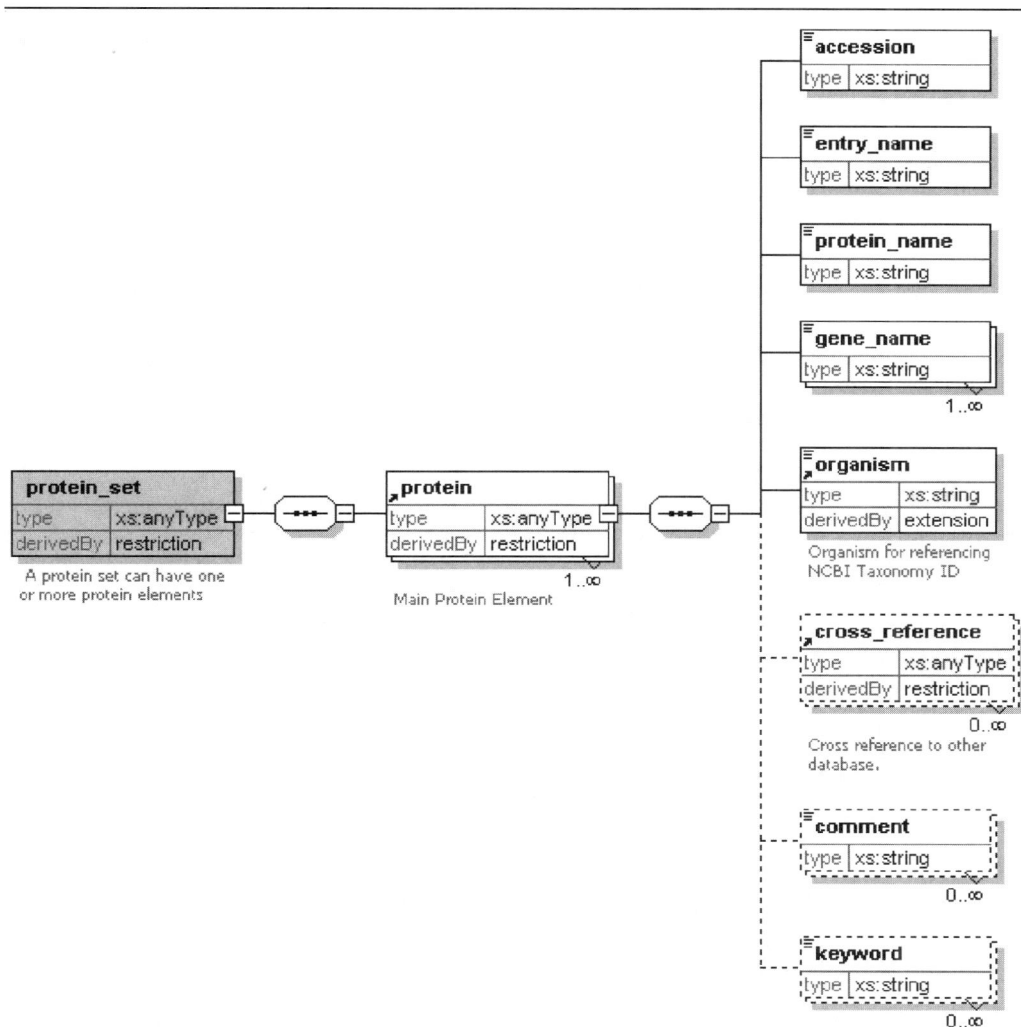

Figure 4.1 Visual overview of the protein.xsd schema. (Diagram was created with XMLSpy®)

In the line above, the `schema` element defines a namespace prefix "`xs`," which references the namespace for XML Schemas. The prefix could be named anything, but most people and applications use "`xs`" or "`xsd`."

By declaring a namespace for XML Schemas, you can easily reference schema specific elements, such as `xs:annotation` and `xs:complexType`. These elements are referenced via qualified names. A qualified name consists of a namespace prefix, followed by a colon, and a local name. For example, the start tag: `<xs:annotation>` references the `annotation` element in the XML Schema namespace.

4.2.2   Schema Documentation

The third line of Listing 4.1 creates an annotation element and is used to document the schema:

```
<xs:annotation>
  <xs:documentation>
     Sample XML Schema for representing Protein data.
  </xs:documentation>
</xs:annotation>
```

In DTDs, any documentation you provide goes into regular XML comments. You are free to use comments within XML Schemas too, but Schema annotations have two important features.

First, annotations can contain two distinct sets of documentation. Specifically, an `annotation` element can contain a `documentation` element, used primarily for human readable information, and/or an `appinfo` element, used primarily for external tools and applications.

Second, annotations can be tied to specific elements. For example, you can place an `annotation` element at the beginning of most schema elements, such as element, attribute, and type declarations. Tools can take advantage of this feature to automatically generate schema documentation. For example, XMLSpy® automatically extracts annotation data and creates beautifully formatted HTML and Microsoft Word documents.

4.2.3   Simple Types vs. Complex Types

In order to understand XML Schemas, you must first understand the differences between simple types and complex types. In a nutshell, a *simple type* contains a single value, such as a string, integer, or date value. By contrast, a *complex type* can contain more than one value, usually in the form of attribute values or element children.

To make this concrete, let's consider a few examples. First, consider the following `gene_name` element:

```
<gene_name>CSF2RB2</gene_name>
```

This element contains a single string value, and is therefore considered a simple type.

Now, consider the following `organism` element:

```
<organism taxonomy_id="10090">Mus musculus</organism>
```

This element contains more than one value; specifically, it contains a string value, e.g., "Mus musculus," and an attribute named *taxonomy_id*. The `organism` element is therefore considered a complex type.

The protein schema contains a mix of simple types and complex types. For example, `accession`, `protein_name`, and `gene_name` are all defined as simple types. By contrast, the root `protein_set` and the main `protein` element can each contain child elements—these are therefore defined as complex types.

We will explore the details of simple and complex types in the sections that follow.

4.2.4   Global Elements vs. Local Elements

In order to understand XML Schemas, you must also understand the differences between global and local elements. A *global element* is any element which is a direct child of the root `schema` element.

For example, in Listing 4.1, `protein_set`, `protein`, `organism`, and `cross_reference` are all direct children of the `schema` element, and are therefore considered global in scope. By contrast, a *local element* is any element which is scoped within another schema construct and is not a direct child of the `schema` element. For example, the `protein` element is global in scope, but it defines a number of local elements, such as `entry_name`, `protein_name`, and `gene_name`.

Global elements can be referenced and reused multiple times throughout your schema. To do so, you simply specify the *ref* attribute and specify the name of the global element. For example, Listing 4.1 defines a global element named `organism`. This element is then referenced within the `protein` element declaration like so:

```
<xs:element ref="organism"/>
```

By using a *ref* attribute, we have declared that the `protein` element will contain an `organism` element, and that the organism structure is defined by the global element declaration. Since the `organism` element is global, other schema constructs can also reference and reuse it. By their very nature, local elements cannot be reused. In fact, local elements are scoped to their direct parent, and cannot be referenced outside this scope.

4.2.5   Creating Instance Documents

Before diving into the details of simple and complex types, let us consider how to create and validate instance documents. As a quick review, an *instance document* is any XML document that purports to adhere to an XML grammar (the grammar may be specified with a DTD or an XML Schema). Instance documents are not required to specifically reference a grammar document, but most do so in practice.

References to external XML Schemas are always specified within the root `schema` element. You have two main options. The first pertains to schemas *without* a declared namespace. In this case, you use the *noNamespaceSchemaLocation* attribute. The second option pertains to schemas *with* a declared namespace. In this case, you use the alternative *schemaLocation* attribute. We explore the first option in detail here and defer the second until later in the chapter (see Section 4.5).

A sample instance document, which adheres to the protein schema, is provided in Listing 4.2. Note in particular the root element:

```
<protein_set xmlns:xsi="http://www.w3.org/2001/XMLSchema-instance"
    xsi:noNamespaceSchemaLocation="protein.xsd">
```

The root element includes an "`xsi`" namespace declaration that references the XML Schema instance specification. Again, the prefix "`xsi`" is just a convention and you are free to use whatever prefix you like.

By declaring the instance namespace, your document is free to reference schema instance constructs. For example, your document can now reference either the *noNamespaceSchemaLocation* attribute or the *schemaLocation* attribute.

Our protein schema has no declared namespace, and therefore Listing 4.2 uses the *noNamespaceSchemaLocation* attribute. The value of this attribute specifies the location of the associated schema. For example, Listing 4.2 specifies the protein.xsd file. This specific file must be located in the same directory as the instance document, but you could just as easily specify an absolute or relative URL and point to any location on the Internet.

Listing 4.2 Sample protein instance document

```xml
<?xml version="1.0" encoding="UTF-8"?>
<protein_set xmlns:xsi="http://www.w3.org/2001/XMLSchema-instance"
    xsi:noNamespaceSchemaLocation="protein.xsd">
    <protein>
        <accession>P26954</accession>
        <entry_name>IL3B_MOUSE</entry_name>
        <protein_name>Interleukin-3 receptor class II
            beta chain [Precursor]
        </protein_name>
        <gene_name>CSF2RB2</gene_name>
        <gene_name>AI2CA</gene_name>
        <gene_name>IL3RB2</gene_name>
        <gene_name>IL3R</gene_name>
        <organism taxonomy_id="10090">Mus musculus</organism>
        <cross_reference database="EMBL" id="M29855"/>
        <cross_reference database="EMBL" id="AAA39295"/>
        <cross_reference database="PIR" id="A40091"/>
        <cross_reference database="MGD" id="MGI:1339760"/>
        <cross_reference database="InterPro" id="IPR002996"/>
        <comment>FUNCTION: IN MOUSE THERE ARE TWO CLASSES OF
            HIGH-AFFINITY IL-3 RECEPTORS. ONE CONTAINS THIS IL-3-
            SPECIFIC BETA CHAIN AND THE OTHER CONTAINS THE BETA
            CHAIN ALSO SHARED BY HIGH-AFFINITY IL-5 AND GM-CSF
            RECEPTORS.</comment>
        <comment>SUBUNIT: Heterodimer of an alpha
        and a beta chain.</comment>
        <comment>SUBCELLULAR LOCATION: Type I
        membrane protein.</comment>
        <comment>SIMILARITY: BELONGS TO THE CYTOKINE
        FAMILY OF RECEPTORS.
        </comment>
        <keyword>Receptor</keyword>
        <keyword>Glycoprotein</keyword>
        <keyword>Signal</keyword>
    </protein>
</protein_set>
```

4.2.6   Validating Instance Documents

Now that we have a sample instance document, and know how to reference its associated schema, we are ready to validate it. The best option for validating instance documents is to pick a good XML editor (see previous chapter for some suggestions). The second best option is to pick a command-line validator. Information on two of the best-known schema validators is provided below:

- xsv is an open source Schema validator created by Henry S. Thompson and Richard Tobin. You can validate documents via a web interface at: *http://www.w3.org/2001/03/webdata/xsv*, or download a Windows command line interface tool at: *ftp://ftp.cogsci.ed.ac.uk/pub/XSV/ XSV14.EXE.*

- The Sun Multi-Schema XML Validator is a free command-line validation tool written in Java. It is capable of validating against several different grammar specifications, including DTDs, Schema, RELAX, and TREX. It is available for download at: *http://wwws.sun.com/software/ xml/developers/multischema*.

Before moving on, make sure you have a good XML editor, or one of the free command line tools. You can therefore experiment with the remaining examples in the chapter.

4.3   Working with Simple Types

As described above, a *simple type* contains a single value, such as a string, integer, or date value. By contrast, a *complex type* can contain more than one value, usually in the form of attribute values or element children. As our next step in exploring XML Schemas, we now move onto an in-depth discussion of simple types. We begin with a discussion of built-in Schema data types, and conclude with a discussion of XML Schema facets.

4.3.1   Built-in Schema Types

As described in the introduction, one of the main benefits of XML Schemas over DTDs is its built-in support for data types. All told, the XML Schemas specification includes support for 44 different data types, including integers, floats, doubles, and dates. Details on the main data types are provided in Table 4.2.

Table 4.2 Main XML Schema data types

Type	Notes
String	A character string
NormalizedString	A character string, which does not contain carriage returns, line feeds, or tabs
Byte	An integer value restricted to the range: –128 to +127
unsignedByte	An integer value restricted to the range: 0 to 255
base64Binary	Binary value encoded in Base64 format
HexBinary	Binary value encoded in Hex format
Integer	An integer value with no minimum/maximum range specified
positiveInteger	A positive integer with no maximum range specified
negativeInteger	A negative integer with no minimum range specified
Int	An integer value restricted to the range: –2,147,483,648 to +2,147,483,647
unsignedInt	An integer value restricted to the range: 0 to 4,294,967,295
Long	An integer value restricted to the range: –9,223,372,036,854,775,808 to +9,223,372,036,854,775,807
unsignedLong	An integer value restricted to the range: 0 to 18,446,744,073,709,551,615
Short	An integer value restricted to the range: –32768 to 32767
unsignedShort	An integer value restricted to the range: 0 to 65,535
Decimal	A decimal value with no minimum/maximum range specified
Float	IEEE 32-bit floating point type
Double	IEEE 64-bit floating point type
Boolean	Boolean value containing the value: true, false, 1, or 0
Time	Time value represented in ISO 8601 format
DateTime	Date–Time value represented in ISO 8601 format
Qname	A namespace qualified value, consisting of a namespace prefix and a local name. For example: `prot:gene_name`
AnyURI	An absolute or relative Uniform Resource Identifier (URI)

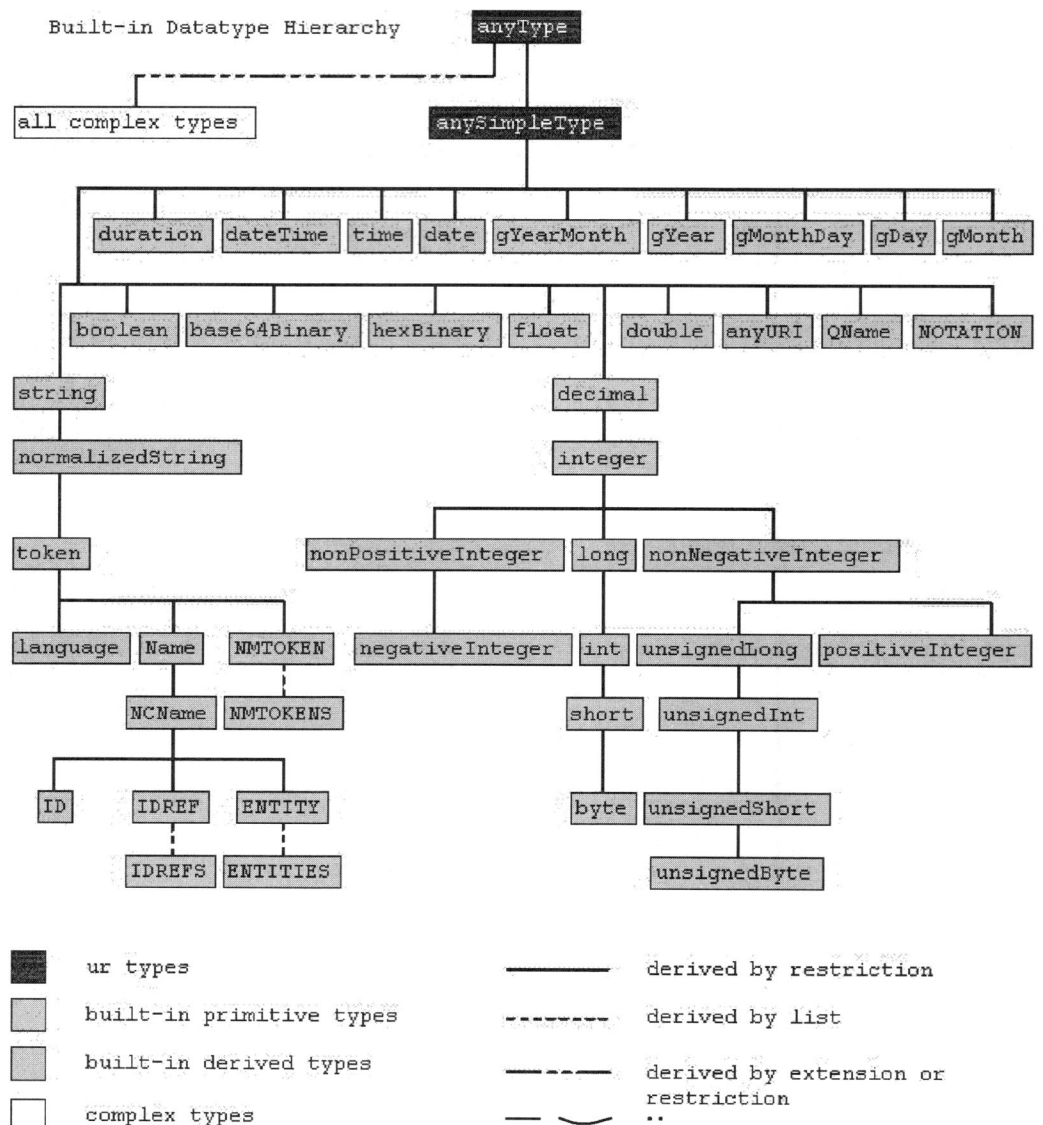

Figure 4.2 The W3C XML Schema type hierarchy. Diagram is copied from the W3C specification, XML Schema Part 2: Datatypes [42], available online at: *http://www.w3.org/TR/xmlschema-2.*

The built-in schema types are organized into a type hierarchy, as shown in Figure 4.2. Note that the root of the type hierarchy is the anyType type. This type places no restrictions on content, and all other types are derived from it.

Creating an element or an attribute with an assigned data type is straightforward. You simply specify the *type* attribute and indicate which type you want to use. For example, the following declaration comes from our protein schema:

```
<xs:element name="protein_name" type="xs:string"/>
```

This declares a `protein_name` element with a `string` data type. Now, consider a second example from our protein schema:

```
<xs:attribute name="taxonomy_id" type="xs:integer" use="required"/>
```

This declares a *taxonomy_id* attribute with an `integer` data type. Your schema validator will ensure that only valid integer values are used. For example, the following example would be considered valid:

```
<organism taxonomy_id="10090">Mus musculus</organism>
```

However, the next two examples would be considered invalid:

```
<organism taxonomy_id="Mus musculus">Mus musculus</organism>
<organism taxonomy_id="10090.10">Mus musculus</organism>
```

4.3.2   Working with Facets

Beyond the 44 primitive data types, XML Schema provides a built-in mechanism for creating new types. All new types are derived from existing types, and this forms the basis of the type hierarchy in Figure 4.2. For example, the `byte` data type is derived from the `short` data type, which is derived from the `int` data type, and so on, up the type hierarchy.

XML Schema provides two primary mechanisms by which you can derive new types. The first is *derivation by extension*. This means that the newly derived type has the same properties of the base type, plus a few additionally specified properties. The second is *derivation by restriction*. This means that the newly derived type has the same properties of the base type, but that additional restrictions are placed on the newly derived type.

Most data types in the type hierarchy are derived via restriction, and most new data types you create are likely to be derived by restriction as well. For simple types, XML Schema supports a concept, called *facets*, which enables you to derive new types via restriction and to place specific restrictions on data values. There are a total of 12 facets, including *length*, *minLength*, *maxLength*, *pattern*, and *enumeration*. For an overview of the main schema facets, refer to Table 4.3.

To use the schema facets, you must create a new data type. This new type must be based on an existing data type in the type hierarchy. For example, Listing 4.3 shows our first schema facet example.

In Listing 4.3, we are declaring a new data type, named `accessionType`. This type will contain string data, but the string data must be between four and eight characters in length.

Let us dissect each line in detail. The first line indicates that we are declaring a new `simpleType`. This is a simple type because it will contain a single value, and will not contain any attributes or child elements. The second line indicates that the `accessionType` will be derived from the built-in `string` data type, and that we will be deriving via restriction. The third and fourth lines specify the *minLength* and *maxLength* facets.

To summarize, `accessionType` is the same as `string` data type, except that additional restrictions on character length have been specified. Because of these additional restrictions, the following element is considered valid:

```
<accession>P26954</accession>
```

Table 4.3 Main XML Schema facets

Facet	Description
length	Specifies the exact number of characters for a string. Can also be applied to other data types, such as `hexBinary` and `base64Binary`
minLength	Specifies the minimum number of characters for a string. Can also be applied to other data types, such as `hexBinary` and `base64Binary`
maxLength	Specifies the maximum number of characters for a string. Can also be applied to other data types, such as `hexBinary` and `base64Binary`
pattern	Restricts content to those values which match a regular expression pattern
enumeration	Restricts content to an enumerated list of valid values
minInclusive	Specifies the minimum value (inclusive) for a numeric type
maxInclusive	Specifies the maximum value (inclusive) for a numeric type
minExclusive	Specifies the minimum value (exclusive) for a numeric type
maxExclusive	Specifies the maximum value (exclusive) for a numeric type
totalDigits	Specifies the maximum number of digits for a decimal derived type
fractionDigits	Specifies the maximum number of fractional digits for a decimal derived type

Listing 4.3 Illustrates basic use of XML Schema facets

```
<xs:simpleType name="accessionType">
  <xs:restriction base="xs:string">
    <xs:minLength value="4"/>
    <xs:maxLength value="8"/>
  </xs:restriction>
</xs:simpleType>
```

However, this element has too many characters, and is therefore considered invalid:

```
<accession>P26954382</accession>
```

Once you have created a new type, you can reference that type throughout your schema. To do so, you simply use the *type* attribute, just like before. However, instead of specifying one of the built-in data types, you specify the name of your newly derived data type.

For example, Listing 4.4 shows a bare bones version of our complete protein schema. Within this example, we have declared the `accessionType` as before. We have also declared a protein `accession` element. However, instead of specifying `accession` with a data type of "`string`", we now reference the `accessionType`. In a nutshell, the example states that the `protein` element contains a single `accession` element and that the `accession` element must follow all the rules specified by `accessionType`.

The Pattern Facet

The pattern facet restricts string values to those that match a regular expression pattern. The pattern syntax adopted by XML Schema is based on the regular expression syntax provided in the Perl programming language.

Let us look at two pattern examples. First, Listing 4.5 shows a new variation on the `accession` `Type`. This `accessionType` must be between four and eight characters and must begin with the

Listing 4.4 Illustrates how to declare and reference a named simple type

```xml
<?xml version="1.0" encoding="UTF-8"?>
<xs:schema xmlns:xs="http://www.w3.org/2001/XMLSchema"
elementFormDefault="qualified">
  <xs:element name="protein_set">
    <xs:complexType>
      <xs:sequence>
       <xs:element ref="protein" minOccurs="1" maxOccurs="unbounded"/>
      </xs:sequence>
    </xs:complexType>
  </xs:element>
  <xs:element name="protein">
    <xs:complexType>
      <xs:sequence>
        <xs:element name="accession" type="accessionType"
          minOccurs="1" maxOccurs="1"/>
      </xs:sequence>
    </xs:complexType>
  </xs:element>
  <xs:simpleType name="accessionType">
    <xs:restriction base="xs:string">
      <xs:minLength value="4"/>
      <xs:maxLength value="8"/>
    </xs:restriction>
  </xs:simpleType>
</xs:schema>
```

Listing 4.5 Using the pattern facet, first example

```xml
<xs:simpleType name="accessionType">
  <xs:restriction base="xs:string">
  <xs:minLength value="4"/>
    <xs:maxLength value="8"/>
    <xs:pattern value="P.*"/>
  </xs:restriction>
</xs:simpleType>
```

letter "P." The following example is therefore considered valid:

```xml
<accession>P26954</accession>
```

However, this example does not begin with a "P," and is therefore considered invalid:

```xml
<accession>YDR26954</accession>
```

The second pattern example is shown in Listing 4.6. This example restricts the sequence alphabet to the letters: ACGT. The following example is therefore considered valid:

```xml
<sequence>CCAAGGGTT</sequence>
```

However, this amino acid sequence is considered invalid:

```xml
<sequence>MPVKGGSKCIK</sequence>
```

Listing 4.6 Using the pattern facet, second example

```
<xs:simpleType name="sequenceType">
  <xs:restriction base="xs:string">
    <xs:pattern value="[ACGT]*"/>
  </xs:restriction>
</xs:simpleType>
```

Listing 4.7 Using the enumeration facet

```
<xs:simpleType name="databaseType">
  <xs:restriction base="xs:string">
    <xs:enumeration value="EMBL"/>
    <xs:enumeration value="PIR"/>
    <xs:enumeration value="MGD"/>
    <xs:enumeration value="InterPro"/>
    <xs:enumeration value="Pfam"/>
    <xs:enumeration value="SMART"/>
    <xs:enumeration value="PROSITE"/>
  </xs:restriction>
</xs:simpleType>
```

The Enumeration Facet

The enumeration facet restricts values to a list of predefined values. It is similar to attribute enumerations provided by DTDs; however, schema enumerations are more powerful, as you can apply them to both attributes and elements.

Listing 4.7 provides a useful illustration of the enumeration facet. In it, we create a new databaseType which is restricted to a specific set of biological databases, e.g., EMBL, Inter-Pro, Pfam, etc. This type might be usefully used within a cross-reference element, so that you can easily restrict cross-references to known biological databases.

4.4 Working with Complex Types

4.4.1 Introduction to Complex Types

As stated earlier in the chapter, complex types are those than can contain attributes and/or child elements. They are complex in the sense that they can contain more pieces of data, and when combined with other complex types, can form elaborate content models and tree hierarchies.

When creating a new type for your schema, the first question to ask is whether the new type will be simple or complex. If the type will contain a single value, such as a string, integer, or date, it will be a simple type. If the type will contain attributes and/or child elements, it will be a complex type.

This much we already know. However, having decided that you want to create a complex type, you must ask yourself a second question: will the complex type contain child elements? If the type will *not* contain child elements, this is formally known as a complex type with simple content.

Listing 4.8 An example of a complex type with simple content

```
<xs:element name="organism">
  <xs:complexType>
    <xs:simpleContent>
      <xs:extension base="xs:string">
      <xs:attribute name="taxonomy_id" type="xs:integer"
          use="required"/>
      </xs:extension>
    </xs:simpleContent>
  </xs:complexType>
</xs:element>
```

Listing 4.9 An example of a complex type with complex content

```
<xs:element name="protein_set">
  <xs:complexType>
    <xs:complexContent>
      <xs:restriction base="xs:anyType">
        <xs:sequence>
          <xs:element ref="protein" maxOccurs="unbounded"/>
        </xs:sequence>
      </xs:restriction>
    </xs:complexContent>
  </xs:complexType>
</xs:element>
```

If the type *will* contain child elements, this is formally known as a complex type with complex content.*

Concrete examples will make the distinctions very clear. First, consider Listing 4.8. This is an excerpt of the `organism` element declaration from our main protein schema. The `organism` element contains a string value and a *taxonomy_id*. It does not, however, contain child elements. It is therefore defined as a complex type with simple content.

Note that `organism` is derived from the `string` data type, and that we are deriving via extension. When deriving via extension, you can add additional properties not present in the base type. In this case, we are adding a *taxonomy_id,* which is not present in the string base type.

Next, consider Listing 4.9. This is an excerpt of the `protein_set` element from our main protein schema. The `protein_set` element contains child elements, namely, one or more `protein` elements. The element is therefore defined as a complex type with complex content.

Note that the `protein_set` element is derived from the `anyType` type, and we are now deriving via restriction. Recall that `anyType` is the root of the XML Schema type hierarchy and places no restrictions on content. The `protein_set` element is therefore derived from `anyType` and restricted to contain `protein` elements only.

---

* The line of schema questioning here is based on an excellent article by Donald Smith, "Understanding W3C Schema Complex Types" [51], available online at: *http://www.xml.com/pub/a/2001/08/22/easyschema.html.* The article includes a handy visual "Schema Type Decision Tree" available as a separate PDF document.

Listing 4.10 An example of the abbreviated `complexType` syntax

```
<xs:element name="protein_set">
  <xs:complexType>
    <xs:sequence>
      <xs:element ref="protein" maxOccurs="unbounded"/>
    </xs:sequence>
  </xs:complexType>
</xs:element>
```

Creating a new complex type with complex content and deriving from `anyType` is such a common occurrence that it is actually the default behavior for new complex types. You can therefore use an abbreviated syntax, as shown in Listing 4.10.

Listing 4.10 is functionally equivalent to Listing 4.9. However, Listing 4.9 uses the long syntax, whereas Listing 4.10 uses the abbreviated syntax. Which syntax you use is a matter of choice. Some people prefer the long syntax because it offers greater conceptual clarity, whereas others prefer the abbreviated syntax, simply because it is more concise.*

4.4.2   Declaring Empty Element Types

Next, let us consider the mechanics of declaring empty elements. An empty element is one that does not contain any text or child elements, but may contain attributes. For example, the `cross_reference` element is considered empty:

```
<cross_reference database="EMBL" id="M29855"/>
```

The easiest option for declaring an empty element is to define a complex type with complex content. This is a bit of a hack, and may even seen counter-intuitical, but recall that complex content is reserved for elements which contain child elements. However, if you neglect to specify any child elements, the element is defined as empty. Using this small loophole, you can therefore declare empty elements.**

Let us take a look at an example. Listing 4.11 provides an excerpt of the `cross_reference` element declaration. As you can see, the element is defined as a complex type with complex content. Within complex content, we define two attributes, but do not define any child elements. That is all there is to it.

If you recall the abbreviated syntax from the previous section, you can create a much more concise, empty element declaration. For example, the following declaration is functionally equivalent to Listing 4.11:

```
<xs:element name="cross_reference">
  <xs:complexType>
    <xs:attribute name="database" type="xs:string" use="required"/>
    <xs:attribute name="id" type="xs:string" use="required"/>
  </xs:complexType>
</xs:element>
```

---

* Again, I am indebted to Donald Smith and his xml.com article, "Understanding W3C Schema Complex Types." Smith argues that the long syntax offers greater conceptual clarity and provides new users with a better window into the inner mechanics of XML Schemas. Once you understand these mechanics, you can more easily transition to the abbreviated syntax.

**This is not the only option for declaring empty elements. A full discussion is provided in Chapter 7 of Van der Vlist, XML Schema (O'Reilly 2002) [53]

Listing 4.11 Declaring an empty element

```
<xs:element name="cross_reference">
  <xs:complexType>
    <xs:complexContent>
      <xs:restriction base="xs:anyType">
      <xs:attribute name="database" type="xs:string" use="required"/>
        <xs:attribute name="id" type="xs:string" use="required"/>
      </xs:restriction>
    </xs:complexContent>
  </xs:complexType>
</xs:element>
```

Listing 4.12 Declaring an element with mixed content

```
<xs:element name="PubmedUpdate">
  <xs:complexType mixed="true">
      <xs:sequence>
        <xs:element name="ArticleTitle" type="xs:string"/>
        <xs:element name="Url" type="xs:anyURI"/>
      </xs:sequence>
  </xs:complexType>
</xs:element>
```

The example above is considerably more concise, and I would argue that in this case, the abbreviated syntax is also more intuitive. Again, it is up to you to choose the long or the abbreviated syntax, and you may want to do so on a case-by-case basis.

4.4.3 Declaring Mixed Element Types

A mixed element type is one that contains text data interspersed with child elements. For example, the following PubmedUpdate element contains mixed type content:

```
<PubmedUpdate>
The article you requested:
<ArticleTitle>Initial sequencing and analysis of the
human genome.</ArticleTitle> is available for download from the
following website: <Url>http://www.nature.com</Url>
</PubmedUpdate>
```

To declare that an element supports mixed type content, you simply set the complexType *mixed* attribute to "true." For example, Listing 4.12 shows the element declaration for PubmedUpdate.

Note that mixed content functionality in XML Schemas is considerably more powerful that that provided in DTDs. In DTDs, you can specify mixed content along with a list of valid child elements. However, you cannot specify the sequence of the elements, nor can you specify the occurrence with which they appear. In XML Schema, you can specify mixed content, along with a detailed content model. For example, in Listing 4.12, we specify that the PubmedUpdate element must contain exactly one ArticleTitle element followed by exactly one Url element. The following document is therefore considered valid:

```
<PubmedUpdate>
  <ArticleTitle>Initial sequencing and analysis of the human genome.
    </ArticleTitle>
  <Url>http://www.nature.com</Url>
</PubmedUpdate>
```

However, this document does not follow the specified sequence and is therefore considered invalid:

```
<PubmedUpdate>
    <Url>http://www.nature.com</Url>
    <ArticleTitle>Initial sequencing and analysis of the human genome.
      </ArticleTitle>
</PubmedUpdate>
```

4.4.4   Occurrence Constraints

When declaring an element which contains child elements, you can specify the exact number of times the child element may appear. Within XML Schema, these are formally known as *occurrence constraints*. Occurrence constraints are set via two attributes: *minOccurs* and *maxOccurs*.

The *minOccurs* attribute specifies the minimum number of times the element must occur, and must be specified with an integer value. For example, a *minOccurs* value of 1 indicates that the element must appear at least once. A *minOccurs* value of 0 indicates that the element may occur zero times and is therefore optional. If the *minOccurs* attribute is not specified, its default value is set to 1.

The *maxOccurs* attribute specifies the maximum number of times the element may occur. Its value must be specified with a positive integer value or the string value, "unbounded." For example, a *maxOccurs* value of 5 indicates that the element can appear at most five times. A *maxOccurs* value of "unbounded" indicates that the element may appear as many times as you like. If the *maxOccurs* attribute is not specified, its default value is set to 1.

As a concrete example, take a look at Listing 4.13. This shows an excerpt of the protein element from our main protein schema. Within the example, you can find numerous examples of occurrence constraints. For example, the `protein_name` element has no explicit occurrence constraints. In this case, the default values kick in, and the element must occur exactly once. By contrast, the `comment` element is entirely optional, and yet there are no limits on the maximum number of times it may occur.

Attributes have entirely different occurrence constraints than elements. For example, you cannot specify that an attribute appear more than once. You can, however, specify whether an attribute is optional or required. To do so, simply specify the *use* attribute. For example, the following attribute is required, and must be specified within all instance documents:

```
<xs:attribute name="database" type="xs:string" use="required"/>
```

To make the attribute optional, redo the example like this:

```
<xs:attribute name="database" type="xs:string" use="optional"/>
```

If you do not specify a use attribute, the default value is set to "optional."

Listing 4.13 Using occurrence constraints

```
<xs:element name="protein">
  <xs:complexType>
    <xs:complexContent>
      <xs:restriction base="xs:anyType">
        <xs:sequence>
          <xs:element name="accession" type="xs:string"/>
          <xs:element name="entry_name" type="xs:string"/>
          <xs:element name="protein_name" type="xs:string"/>
          <xs:element name="gene_name" type="xs:string"
            maxOccurs="unbounded"/>
          <xs:element ref="organism"/>
          <xs:element ref="cross_reference" minOccurs="0"
            maxOccurs="unbounded"/>
          <xs:element name="comment" type="xs:string" minOccurs="0"
            maxOccurs="unbounded"/>
          <xs:element name="keyword" type="xs:string" minOccurs="0"
            maxOccurs="unbounded"/>
        </xs:sequence>
      </xs:restriction>
    </xs:complexContent>
  </xs:complexType>
</xs:element>
```

Listing 4.14 Declaring a default value

```
<xs:element name="source">
  <xs:complexType>
    <xs:sequence>
      <xs:element name="organization" type="xs:string"
        default="Memorial Sloan-Kettering Cancer Center"/>
    </xs:sequence>
  </xs:complexType>
</xs:element>
```

4.4.5   Declaring Default Values

In addition to specifying occurrence constraints, you can also specify default values for your elements and attributes. To do so, simply specify the *default* attribute.

For example, Listing 4.14 shows an `organization` element with a default value.

If you have an organization element with no content, your XML parser will automatically insert the default text. For example, consider the following instance document:

```
<source>
    <organization/>
</source>
```

Your XML parser will automatically insert "Memorial Sloan-Kettering Cancer Center" within the `organization` element. After parsing, the document will therefore look like

Listing 4.15 Declaring a fixed value

```
<xs:element name="source">
  <xs:complexType>
  <xs:sequence>
    <xs:element name="organization" type="xs:string"
       fixed="Memorial Sloan-Kettering Cancer Center"/>
  </xs:sequence>
  </xs:complexType>
</xs:element>
```

this:

```
<source>
  <organization>Memorial Sloan-Kettering Cancer Center </organization>
</source>
```

Although used less commonly, you can also specify fixed values for your elements and attributes. In this case, if the instance value is specified, it must exactly match that specified by the fixed value. Otherwise, it is considered a validation error. If the instance value is not specified, the fixed value is automatically inserted, just like a regular default value.

For example, Listing 4.15 shows a slight variation on our `organization` element.

This time, `organization` is defined with a fixed value. Since it has a fixed value, the following instance data is considered invalid:

```
<source>
    <organization>MIT</organization>
</source>
```

However, this element is valid:

```
<source>
  <organization>Memorial Sloan-Kettering Cancer Center </organization>
</source>
```

This element is also considered valid:

```
<source>
    <organization/>
</source>
```

After parsing, the fixed value is automatically inserted and the element becomes:

```
<source>
  <organization>Memorial Sloan-Kettering Cancer Center </organization>
</source>
```

4.4.6   Compositors: Sequence and Choice

Elements which contain child elements can organize those elements into specific content models. Within XML Schema, elements can be organized via schema *compositors*. We will examine two

Listing 4.16 Using the sequence compositor

```
<xs:element name="PubmedArticle">
  <xs:complexType>
    <xs:sequence>
      <xs:element name="MedlineID" type="xs:long"/>
      <xs:element name="PMID" type="xs:long"/>
    </xs:sequence>
  </xs:complexType>
</xs:element>
```

Listing 4.17 Using the choice compositor

```
<xs:element name="PubmedArticle">
  <xs:complexType>
    <xs:choice>
      <xs:element name="MedlineID" type="xs:long"/>
      <xs:element name="PMID" type="xs:long"/>
    </xs:choice>
  </xs:complexType>
</xs:element>
```

of the main compositors: *sequence* and *choice*. To make the concepts clear, we examine several example schemas for storing PubMed scientific literature data.

First up is the sequence compositor. The sequence compositor specifies an exact sequence of child elements. An instance document must follow the exact sequence specified—otherwise, the document is considered invalid.

Listing 4.16 shows a sample schema for storing PubMed data. Note that we are using the sequence compositor to specify that the MedlineID element must occur *before* the PMID element. The following is therefore considered valid:

```
<PubmedArticle>
  <MedlineID>21131739</MedlineID>
  <PMID>11237011</PMID>
</PubmedArticle>
```

However, this example uses the wrong sequence and is therefore considered invalid:

```
<PubmedArticle>
  <PMID>11237011</PMID>
  <MedlineID>21131739</MedlineID>
</PubmedArticle>
```

Next up is the choice compositor. The choice compositor is used to indicate that an instance document can select from one of several element options. For example, Listing 4.17 shows a new variation on our PubMed schema. The MedlineID and PMID elements are now defined within a choice compositor. A valid instance document must therefore include either a MedlineID or a PMID, but not both. For example, the following is considered valid:

```
<PubmedArticle>
  <MedlineID>21131739</MedlineID>
</PubmedArticle>
```

Listing 4.18 Combining compositors

```
<xs:element name="PubmedArticle">
  <xs:complexType>
    <xs:sequence>
      <xs:choice>
        <xs:element name="MedlineID" type="xs:long"/>
        <xs:element name="PMID" type="xs:long"/>
      </xs:choice>
      <xs:element name="ArticleTitle" type="xs:string"/>
      <xs:element name="AbstractText" type="xs:string"/>
    </xs:sequence>
  </xs:complexType>
</xs:element>
```

However, this example includes both choices and is therefore considered invalid:

```
<PubmedArticle>
  <MedlineID>21131739</MedlineID>
  <PMID>11237011</PMID>
</PubmedArticle>
```

Of course, you can combine compositors to create more complicated content models. For example, Listing 4.18 uses a choice element within a sequence element. In plain English, this schema states that a `PubmedArticle` must contain a `MedlineID` or a `PMID`, followed by an `ArticleTitle` and an `AbstractText`.

4.4.7   Defining Named Complex Types

Up until now, our schema examples have used a combination of global and local elements. As a quick recap, global elements are direct children of the `schema` element and can be reused throughout a schema via the *ref* attribute. Global elements represent one viable option for code reuse, and just like programming code reuse is almost always a good thing. However, global elements are not the only reusable schema components. As we have already seen with simple types, it is possible to create named simple types, and to reuse these types via the *type* attribute. The same mechanism exists for named complex types. We now turn to this alternative.

To create a reusable, named complex type, you make it a direct child of the `schema` element and specify its *name* attribute. If you do not specify a name, you are creating an anonymous complex type (so far, all of our complex types have been anonymous). Given a unique name, you can then reuse the complex type via the *type* attribute—this is the same mechanism used to reference built-in or user derived simple types. For example, the following creates an organism type:

```
<xs:complexType name="organismType">
  <xs:simpleContent>
    <xs:extension base="xs:string">
      <xs:attribute name="taxonomy_id" type="xs:integer"
        use="required"/>
    </xs:extension>
  </xs:simpleContent>
</xs:complexType>
```

This named complex type represents a blueprint for an element, and this blueprint can be reused throughout your schema. For example, consider the following declaration:

```
<xs:element name="organism" type="organismType"/>
```

This declares a new `organism` element, which will use the blueprint defined by `organismType`. If you are used to thinking in terms of object-oriented design, you can think of the named complex type as a class, and the element declaration as an instantiation of that class.*

4.4.8 All Together Now!

By now, you should be confident in creating complex types, setting occurrence constraints, using schema compositors, and creating named complex types. To tie all these concepts together, let us take a look at one more example. Listing 4.19 shows a revised version of our protein schema.

The new schema is functionally equivalent to the first protein schema in Listing 4.1. However, there are several important differences in schema design. First, the new schema uses the abbreviated syntax for complex types, making for a much more concise example. Second, the new schema contains only one global element and makes extensive use of named complex types. These named complex types are referenced throughout the schema via the *type* attribute.

Take a few moments now to review the complete schema in Listing 4.19. If everything makes sense, you are in good shape!

This final complex type example touches lightly upon two schema design patterns. Our first protein schema uses global elements throughout, and this design pattern is sometimes referred to as the "salami slice" pattern. The second protein schema uses named complex types throughout, and is sometimes called the "Venetian blind" pattern. For an in-depth discussion of schema design patterns, and the pros and cons of each approach, see Ayesha Malik's article, "Create flexible and extensible XML schemas: Building XML schemas in an object-oriented framework" [48]. The article is available at the IBM Developer Works web site: *http://www-106.ibm.com/developerworks/*.

Do complex types seem too complicated? If so, you might want to consider an alternative to XML Schemas, called RELAX NG. For more information see *http://relaxng.org* or Eric van der Vlist's "RELAX NG" (O'Reilly, 2003). The RELAX NG web site includes a tutorial, a list of RELAX NG validators, and a list of conversion tools for converting RELAX NG to XML Schemas.

4.5 Basic Namespace Issues

When creating a new XML Schema, you can define an associated namespace. In XML Schema, this associated namespace is formally known as the *target namespace*. All newly defined elements

---

* I first came across this specific analogy from the writings of Eric van der Vlist. For additional details on the connection between XML Schemas and object-oriented design, see van der Vlist's xml.com article, "Using XML Schema," available online at: *http://www.xml.com/pub/a/2000/11/29/schemas/part1.html* .

Listing 4.19 Our final complex type example

```xml
<?xml version="1.0" encoding="UTF-8"?>
<xs:schema xmlns:xs="http://www.w3.org/2001/XMLSchema">
  <xs:element name="protein_set">
    <xs:complexType>
      <xs:sequence>
        <xs:element name="protein" type="proteinType"
          maxOccurs="unbounded"/>
      </xs:sequence>
    </xs:complexType>
  </xs:element>
  <xs:complexType name="proteinType">
    <xs:sequence>
      <xs:element name="accession" type="xs:string"/>
      <xs:element name="entry_name" type="xs:string"/>
      <xs:element name="protein_name" type="xs:string"/>
      <xs:element name="gene_name" type="xs:string"
        maxOccurs="unbounded"/>
      <xs:element name="organism" type="organismType"/>
      <xs:element name="cross_reference" type="crossReferenceType"
        minOccurs="0"  maxOccurs="unbounded"/>
      <xs:element name="comment"  type="xs:string" minOccurs="0"
        maxOccurs="unbounded"/>
      <xs:element name="keyword" type="xs:string" minOccurs="0"
        maxOccurs="unbounded"/>
    </xs:sequence>
  </xs:complexType>
  <xs:complexType name="organismType">
    <xs:simpleContent>
      <xs:extension base="xs:string">
        <xs:attribute name="taxonomy_id" type="xs:integer"
          use="required"/>
      </xs:extension>
    </xs:simpleContent>
  </xs:complexType>
  <xs:complexType name="crossReferenceType">
    <xs:complexContent>
      <xs:restriction base="xs:anyType">
        <xs:attribute name="database" type="xs:string"
          use="required"/>
        <xs:attribute name="id" type="xs:string" use="required"/>
      </xs:restriction>
    </xs:complexContent>
  </xs:complexType>
</xs:schema>
```

and types within your schema will belong to this target namespace. If you do not specify a target namespace, schema constructs will not be associated with any namespace.

Associating a schema with a namespace is accomplished via the *targetNamespace* attribute, which is specified in the root schema element. Using a target namespace is best understood in the context of real examples. The full implications are also best understood when comparing schemas with target namespaces, and instance documents which adhere to those schemas.

First, let us take a look at the newly revised version of our protein schema (see Listing 4.20). The content model has been significantly simplified to keep the example short, and to enable us to concentrate solely on namespace issues. Most of the new information is also contained in the first three lines of the example, all within the root `schema` element.

As you examine Listing 4.20, first note that the root `schema` element now contains a *target-Namespace* attribute. This target namespace is set to the value: *http://www.xmlbio.org/protein*. Recall that namespaces are used to disambiguate elements with the same names, and that namespace values are merely identifiers. Namespace values therefore do not necessarily need to actually point to meaningful resources. For example, if you type "*http://www.xmlbio.org/protein*" into a web browser, you get a 404 Not Found error. Nonetheless, our namespace value is still a useful, unique identifier and newly defined elements, such as `protein` and `organism`, now belong to this unique namespace.

One consequence of using a target namespace is that all references to newly defined schema components must be namespace qualified. For example, consider the following variation for the

Listing 4.20 Using a target namespace

```
<?xml version="1.0" encoding="UTF-8"?>
<xs:schema targetNamespace="http://www.xmlbio.org/protein"
    xmlns:prot="http://www.xmlbio.org/protein"
    xmlns:xs="http://www.w3.org/2001/XMLSchema"
    elementFormDefault="qualified">
<xs:annotation>
    <xs:documentation>
    Simplified Protein Schema that illustrates basic namespace
      issues.
    </xs:documentation>
</xs:annotation>
<xs:element name="protein">
    <xs:complexType>
        <xs:sequence>
            <xs:element name="entry_name" type="xs:string"/>
            <xs:element ref="prot:organism"/>
        </xs:sequence>
    </xs:complexType>
</xs:element>
<xs:element name="organism">
    <xs:complexType>
        <xs:simpleContent>
            <xs:extension base="xs:string">
                <xs:attribute name="taxonomy_id"
                    type="xs:integer" use="required"/>
            </xs:extension>
        </xs:simpleContent>
    </xs:complexType>
</xs:element>
</xs:schema>
```

`protein` element declaration:

```
<xs:element name="protein">
  <xs:complexType>
    <xs:sequence>
    <xs:element name="entry_name" type="xs:string"/>
      <xs:element ref="organism"/>
    </xs:sequence>
  </xs:complexType>
</xs:element>
```

If you look closely, you will note that the second local element references the global `organism`
element. However, `organism` is actually in the target namespace and must therefore be namespace
qualified.

To address this issue, the example in Listing 4.20 actually includes two namespace declarations.
The first is the namespace for the XML Schema specification. The second is named "prot" and
specifies the value: "*http://www.xmlbio.org/protein.*" This is the same value as our target namespace. The "prot" namespace prefix can therefore reference any items within the target namespace.
In fact, the `protein` element does just that, and specifically references the `organism` element like
the following:

```
<xs:element name="protein">
  <xs:complexType>
    <xs:sequence>
      <xs:element name="entry_name" type="xs:string"/>
      <xs:element ref="prot:organism"/>
    </xs:sequence>
  </xs:complexType>
</xs:element>
```

The value of the *ref* attribute is now specified as a *qualified name*. The first part specifies the
namespace prefix, and the second part specifies the local element name.

What about instance documents? If an instance document adheres to a schema, and that
schema does not have a target namespace, you use the *noNamespaceSchemaLocation* attribute.
We illustrated this approach early in the chapter in Listing 4.2. However, if an instance document adheres to a schema and that schema does have a target namespace, you must use the
schemaLocation attribute. The peculiar nature of this attribute is that it contains two values. The
first value is the namespace of the schema, and the second value is the location of the schema
document itself. In other words, it associates a namespace identifier with a concrete schema
document.

Listing 4.21 shows one such instance document. In particular, note that the example uses the
schemaLocation attribute, and that this value references the protein namespace and the schema
document itself. Note also that all protein schema elements must now be namespace qualified. This
is accomplished via a namespace declaration for the namespace prefix, "prot." All protein schema
elements, including `protein`, `entry_name`, and `organism`, are then qualified with the "prot"
namespace prefix.

If you do not feel like providing a namespace prefix for each element, you can declare a default
namespace. This approach is shown in Listing 4.22.

Listing 4.21 Using the schemaLocation attribute

```
<?xml version="1.0" encoding="UTF-8"?>
<prot:protein xmlns:prot="http://www.xmlbio.org/protein"
    xmlns:xsi="http://www.w3.org/2001/XMLSchema-instance"
    xsi:schemaLocation="http://www.xmlbio.org/protein protein_ns1.xsd">
        <prot:entry_name>IL3B_MOUSE</prot:entry_name>
        <prot:organism taxonomy_id="10090">Mus musculus</prot:organism>
</prot:protein>
```

Listing 4.22 Using the schemaLocation attribute with a default namespace

```
<?xml version="1.0" encoding="UTF-8"?>
<protein xmlns:prot="http://www.xmlbio.org/protein"
    xmlns:xsi="http://www.w3.org/2001/XMLSchema-instance"
    xsi:schemaLocation="http://www.xmlbio.org/protein protein_ns1.xsd"
    xmlns="http://www.xmlbio.org/protein">
        <entry_name>IL3B_MOUSE</entry_name>
        <organism taxonomy_id="10090">Mus musculus</organism>
</protein>
```

This section only presents the tip of the iceberg for namespace issues, and we have intentionally glossed over some of the more advanced concepts. Nonetheless, it does provide enough information to get you started, and to work with most basic and intermediate examples. If you want to dig deeper, Priscilla Walmsley provides an in-depth discussion in Chapter 3, "Namespaces," of her book, *Definitive XML Schema* (Prentice–Hall, 2002) [54].

4.6   Case Study: The HUPO PSI Molecular Interaction Format

Now that you have a firm grasp of XML Schemas, let us turn to a specific case study in bioinformatics. The HUPO PSI Molecular Interaction (PSI-MI) [46; 50] format is a standard for representing and exchanging protein–protein interactions. It is one of the first XML standards in the emerging field of proteomics, and aims to provide a common format for several heterogeneous interaction databases. With recent advances in proteomics, high-throughput techniques are now available for detecting large-scale protein–protein networks. However, much of this data remains dispersed among many databases, and each of these databases has its own specific database platform and data format. With a common format for interaction data, applications will be able to more easily aggregate data from diverse databases and import this data into modeling tools and visualization applications. Interaction databases will also be able to more easily exchange and share data.

PSI-MI was developed under the auspices of the Proteomics Standards Initiative (PSI) of the Human Proteome Organization (HUPO), and is currently endorsed by a number of major biological interaction databases. These include: the Biomolecular Interaction Network Database (BIND), the Database of Interacting Proteins (DIP), IntAct, the Molecular Interactions Database (MINT), and the Human Protein Reference Database (HPRD).* Each of these databases plans to provide its data in PSI-MI format, and by the time this book goes to press, most will have already completed the process. Check each database web site for complete details.

PSI-MI is maintained as an open source project and is hosted on Source Forge at: *http://psidev.sourceforge.net.*

The PSI-MI standard consists of an XML Schema, and a set of externally controlled vocabulary terms (more on the controlled vocabulary in Section 4.6.3). The standard is currently being developed among several working groups with a multilevel approach. Level 1 represents the first level of standardization and focuses exclusively on protein–protein interactions. As the standard evolves, incremental levels will be developed and will provide additional layers of complexity and detail.

As this book goes to press, work on PSI-MI Level 2 is in progress. For full details, visit the PSI-MI web site at: *http://psidev.sourceforge.net.*

The XML Schema for PSI-MI is approximately 14 pages long. Rather than examining the entire schema, we will instead give you an overview of the main structure, examine a few excerpts, and take a look at a sample document. We will also briefly explore the mechanics of using the controlled vocabulary terms. If you want to view the full schema, you can download a copy from the PSI web site at: *http://psidev.sourceforge.net.*

4.6.1   PSI-MI Schema Overview

To get started, let us consider the main structure of the PSI-MI schema. Figure 4.3 shows an overview of the top-level elements, as generated by XMLSpy®. The root element is named `entrySet`, and it can contain one or more `entry` elements. The `entry` element is the main record type and is defined to contain six elements, which are defined as follows:

- `source` (optional): describes the source of the data. This is usually the name of a specific database or a specific institution.
- `availability` (optional): describes the availability of the data. For example, the data may be freely available to the public or may contain copyright statements.
- `experimentList` (optional): describes a list of experiments used to identify interactions. This includes a literature reference that fully describes the experiment, and one or more controlled vocabulary terms that describe the experimental technique.
- `interactorList` (optional): describes a list of interactors, which participate in interactions. Currently, this consists of protein interactors only, but may be expanded in the future to include other entities, such as small molecules.

---

*Information on each of these databases is available online. BIND: *http://www.bind.ca*. DIP: *http://dip.doe-mbi.ucla.edu*. IntAct: *http://www.ebi.ac.uk/intact*. MINT: *http://cbm.bio.uniroma2.it/mint*. HPRD: *http://www.hprd.org*.

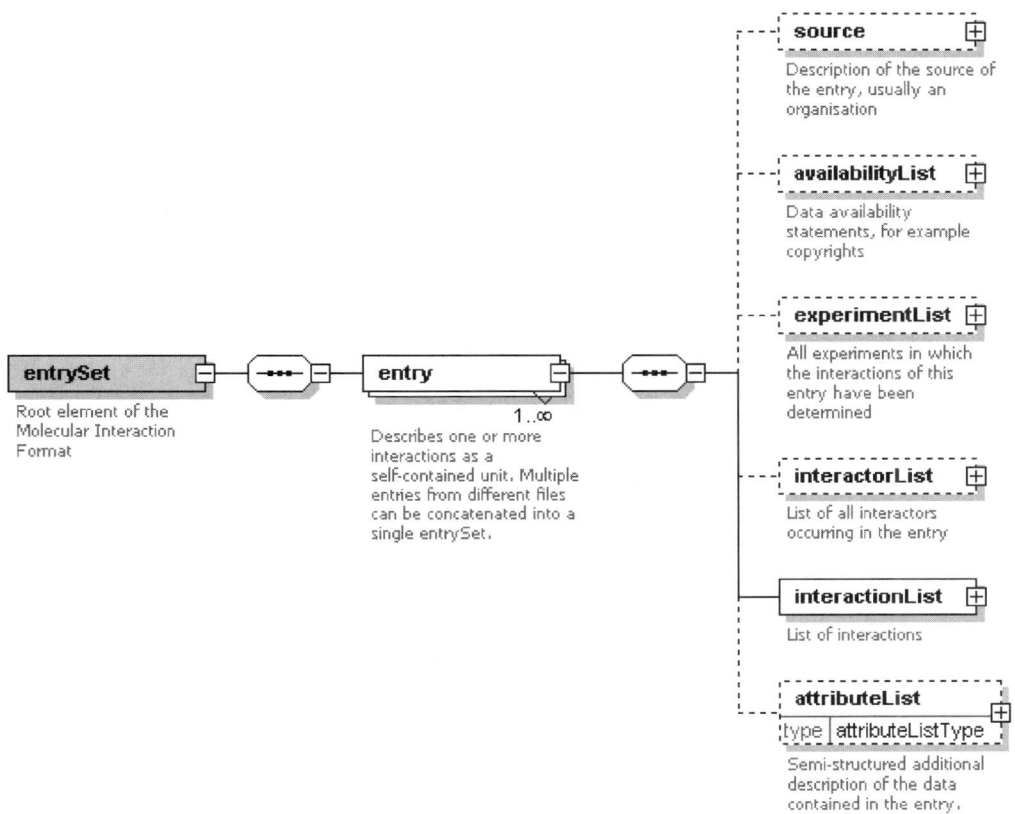

Figure 4.3 Visual overview of the PSI-MI XML Schema.

- `interactionList` (required): this is the heart of the exchange format, as it represents the actual protein–protein interactions. Interactions minimally consist of an experimental description and two or more protein participants.
- `attributeList` (optional): describes additional details about the entry. Attributes are designed to be semistructured in order to accommodate a wide range of additional details.

4.6.2   A Sample PSI-MI Instance Document

To explore PSI-MI further, let us examine a sample instance document. Listing 4.23 shows a sample file that includes exactly one interaction. The specified interaction occurs between two yeast proteins: YER168C and YHR174W.

First, note that the instance document references the PSI-MI schema via the *schemaLocation* attribute. The PSI-MI namespace is "net:sf:psidev:mi" and we are referencing a local copy of the schema. Also, note that we have declared a default namespace, *xmlns* ="net:sf:psidev:mi." All unqualified elements, such as `entry`, `interactorList`, etc., use this default namespace, and are therefore within the PSI namespace.

Listing 4.23 Sample PSI-MI instance document

```xml
<?xml version="1.0" encoding="UTF-8"?>
<entrySet level="1" version="1" xmlns="net:sf:psidev:mi"
xmlns:xsi="http://www.w3.org/2001/XMLSchema-instance"
xsi:schemaLocation="net:sf:psidev:mi MIF.xsd">
  <entry>
    <interactorList>
      <proteinInteractor id="YER168C">
        <names>
          <shortLabel>YER168C</shortLabel>
          <fullName>tRNA nucleotidyltransferase tRNA
            CCA-pyrophosphorylase</fullName>
        </names>
      </proteinInteractor>
      <proteinInteractor id="YHR174W">
        <names>
          <shortLabel>YHR174W</shortLabel>
          <fullName>enolase</fullName>
        </names>
      </proteinInteractor>
    </interactorList>
    <interactionList>
      <interaction>
        <experimentList>
          <experimentDescription id="exp1">
            <bibref>
              <xref>
                <primaryRef db="pubmed" id="11805826"/>
              </xref>
            </bibref>
            <interactionDetection>
              <names>
                <shortLabel>affinity chromatography
                  technologies</shortLabel>
              </names>
              <xref>
                <primaryRef db="PSI-MI" id="MI:0004"/>
              </xref>
            </interactionDetection>
          </experimentDescription>
        </experimentList>
        <participantList>
          <proteinParticipant>
            <proteinInteractorRef ref="YER168C"/>
          </proteinParticipant>
          <proteinParticipant>
            <proteinInteractorRef ref="YHR174W"/>
          </proteinParticipant>
        </participantList>
      </interaction>
    </interactionList>
  </entry>
</entrySet>
```

Second, note that the `interactorList` consists of two protein interactors. Each of these interactors includes a required *id* attribute. This attribute value must be unique, and serve as a global reference throughout the instance document. You can then reference a specific protein later in the instance document by simply referencing its unique ID. For example, the `interaction` element includes two protein participants, each of which references the interactors defined earlier in the file. In PSI-MI, this technique is formally known as the *canonical* or compact form. However, PSI-MI also supports a noncanonical or expanded form. In the expanded form, interactors are not referenced via IDs; rather, interactors are defined directly within interaction elements and can be repeated as many times as necessary.

In Listing 4.23, we have intentionally defined the protein interactors with a minimal set of data. However, interactors can also include more detailed information, such as cross-reference links to other databases, organism information, and sequence data.

The PSI-MI schema enforces that interactions only reference existing protein IDs. Within XML Schemas, this is referred to as referential integrity. Referential integrity can be maintained via Schema key/keyref constraints. Full details are available in Chapter 9 of Eric van der Vlist's book, *XML Schema* (O'Reilly, 2002) [53].

Lastly, note that the `interaction` element includes an `interactionDetection` element. This element is defined to be of type `cvType`. A `cvType` is a complex type used to reference externally controlled vocabulary terms. To dig a little deeper into how this works, let us take a look at the XML Schema excerpt for `cvType` and a few other related types (see Listing 4.24).

Listing 4.24 Excerpts from the PSI-MI XML Schema

```
PSI-MI cvType:

<xs:complexType name="cvType">
    <xs:annotation>
        <xs:documentation>
        Reference to an external controlled vocabulary.
        </xs:documentation>
    </xs:annotation>
    <xs:sequence>
        <xs:element name="names" type="namesType"/>
        <xs:element name="xref" type="xrefType"/>
    </xs:sequence>
</xs:complexType>

PSI-MI namesType:

<xs:complexType name="namesType">
    <xs:annotation>
        <xs:documentation>Names for an object.</xs:documentation>
    </xs:annotation>
    <xs:sequence>
```

Listing 4.24 (*cont.*)

```
        <xs:element name="shortLabel" type="xs:string"/>
        <xs:element name="fullName" type="xs:string" minOccurs="0"/>
    </xs:sequence>
</xs:complexType>


 PSI-MI xrefType:


<xs:complexType name="xrefType">
    <xs:annotation>
        <xs:documentation>
        Crossreference to an external database.
        </xs:documentation>
    </xs:annotation>
    <xs:sequence>
        <xs:element name="primaryRef" type="dbReferenceType">
            <xs:annotation>
                <xs:documentation>Primary reference to an
                external database.</xs:documentation>
            </xs:annotation>
        </xs:element>
        <xs:element name="secondaryRef" type="dbReferenceType"
            minOccurs="0" maxOccurs="unbounded">
            <xs:annotation>
                <xs:documentation>Further external objects
                describing the object.</xs:documentation>
            </xs:annotation>
        </xs:element>
    </xs:sequence>
</xs:complexType>


 PSI-MI dbReferenceType:


<xs:complexType name="dbReferenceType">
    <xs:annotation>
        <xs:documentation>Refers to a unique object in
        an external database.</xs:documentation>
    </xs:annotation>
    <xs:attribute name="db" type="xs:string" use="required"/>
    <xs:attribute name="id" type="xs:string" use="required"/>
    <xs:attribute name="secondary" type="xs:string" use="optional"/>
    <xs:attribute name="version" type="xs:string" use="optional"/>
</xs:complexType>
```

The cvType is defined as a complex type with two child elements. These child elements must appear within the sequence specified, and each element uses the *type* attribute to reference another complex type. For example, the names element is defined to be of type: namesType, and the xref element is defined to be of type: xrefType. These additional complex types are also included in Listing 4.24. For example, you can see that the xrefType includes a `primaryRef` and any number of `secondaryRefs`. Each of these is defined to be of type: `dbReferenceType`.

Hopefully, this gives you a flavor for the inner mechanics of the PSI-MI Schema. If you have been following this chapter closely, none of the concepts should be new. In fact, these are all the same concepts we have been discussing throughout the chapter. The main difference is that PSI-MI simply has more layers of data than our earlier examples. For example, our protein schema has just a few complex types, whereas the PSI-MI schema has dozens of complex types, and each of these complex types are layered on top of one another. Nonetheless, the concepts remain the same, and you are well equipped to understand the full PSI-MI schema in its entirety.

4.6.3   Working with the PSI-MI Controlled Vocabulary

Several portions of the PSI-MI schema require reference to externally controlled vocabulary terms. These terms are defined externally to the XML schema and are not actually enforced by the schema. However, the controlled vocabulary is a vital component of the PSI-MI standard. In general, controlled vocabularies are also a vital component in sharing and exchanging data, particularly in the realm of bioinformatics.

Controlled vocabularies provide a common set of terms and a set of relationships between those terms. The best known example in bioinformatics is the Gene Ontology (GO) [41; 45]. GO provides a comprehensive set of terms for annotating genes, including three categories of data: biological process, molecular function, and cellular component. The terms are defined within two sets of files. The first file defines relationships between terms, and the second file defines the terms themselves, including a short description and a literature reference.

The PSI-MI controlled vocabulary is based on GO, and uses the GO file format. The evolving list of PSI terms is available at the Open Biological Ontologies (OBO) web site: *http://obo.sourceforge.net*. Here, you will find terms for several categories of data. For example, you will find terms for interaction detection—this is a catalog of experimental techniques that are used to determine physical interactions between proteins. You will also find terms for participant detection—this is a catalog of experimental techniques used to identify specific participants within an interaction. By sticking with these defined terms, it is much easier to conclusively identify which experimental techniques were used, and it is therefore much easier to share data with others and integrate data from heterogeneous data sources.

Within the PSI ontology, you will also find terms that describe specific protein features. For example, the binding site term (MI:0117) specifies a binding site where two proteins physically interact. Again, sticking with the defined terms makes it easier to conclusively identify features. If each database specified binding site features with different values, e.g., "binding," "binding-site," "binding site," it would be more difficult to aggregate and compare data.

Note that PSI-MI terms are always specified with a *db* attribute value equal to "PSI-MI," and an *id* attribute value beginning with the prefix "MI." For example, in our sample instance document, we have specified a single controlled vocabulary term. Specifically, we have declared that the interaction between YER168C and YHR174W was determined via "affinity chromatography technologies" (MI:0004).

If you would like to browse the entire PSI-MI ontology, it is best to use one of the many useful GO ontology browsers. For example, the DAG-Edit Java application lets you navigate through a set of terms, view relationships between terms, and view term definitions. It also enables you to create new ontologies or edit existing ontologies. DAG-Edit is freely available for download. Check the Gene Ontology web site (*http://www.geneontology.org/doc/GO.tools.html*) for details.

If you want to explore PSI-MI further, you can download the complete schema file from the PSI web site on SourceForge. You should now be well equipped to understand its complete details. You should also be well equipped to read any other schema files that come your way and even start building your own.

Parsing NCBI XML in Perl 5

Perl remains the programming language of choice for many in bioinformatics. Perl has excellent support for processing and manipulating text, finding regular expression patterns, retrieving files via the Internet, and connecting to a wide variety of relational databases. This makes it an ideal language for parsing flat text files, such as GenBank Flat File records, integrating biological data from multiple sources, and performing sequence analysis. Building on these strengths, the bioinformatics community has developed BioPerl [72], a very successful open source module that includes numerous features, including the ability to retrieve biological data from remote data sources, run BLAST searches, and manipulate sequence data.

Perl also has excellent support for XML, and is supported by a wide variety of third-party open source XML modules. This chapter provides an introduction to XML parsing in Perl, and introduces two standard interfaces: the Simple API for XML (SAX) and the Document Object Model (DOM). To explore SAX, we focus on the XML::SAX module, and to explore the DOM, we focus on the XML::LibXML module. To illustrate basic concepts, the chapter includes numerous examples for parsing XML documents from the National Center for Biotechnology Information (NCBI) at the U.S. National Institutes of Health. We also explore the NCBI EFetch service, and illustrate how to dynamically retrieve and parse sequence records from NCBI.

This chapter assumes that you have a basic familiarity with Perl programming, and understand the fundamentals of object-oriented programming in Perl. If you do not have such background, you may want to check out one of the recommended Perl references [56; 66; 67; 73; 74].

5.1   Introduction to XML Parsing in Perl

A quick search of CPAN, the Comprehensive Perl Archive Network, will reveal several hundred modules pertaining to XML. Given the sheer scope of XML-related modules, it is difficult to figure out where to begin. For example, both XML::Simple [59] and XML::Twig [63] provide "perlish" interfaces for parsing XML documents, and have proven to be quite successful. The original XML::Parser [75] module, originally written by Larry Wall himself, also remains quite popular. Given the multitude of options, we have chosen to focus exclusively on Perl modules, which adhere to specific well-known and well-documented standards. The chapter therefore focuses on Perl implementations of the Simple API for XML (SAX) and the Document Object Model (DOM). If you are interested in some of the other more "perlish" interfaces, such as XML::Simple or XML::Twig, check out some of the recommend references [59; 62].

For general questions about XML and Perl, check out the excellent Perl-XML FAQ [58], available at: *http://perl-xml.sourceforge.net/faq/*.

5.1.1 Tree-Based vs. Event-Based XML Parsers

XML parser interfaces are broadly divided into two categories: *tree-based* and *event-based*. A tree-based interface, such as the DOM, will parse an XML document and build an in-memory tree of all its XML elements. For example, consider the XML document in Listing 5.1. This is a sample GBSeq XML document, retrieved from NCBI. If you send this document through a tree-based interface, the parser will create a tree like that displayed in Figure 5.1. The root element is specified as the GBSet element and your application can navigate through the tree one node at a time. As your application traverses the tree, it can extract any and all data that it needs.

Contrast this with an event-based parser, such as SAX. An event-based interface will read the document one line at a time. Each time the parser encounters an important piece of data, it will immediately fire off an event. For example, when the parser reaches the start <GBSeq_locus> tag, it fires off a start element event. When it sees the text, "BC034957," it immediately fires off one or more character events. The same GBSeq example in Listing 5.1 will therefore trigger a very

Listing 5.1 Excerpt of a sample GBSeq document from NCBI

```
<?xml version="1.0"?>
<!DOCTYPE GBSet PUBLIC "-//NCBI//NCBI GBSeq/EN"
"http://www.ncbi.nlm.nih.gov/dtd/NCBI_GBSeq.dtd">
<GBSet>
 <GBSeq>
  <GBSeq_locus>BC034957</GBSeq_locus>
  <GBSeq_length>2547</GBSeq_length>
  <GBSeq_strandedness value="not-set">0</GBSeq_strandedness>
  <GBSeq_moltype value="mrna">5</GBSeq_moltype>
  <GBSeq_topology value="linear">1</GBSeq_topology>
  <GBSeq_division>PRI</GBSeq_division>
  <GBSeq_update-date>04-OCT-2003</GBSeq_update-date>
  <GBSeq_create-date>15-OCT-2002</GBSeq_create-date>
  <GBSeq_definition>Homo sapiens a disintegrin and metalloproteinase
  domain 2 (fertilin beta), mRNA (cDNA clone MGC:26432 IMAGE:4826530),
  complete cds</GBSeq_definition>
  <GBSeq_primary-accession>BC034957</GBSeq_primary-accession>
  <GBSeq_accession-version>BC034957.2</GBSeq_accession-version>
  <GBSeq_other-seqids>
    <GBSeqid>gb|BC034957.2|</GBSeqid>
    <GBSeqid>gi|34783181</GBSeqid>
  </GBSeq_other-seqids>
  ...
 </GBSeq>
</GBSet>
```

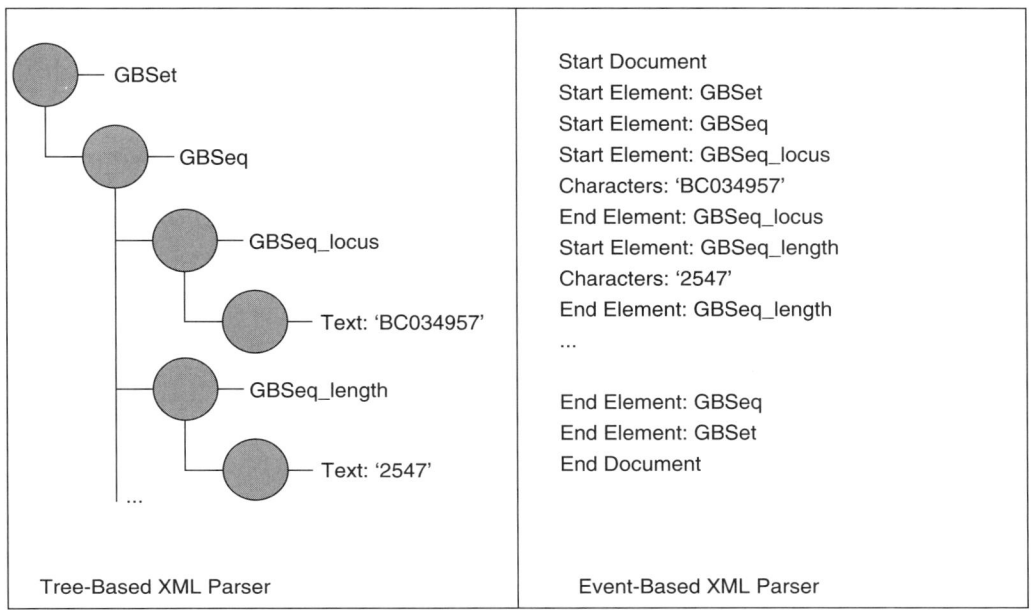

Tree-Based XML Parser	Event-Based XML Parser
GBSet GBSeq GBSeq_locus Text: 'BC034957' GBSeq_length Text: '2547' ...	Start Document Start Element: GBSet Start Element: GBSeq Start Element: GBSeq_locus Characters: 'BC034957' End Element: GBSeq_locus Start Element: GBSeq_length Characters: '2547' End Element: GBSeq_length ... End Element: GBSeq End Element: GBSet End Document

Figure 5.1 Tree-based vs. event-based XML parsing.

specific sequence of events (see Figure 5.1). To extract the XML data, your application must be registered to receive parsing events and act upon them appropriately.

In event-based parsing, the XML parser is typically referred to as an event *producer*, and your application handler is referred to as an application *consumer* [61]. As the XML parser reads in an XML document, the parser will "push" events to the application consumer, and the consumer can choose to record these events or ignore them. Note that event-based parsers are always sequential, and therefore do not provide random access to the XML document content.

5.1.2   Installing Modules via CPAN

All of the modules discussed in this chapter are available from CPAN, the Comprehensive Perl Archive Network. The easiest way to install them is via the interactive CPAN shell. For example, the following command starts the interactive CPAN shell and installs XML::SAX:

```
perl -MCPAN -e shell

cpan shell -- CPAN exploration and  modules installation (v1.61)
ReadLine support available (try  'install Bundle::CPAN')

cpan> install XML::SAX
```

Note, however, that the XML::LibXML module (discussed in the second half of the chapter) requires platform specific binary files, and may present specific installation challenges. If you are using a Windows platform, you may be able to use the Active State Programmer's Package Module (PPM), to automatically install XML::LibXML. For up-to-date information about platform-specific installation issues, refer to the Perl-XML FAQ [58].

The CPAN Shell can sometimes be a bit daunting, even to those with considerable Perl experience. If you find yourself having difficulties, check out the official CPAN documentation at: *http://www.perl.com/doc/manual/html/lib/CPAN.html*. Alternatively, check out Section 2.4, "Getting and Installing Modules," in *Perl in a Nutshell, 2nd edition* (O'Reilly, 2002).

5.2   The Simple API for XML (SAX)

5.2.1   Introduction to SAX

The Simple API for XML (SAX) is a standard event-based interface for parsing XML documents [71]. Unlike XML itself or the Document Object Model (DOM), SAX is not an official standard of any organization, such as the World Wide Web Consortium (W3C). Rather, SAX is a de facto standard, developed by a group of volunteers, freely available to the public, and widely implemented by dozens of XML parsers. SAX was originally designed for Java, but SAX and SAX-inspired implementations are now available for other languages, including Perl, Python, C++, Visual Basic, and Pascal (for a complete listing of SAX implementations see [71]).

The official SAX web site is available at: *http://www.saxproject.org*. Continued SAX development is now hosted at SourceForge.net.

5.2.2   SAX and Bioinformatics Applications

Using an event-based XML parsing interface like SAX provides a number of advantages, particularly when used for bioinformatics applications. First, SAX is a de facto standard and has wide support within the industry. Second, you can learn the SAX interface in one language and immediately apply it to a second language; for example, you can learn the Perl XML::SAX interface and apply your knowledge to the Java SAX interface.

Third, SAX is very fast and requires little memory. Unlike a tree-based interface, SAX will not store a complete representation of the XML document within memory. After an event has been reported to the application handler, the parser will immediately discard the event. This saves memory and is particularly important when parsing very large XML documents. For example, consider the implications of parsing the complete contents of UniProt [55], a comprehensive database of protein sequences and annotations. From the UniProt web site, you can download a complete snapshot of the database in UnitProt XML format. However, the complete snapshot consists of a single XML document, which is several hundreds of megabytes long. To parse this document via a tree-based interface, you need enough memory to hold the entire tree, and you always run the risk of receiving "out of memory" errors. In contrast, SAX requires very little overhead and is capable of parsing arbitrarily large documents.

Although SAX itself is very memory efficient, keep in mind that your application handler can choose to store any and all XML events within its own internal data structures, and this will require its own set of memory requirements. You therefore need to carefully evaluate the specific requirements for your application. For example, if you want to parse the UnitProt XML file, and locate only a specific subset of proteins, your application handler can choose to record events specific to the target set, and discard the rest. Alternatively, if you want to import the UnitProt XML file into a set of relational database records, you only need to store one protein record at a time. After each record is committed to the database, you can purge your internal data structures, and move onto the next record.

The main disadvantage of SAX is that it does not provide facilities for easily modifying existing XML documents, or easily creating new XML documents from scratch. Some programmers also find the capturing and processing of individual SAX events counterintuitive, and prefer the simplicity of a tree traversal API.

5.2.3 SAX 2.0

The SAX API was originally developed by a group of volunteers, coordinating over the xml-dev mailing list. SAX 1.0 was originally released in May 1998 [65]. SAX 2.0.1 was officially released in January 2002. Several major changes were introduced in SAX 2.0, including complete support for XML Namespaces, standard methods for configuring XML parser features and properties, and support for SAX filters, enabling you to chain XML applications together [57; 64]. This book focuses on SAX 2.0 only. For details regarding SAX 1.0, refer to the SAX web site at: *http://www.saxproject.org/?selected=sax1*.

5.2.4 Introduction to XML::SAX

The Perl XML::SAX [70] module provides a complete implementation of the SAX 1.0 and 2.0 interfaces. The module is available for download from CPAN, and includes an XML parser written in Perl, called, XML::SAX::PurePerl [68]. PurePerl is considered quite slow, but it enables XML::SAX to work right out of the box on all platforms. XML::SAX also works with other faster SAX-compliant XML parsers, such as XML::LibXML [69], and provides a simple interface for selecting an XML parser at runtime.

To get started with XML::SAX, follow these three steps:

- Create a SAX event handler. As the XML parser reads a document, it will encounter specific XML constructs and notify your handler via callback methods. For example, when the XML parser encounters a new start element tag, it will call the event handler `start_element()` method, and pass information about the element. As a convenience, your event handler can extend `XML::SAX::Base`, and provide implementations of only those call-back methods that are of interest.
- Obtain an XML parser and register your event handler with the parser. XML::SAX provides a factory class, called `ParserFactory`, which will locate, instantiate, and initialize an XML

parser of your choosing. Later in this section, we will also explore options for selecting an XML parser at runtime.

- Initiate parsing by calling one of the `parse_xxx()` methods. For example, you can parse a local document by calling the `parse_uri()` method and supplying an absolute or relative path to the file. Alternatively, you can use the `parse_file()` method and pass in a stream or file handle, such as an `IO::File`. You can also use the `parse_string()` method and pass in the complete XML document as one string. In all cases, the XML parser will immediately start parsing the document of your choosing, and report each XML event to the registered handler.

To illustrate the basic concepts of XML::SAX, consider the sample Perl code in Listing 5.2.

The source code in Listing 5.2 instantiates a parser object via the `XML::SAX::ParserFactory` class. By default, the `ParserFactory` will look for a package variable to determine which parser to instantiate. For example, the following package variable will load the PurePerl SAX parser:

```
$XML::SAX::ParserPackage="XML::SAX::PurePerl";
```

Alternatively, the following package variable will load the LibXML SAX parser:

```
$XML::SAX::ParserPackage="XML::LibXML::SAX::Parser";
```

If no package variable is set, XML::SAX will automatically search all the directories in @INC in search of a SAX.ini file. The SAX.ini file uses a simple key=value format. For example, the following file specifies the LibXML SAX parser:

```
ParserPackage = XML::LibXML::SAX::Parser
```

If XML::SAX is unable to find a package variable or a SAX.ini file, it will automatically default to the PurePerl SAX parser.

Listing 5.2 First XML::SAX application

```perl
#!/usr/bin/perl
# Basic SAX Example.
# Author: Ethan Cerami
use strict;
use XML::SAX;
use ContentReporter;

# Create ElementReporter Instance
my $handler = ContentReporter->new;

# Obtain SAX Parser via ParserFactory
my $parser  = XML::SAX::ParserFactory->parser(Handler =>$handler);

# Parse TinySeq XML Document
my $element_counter = $parser->parse_uri("sample/ncbi.xml");

print "Total Number of Elements: $element_counter\n";
```

Tip: If you want to confirm which SAX parser is currently in use, you can use the very helpful `Data::Dumper` Perl module. Just obtain an XML parser via the `ParserFactory` and dump out its data. For example:

```
use Data::Dumper;
...
my $parser = XML::SAX::ParserFactory->parser(
        Handler =>$handler);
print Dumper ($parser);
```

The `Data::Dumper` will display all information about the parser, including the package name of the selected parser, and a list of all features and properties.

The bulk of the work in using XML::SAX goes into creating a custom SAX event handler. A sample event handler, called `ContentReporter`, is shown in Listing 5.3. Examine the code now, and we will survey its components below.

The `ContentReporter` in Listing 5.3 extends `XML::SAX:Base`, and selectively listens for four types of events: start_document, end_document, start_element, and end_element. The `ContentReporter` also keeps a running count of the number of elements encountered, and returns the total count to the main calling application. When we use this event handler on our sample GBSeq document in Listing 5.1, we get the following output:

```
Start Document
Start Element: GBSet
Start Element: GBSeq
Start Element: GBSeq_locus
End Element: GBSeq_locus
Start Element: GBSeq_length
End Element: GBSeq_length
Start Element: GBSeq_strandedness
...
End Document
Total Number of Elements: 320
```

Depending on the call-back method, you may or may not receive additional information about the event. For example, when the parser encounters the beginning of an XML document, it will call the `start_document()` method, but will not pass any additional event details. By contrast, when the parser encounters a new element, it will call the `start_element()` method, and pass along specific element details. Event details are specified as a hash reference with specific predefined keys. For example, the `start_element()` method receives a hash reference with several element specific keys, including "Name," "LocalName," and "Prefix." You can then reference these keys to display additional information about the event. For example:

```
sub start_element {
    my ($self, $element) = @_;
    my $name = $element->{"Name"};
    print "Start Element: $name\n";
}
```

The element hash reference also contains an optional "Attributes" key, which contains all attribute data associated with the element. For example, the following code will extract all attribute data and display it to the console:

```perl
# Extract All Attributes
my $attributes = $element->{"Attributes"};
foreach my $key (keys %$attributes) {
    my $attribute = $attributes->{$key};
    my $name = $attribute->{"Name"};
    my $value = $attribute->{"Value"};
    print "Attribute: $name --> $value\n";
}
```

Listing 5.3 ContentReporter.pm

```perl
#!/usr/bin/perl
# Basic SAX Handler
# Reports Basic Content Events
use strict;
package ContentReporter;

# Extend XML::SAX::Base
use base qw (XML::SAX::Base);

my $element_counter = 0;

# Report Start Document Event.
sub start_document {
    my ($self, $doc) = @_;
    print "Start Document\n";
}

# Report Start Element Events.
sub start_element {
    my ($self, $element) = @_;
    my $name = $element->{"Name"};
    print "Start Element: $name\n";
    $element_counter++;
}

# Report End Element Events.
sub end_element {
    my ($self, $element) = @_;
    my $name = $element->{"Name"};
    print "End Element: $name\n";
}

# Report End Document.
sub end_document {
    my ($self, $doc) = @_;
    print "End Document\n";
    return $element_counter;
}
1;
```

Table 5.1 provides a listing of the main methods in `XML::SAX::Base`. Note in particular that the `end_document()` method is the final method called, and that its return value is returned by the `parse_xxx()` methods. This provides a convenient mechanism to propagate data from the event handler back to the main calling application.

Table 5.1 Main methods of `XML::SAX::Base`

Method	Description
attribute_decl ($self, $attribute_info)	Indicates a DTD Attribute Declaration. The $attribute_info parameter is a hash reference containing the following keys: • Type: the attribute type, e.g. CDATA or ID • eName: the element name • aName: the attribute name • ValueDefault: attribute default flag, e.g. #REQUIRED or #IMPLIED • Value: default value, or undef if a default value is not supplied
characters ($self, $text)	Indicates a character event. The $text parameter is a hash reference containing a single key: • Data: contains the character content. Given a string of XML text, each SAX parser is free to report character events as it sees fit. For example, one parser may report the entire text string via a single call to characters(); a second parser may split the string and call characters() twice. Given this flexibility, it is important that your handler provide some type of character buffering
comment ($self, $comment)	Indicates an XML comment. The $comment parameter is a hash reference containing a single key: • Data: contains the comment text
element_decl ($self, $element_info)	Indicates a DTD Element Declaration. The $element_info parameter is a hash reference containing the following keys: • Name: name of the element • Model: content model of the element
end_cdata ($self)	Indicates the end of a CDATA section
end_document ($self)	Indicates the end of an XML document. The return value of end_document() is returned by the parse_xxx() methods, and therefore provides a convenient mechanism for propagating data from the event handler back to the main calling application
end_element ($self, $element)	Indicates the end of an XML element. The $element parameter is a hash reference containing the same keys as those defined in start_element(). See start_element() for details
end_prefix_mapping ($self, $namespace_info)	Indicates the end scope of a namespace declaration. The $namespace_info parameter is a hash reference containing the following keys: • Prefix: the namespace prefix, e.g., "psi" • NamespaceURI: the namespace URI, e.g., "net:sf:psidev:mi"
processing_instruction ($self, $pi)	Indicates a processing instruction. The $pi parameter is a hash reference containing the following keys: • Target: the target of the processing instruction • Data: the complete text of the processing instruction
set_document_locator ($self, $doc_locator)	Sets a document locator object. This is usually the very first method called, directly before a call to start_document(). If you want to determine the location of all subsequent SAX events, store the $doc_locator object locally and reference it within other SAX call-back methods. The $doc_locator parameter is a hash reference, containing the following keys: • ColumnNumber: column number where the event occurred • LineNumber: line number where the event occurred

Table 5.1 (*cont.*)

Method	Description
	• SystemId: the system identifier of the current document or undef, if it is not defined
	• PublicId: the public identifier of the current document or undef, if it is not defined
	Note that SAX parsers are strongly encouraged to supply a document locator, but are not required to do so
`start_cdata ($self)`	Indicates the start of a CDATA section. The actual content of the CDATA section is subsequently reported via the `characters()` method
`start_document ($self)`	Indicates the start of an XML document
`start_element` `($self, $element)`	Indicates the start of an XML element. The `$element` parameter is a hash reference containing the following keys: • Prefix: the namespace prefix of the element, e.g., "psi" • LocalName: the local name of the element. This is the name of the element, without its namespace prefix, e.g., "interaction" • Name: a fully qualified element name. This is the name of the element with its namespace prefix, e.g., "psi:interaction" • NamespaceURI: the namespace URI of this element • Attributes: a hash reference containing all the element's attributes If the element has attributes, the Attributes hash reference will contain one key for each attribute. The key is specified in the following form: "{NamespaceURI}LocalName". If the attribute is not associated with any namespace, it will have an empty NamespaceURI, e.g., "{ }LocalName". Each attribute will in turn contain the following keys: • Prefix: the namespace prefix of the attribute • LocalName: the local name of the attribute • Value: the attribute value • Name: a fully qualified attribute name • NamespaceURI: the namespace URI of the attribute
`start_prefix_mapping` `($self, $namespace_info)`	Indicates the beginning scope of a namespace declaration. The `$namespace_info` parameter is a hash reference containing the following keys: • Prefix: the namespace prefix, e.g., "psi" • NamespaceURI: the namespace URI, e.g., "net:sf:psidev:mi"
`xml_decl` `($self, $declaration)`	Indicates an XML declaration. The `$declaration` parameter is a hash reference containing the following keys: • Version: XML version, e.g., "1.0" • Encoding: character encoding, e.g., "UTF-8"

Error Handling

It is important to note that if your XML parser encounters an error in well-formedness, the parser will consider this a fatal error and stop program execution via a call to `die()`. (Recall from Chapter 2 that an XML document is considered well-formed if it follows the basic rules of document construction, e.g., all start tags must have matching end tags, elements must be properly nested, attributes must appear in quotes, etc.) If you want to prevent your program from dying, you can wrap your call to `parse_xxx()` inside an eval block. For example:

```
eval {
    $parser->parse_uri("sample/ncbi.xml");
};
```

```
if ($@) {
    my $message = $@->{"Message"};
    my $line_number = $@->{"LineNumber"};
    print "Error! -->  $message\n";
    print "Error Occurred at line number: $line_number\n";
}
```

Note that the $@ hash reference contains a number of predefined keys, such as "Message," "LineNumber," and "ColumnNumber," allowing you to access specific details about the error.

5.2.5   Using NCBI EFetch and XML::SAX

Now that you understand the basics of XML::SAX , you can apply this knowledge to dynamically retrieve and parse sequence data from NCBI. Fortunately for us, NCBI provides a web service, called EFetch that simplifies the process of retrieving sequence records. EFetch is actually an example of a REST-based web service (for details on REST-based web services, refer to Chapter 9). In a nutshell, client applications connect to EFetch via HTTP and specify search criteria with a set of URL parameters. Based on the search criteria, the EFetch service will connect to the NCBI Entrez back-end database system, find a matching record, and return the requested record in the format of your choosing. EFetch currently provides access to several NCBI Entrez databases, including sequence, literature, and taxonomy databases; and can return data in several data formats, including text, HTML, ASN.1 and XML. If you are eager to try out a few sample EFetch requests, refer to Table 5.2.

As of this writing, the base URL for connecting to EFetch is:

http://eutils.ncbi.nlm.nih.gov/entrez/eutils/efetch.fcgi

To retrieve a specific nucleotide sequence record, you must append a database parameter and an ID parameter, which uniquely identifies the record. For example, the following URL retrieves the complete genome record for the SARS coronavirus, formatted in the GenBank flat file format:

Table 5.2 Example EFetch queries

How to Retrieve a Nucleotide Record:
Example #1: Retrieves information regarding the BRCA2 gene in Human, and formats the results in TinySeq XML:
http://eutils.ncbi.nlm.nih.gov/entrez/eutils/efetch.fcgi?db= nucleotide&id=U43746&rettype=fasta&retmode=xml
Example #2: Retrieves information regarding the BRCA2 gene in Human, and formats the results in GenBank XML:
http://eutils.ncbi.nlm.nih.gov/entrez/eutils/efetch.fcgi?db=nucleotide&id=U43746&rettype=gb&retmode=xml

How to Retrieve a Protein Record:
Example: Retrieves information regarding the BRCA2 protein in Human, and formats the results in GenPept XML:
http://eutils.ncbi.nlm.nih.gov/entrez/eutils/efetch.fcgi?db=protein&id=AAB07223.1&rettype=gp&retmode=xml

How to Retrieve a Literature Record:
Example: Retrieves citation and abstract information regarding PMID: 14597658, and formats the results in XML:
http://eutils.ncbi.nlm.nih.gov/entrez/eutils/efetch.fcgi?db=pubmed&id=14597658&retmode=xml

How to Retrieve a Taxonomy Record:
Example: Retrieves the species name for NCBI Taxonomy ID: 7227. In this case, E-Fetch returns a single string: "Drosophila melanogaster".
http://eutils.ncbi.nlm.nih.gov/entrez/eutils/efetch.fcgi?db=taxonomy&id=7227&report=brief

http://eutils.ncbi.nlm.nih.gov/entrez/eutils/efetch.fcgi?db=nucleotide&rettype=gb&retmode=
text&id=30271926

In the URL above, the *db* parameter specifies the NCBI nucleotide database, *rettype* specifies the GenBank flat file format, *retmode* specifies text content, and *id* specifies the NCBI GI number for the SARS virus. Conveniently, the *id* parameter accepts both NCBI GI numbers and NCBI accession numbers.

For XML content, set the *retmode* parameter to "xml." For example, to retrieve data in the NCBI TinySeq XML format, set *rettype=fasta* and *retmode=xml*. To retrieve data in the more comprehensive NCBI GBSeq XML, set *rettype=gb* and *retmode=xml*. For example, the following URL retrieves the same SARS virus record, but this time it is formatted in GBSeq XML:

http://eutils.ncbi.nlm.nih.gov/entrez/eutils/efetch.fcgi?db=nucleotide&rettype=gb&retmode=
xml&id=30271926

Complete details regarding NCBI EFetch are available online at:
http://eutils.ncbi.nlm.nih.gov/entrez/query/static/efetchseq_help.html.

Our goal is to write a Perl program capable of automatically retrieving sequence data from EFetch and extracting a small subset of the XML content for display to the console. The program expects a single command line argument, indicating an NCBI GI number or accession number. A sample run of the application is shown below:

```
>fetch.pl NC_004718
Downloading XML from NCBI E-Fetch
Using URL: http://eutils.ncbi.nlm.nih.gov/entrez/eutils/efetch.
fcgi?db=nucleotide&rettype=gb&retmode=xml&id=30271926
Definition: SARS coronavirus, complete genome
Accession: NC_004718
Locus: NC_004718
Organism: SARS coronavirus
Sequence (0..20):  ATATTAGGTTTTTACCTACC...
```

Source code for the Perl fetcher is shown in Listings 5.4 and 5.5. Examine the code now and we will describe its main components below.

As in our first SAX application, the fetch application consists of two parts: a main application, which initiates parsing (Listing 5.4), and a SAX event handler (Listing 5.5). The main application uses the World Wide Web library for Perl (LWP) [60] to connect to NCBI EFetch and retrieve the specified sequence record. It also obtains an XML parser via the SAX factory, and initiates parsing via the parse_string() method. The parse_string() method returns an associative array, which we then print to the console.

The NcbiHandler.pm module listens for specific SAX events, and selectively stores specific GBSeq elements in an internal associative array. There are a few important items to note. First, the characters() method uses a character buffer. This is important because SAX parsers are free to perform character "chunking"—for example, one SAX parser may report a line of text via a single call to characters(), whereas a second SAX parser may break the line into two "chunks" and report it via two calls to characters(). Since there is no way to know ahead of time which chunking method the parser will use, it is always safest to assume multiple calls to characters() and to append to a character buffer each time. Second, the end_element() method is used to

Listing 5.4 Parsing NCBI EFetch data via the SAX API

```perl
#!/usr/bin/perl
# Fetches NCBI XML from the NCBI E-Fetch Utility.
# Author: Ethan Cerami
use XML::SAX;
use LWP::Simple;
use NcbiHandler;
use strict;

# Display Command Line Usage
if (@ARGV == 0) {
    print "Usage: fetch.pl ncbi_identifier (NCBI GI or Accession
        Number)\n";
die "Example: fetch.pl 30271926\n";
}

# Download File from NCBI e-Fetch; uses LWP Module
my $ncbi_url = get_ncbi_url($ARGV[0]);
print "Downloading XML from NCBI E-Fetch\n";
print "Using URL: $ncbi_url\n";
my $xml_doc = LWP::Simple::get($ncbi_url);

# Parse XML Document
my $handler = NcbiHandler->new;
my $parser = XML::SAX::ParserFactory->parser(Handler =>$handler);
my %data = $parser->parse_string($xml_doc);

# Output Results of Parsing
my $sequence = $data{"GBSeq_sequence"};
$sequence = substr ($sequence, 0, 20);
print "Definition: ",  $data{"GBSeq_definition"};
print "\nAccession: ",  $data{"GBSeq_primary-accession"};
print "\nLocus: ", $data{"GBSeq_locus"};
print "\nOrganism: ", $data{"GBSeq_organism"};
print "\nSequence (0..20): $sequence...\n";

# Gets NCBI Identifier from user, and returns an absolute URL
# to the NCBI E-Fetch Utility.
sub get_ncbi_url {
    my $id = $_[0];

    # Set Base URL for NCBI E-Fetch
    my $baseurl = "http://eutils.ncbi.nlm.nih.gov/entrez/eutils/"
    ."efetch.fcgi?db=nucleotide&rettype=gb&retmode=xml&id=";

    return ($baseurl . $id);
}
```

Listing 5.5 NcbiHandler.pm

```perl
#!/usr/bin/perl -w
# Parses NCBI GBSeq XML Documents, and extracts only
# selected elements.
package NcbiHandler;
use strict;

# Extend XML::SAX::Base
use base qw (XML::SAX::Base);

my ($current_text, %data);

# Report Start Element Events.
# Each time we get a start element event,
# reset the character buffer.
sub start_element {
    my ($self, $element) = @_;
    $current_text = "";
}

# Selectively store element information.
sub end_element {
    my ($self, $element) = @_;
    my $name = $element->{"Name"};
    if ($name eq "GBSeq_locus"
            || $name eq "GBSeq_primary-accession"
            || $name eq "GBSeq_definition"
            || $name eq "GBSeq_organism"
            || $name eq "GBSeq_sequence" ) {
        $data{$name} = $current_text;
    }
}

# Keep Character Buffer.
sub characters {
    my ($self, $characters) = @_;
    $current_text .= $characters->{"Data"};
}

# Return Associative Array to main application.
sub end_document {
    my ($self, $doc) = @_;
    return %data;
}
1;
```

selectively filter for specific GBSeq elements. For those specific elements of interest, we store the current character buffer into an associative array and use the element name as a hash key. We subsequently return the associative array to the main calling application by returning it from the end_document() method.

5.3 The Document Object Model (DOM)

The Document Object Model (DOM) is a standard tree-based interface for reading, modifying, and creating XML documents. The DOM is an official recommendation of the W3C, and the DOM API is specified in the Object Management Group Interface Definition Language (OMG IDL). This enables the DOM API to be both platform and language independent. DOM implementations are available in numerous programming languages, including Perl, Java, JavaScript, VBScript, C/C++, and Python. In this section, we provide an introduction to the Perl `XML::LibXML` module, and illustrate its support for the DOM standard. We also revisit the NCBI EFetch service, and re-create the same functionality as our earlier SAX application. This enables you to directly compare and contrast the DOM approach with the SAX approach.

5.3.1 DOM Traversal with XML::LibXML

The `XML::LibXML` Perl module provides an interface to the very popular libxml parser, an XML parser written in C and developed for the Linux Gnome project. The libxml parser itself is packed with numerous features, including support for XML Namespaces, SAX, DOM, XPath, XPointer, and XInclude. For our purposes, we will be focusing exclusively on the DOM implementation provided by libxml. If you are interested in the other features of libxml, check out the main libxml web site at: *http://xmlsoft.org*.

Let us jump right in with our first DOM example. Listing 5.6 provides an example application, which will traverse through all the elements in our sample NCBI XML document. Examine the code now, and we will describe its main components below.

There are several important elements to note about the code in Listing 5.6. First, we instantiate a new `LibXML` parser object and direct the parser to parse a local file:

```
# Instantiate LibXML Parser
my $parser = XML::LibXML->new();

# Parse Sample Document
my $doc = $parser->parse_file("sample/ncbi.xml");
```

The `parse_file()` method returns a DOM `Document` object. This document object contains a complete tree representation of our XML document. To begin tree traversal, we request the root document element:

```
my $root = $doc->getDocumentElement();
```

We then pass this root element to the recursive `traverse_node()` method. In the DOM data model, all XML constructs, e.g., elements, attributes, text data, and processing instructions, are represented as DOM nodes, and all nodes provide a number of very useful attributes/methods. For example, you can retrieve the node name or node type:

```
my $node_name = $node->nodeName;
my $node_type = $node->nodeType;
```

Depending on the node type, our code takes different actions. For example, if we encounter a text node, we extract the embedded text content. If we encounter an element node, we retrieve a list of all its child nodes and pass these nodes recursively to the `traverse_node()` method:

```perl
my @children = $node->childNodes();
foreach my $child (@children) {
    traverse_node ($child, $indent+1);
}
```

Listing 5.6 First DOM example

```perl
# DOM Traversal Example
# Illustrates Basic DOM Functionality

use XML::LibXML;
use strict;

# Instantiate LibXML Parser
my $parser = XML::LibXML->new();

# Parse Sample Document
my $doc = $parser->parse_file("sample/ncbi.xml");

# Get Document Root Element
my $root = $doc->getDocumentElement();

# Initiate DOM Traversal at root
traverse_node ($root, 0);

# Recursive Function to Traverse DOM Tree
sub traverse_node {
    my ($node, $indent) = @_;

    # Indent X Characters
    my $line = "." x ($indent * 2);
    print "$line";

    # Get Node Name and Type
    my $node_name = $node->nodeName;
    my $node_type = $node->nodeType;

    # Take Different Actions depending on node type
    if ($node_type == 3) {
            # This is a text node
            my $text_content = $node->textContent;
            $text_content =~ s/\n/[new line]/;
            print "$node_name: $text_content\n";
    } elsif ($node_type == 1) {
            # This is an Element Node
            print "Element: $node_name\n";
            # Iterate through all Child Nodes
            my @children = $node->childNodes();
            foreach my $child (@children) {
                    traverse_node ($child, $indent+1);
            }
    }
}
```

By examining each element, and recursively exploring each of its child nodes, our sample application is capable of traversing through the entire XML document tree. An excerpt from this traversal is shown below:

```
Element: GBSet
..text: [new line]
..Element: GBSeq
....text: [new line]
....Element: GBSeq_locus
......text: BC034957
....text: [new line]
....Element: GBSeq_length
......text: 2547
[output continues...]
```

If you want to determine if an element has attributes, you can use the `hasAttributes` query method, and then retrieve those attributes via the `attributes` property. For example, the following code excerpt outputs all element attributes to the console:

```
if ($node->hasAttributes) {
    my @attributes = $node->attributes;
    foreach my $attribute (@attributes) {
            my $name = $attribute->nodeName;
            my $value = $attribute->value;
            print "Attribute Name: $name --> Value: $value\n";
    }
}
```

The DOM API also supports several methods for finding specific subelements. For example, the `getChildrenByTagName()` method will find all direct children with the specified tag name. By contrast, the `getElementsByTagName()` method will recursively search all descendants of the current node, and return all descendants with the specified tag name. The LibXML module also provides support for XPath, a W3C specification that enables you to pinpoint specific elements or sets of elements within an XML document. For example, the code snippet below uses the XPath `find` feature to extract two specific GBSeq elements:

```
my $doc = $parser->parse_file("sample/ncbi.xml");
my $locus = $doc->find("/GBSet/GBSeq/GBSeq_locus");
my $def = $doc->find("/GBSet/GBSeq/GBSeq_definition");
print "Locus: $locus\n";
print "Definition: $def\n";
```

The complete LibXML DOM API is quite large, and we could not hope to cover it in its entirety in this chapter. For complete documentation on all relevant classes and methods, refer to the LibXML documentation, available online at: *http://search.cpan.org/dist/XML-LibXML*. If you are working extensively with LibXML, you may find it useful to print out the documented API for `XML::LibXML::Node`, the base class used to represent all DOM nodes, and `XML::LibXML::Element`, the class used to represent element nodes.

5.3.2   Validating XML Documents with XML::LibXML

The LibXML Perl module provides built-in support for validating XML documents against DTDs. To turn XML validation on, simply pass a true value to the parser `validation()` method, and then initiate parsing. If LibXML encounters an error in well-formedness or validity, it will immediately `die()` and report the error to the console. If you want to prevent your program from exiting, you can wrap the parse call in an eval block. For example:

```
my $parser = XML::LibXML->new();
$parser->validation(1);

eval {
    my $doc = $parser->parse_file($file_name);
};

if ($@) {
    print "Error --> $@\n";
} else {
    print "----> OK\n";
}
```

5.3.3   Creating New Documents with XML::LibXML

In addition to reading in XML documents, the Document Object Model also provides convenient mechanisms for modifying existing documents or creating new XML documents from scratch. For example, the code in Listing 5.7 uses the DOM API to create an entirely new document in the NCBI TinySeq XML format.

To create a new XML document, you must first instantiate a `Document` object:

```
my $document = XML::LibXML::Document->new ();
```

You then need to create root element, and add this to the document:

```
my $root = $document->createElement ("TSeq");
$document->addChild ($root);
```

You can then proceed to create new elements, and add these to the root element. Note that if you want to add text to an element, you must first create a text node, and then add this to the element node. This operation is automatically performed by the `appendTextNode()` method. In the same manner, to create an element with attributes, you must first create one or more attribute objects, and then add each attribute to the element node. The DOM API provides dozens of other methods for creating new nodes, removing nodes, and copying nodes. However, the scope of the complete API is beyond the scope of what we hope to cover here. For the complete API, refer to the LibXML API documentation, available online at: *http://search.cpan.org/dist/XML-LibXML.*

5.3.4   Using NCBI EFetch and XML::LibXML

As our final topic, we revisit the NCBI EFetch service. Our goal is to retain the exact same functionality as our first EFetch application (see Section 5.2.5), but to replace the SAX code with

Listing 5.7 Creating a TinySeq XML document via the DOM API

```perl
# Creates a new TinySeq XML Document via the DOM API.
use XML::LibXML;
use strict;

# Instantiate New Document Object
my $document = XML::LibXML::Document->new ();

# Create Root Element
my $root = $document->createElement ("TSeq");
$document->addChild ($root);

# Add Sequence Type with Attribute
my $seq_type = $document->createElement ("TSeq_seqtype");
my $attribute = $document->createAttribute("value", "nucleotide");
$seq_type->addChild ($attribute);
$root->addChild ($seq_type);

# Add Other Sub-elements
add_element ($document, $root, "TSeq_gi", "11497606");
add_element ($document, $root, "TSeq_sid", "ref| NM_001464.2| ");
add_element ($document, $root, "TSeq_taxid", "9606");
add_element ($document, $root, "TSeq_orgname", "Homo sapiens");
add_element ($document, $root, "TSeq_defline",
    "Homo sapiens a disintegrin and metalloproteinase domain 2 "
    . "(fertilin beta) (ADAM2), mRNA");
add_element ($document, $root, "TSeq_length", "2659");
add_element ($document, $root, "TSeq_sequence", "CATCTCGCACTTC...");

# Convert to XML String, with indentation
my $xml = $document->toString(1);
print "XML Document:\n$xml";

# Adds New Element with Single Text Value
sub add_element {
    my ($document, $parent, $element_name, $text_str) = @_;
    my $child = $document->createElement ($element_name);
    $child->appendTextNode ($text_str);
    $parent->addChild ($child);
}
```

DOM code. By comparing the two examples, you can therefore directly compare the SAX and DOM interfaces and gain insight into both approaches.

The complete source code for our new fetch application is shown in Listing 5.8.

As in our first DOM example, we parse an NCBI XML document and immediately extract its root element. However, instead of traversing the entire XML object tree, we now selectively traverse the tree in search of five specific GBSeq elements. To do so, we first obtain the GBSeq element:

```perl
my @seq_children = $root->getElementsByTagName("GBSeq");
my $seq_node = $seq_children[0];
```

Listing 5.8 Parsing NCBI EFetch data via the DOM API

```perl
# Fetches NCBI XML from the NCBI E-Fetch Utility.
# Author: Ethan Cerami
use XML::LibXML;
use LWP::Simple;
use strict;

# Display Command Line Usage
if (@ARGV == 0) {
    print "Usage: fetch.pl ncbi_identifier (NCBI GI or Accession
    Number)\n";
    die "Example: fetch.pl 30271926\n";
}

# Download File from NCBI e-Fetch; uses LWP Module
my $ncbi_url = get_ncbi_url($ARGV[0]);
print "Downloading XML from NCBI E-Fetch\n";
print "Using URL: $ncbi_url\n";
my $xml_doc = LWP::Simple::get($ncbi_url);

# Instantiate LibXML Parser and Parse Document
my $parser = XML::LibXML->new();
my $doc = $parser->parse_string($xml_doc);

# Get Document Root Element
my $root = $doc->getDocumentElement();

# Get First GBSeq Element
my @seq_children = $root->getElementsByTagName("GBSeq");
my $seq_node = $seq_children[0];

# Extract Individual Elements
my $def_line = get_element_text ($seq_node, "GBSeq_definition");
my $acc = get_element_text ($seq_node, "GBSeq_primary-accession");
my $locus = get_element_text ($seq_node, "GBSeq_locus");
my $organism = get_element_text ($seq_node, "GBSeq_organism");
my $sequence = get_element_text ($seq_node, "GBSeq_sequence");
$sequence = substr ($sequence, 0, 20);

print "Definition: $def_line\n";
print "Accession: $acc\n";
print "Locus: $locus\n";
print "Organism: $organism\n";
print "Sequence (0..20): $sequence...\n";

# Gets Element Text
sub get_element_text {
    my ($node, $target_name) = @_;

    # getChildrenByTagName gets direct children only.
    # getElementsByTagName gets all descendants.
```

Listing 5.8 *(cont.)*

```
    my @elements = $node->getChildrenByTagName($target_name);

    if (@elements) {
            my $element = $elements[0];
            my $text = $element->textContent;
    } else {
            return "Not Available";
    }
}


# Gets NCBI Identifier from user, and returns an absolute URL
# to the NCBI E-Fetch Utility.
sub get_ncbi_url {
    my $id = $_[0];

    # Set Base URL for NCBI E-Fetch
    my $baseurl = "http://eutils.ncbi.nlm.nih.gov/entrez/eutils/"
    . "efetch.fcgi?db=nucleotide&rettype=gb&retmode=xml&id=";

    return ($baseurl . $id);
}
```

We then selectively search for direct children with specific tag names. To do so, the local `get_element_text()` method uses the DOM `getChildrenByTagName()` method to find direct children with the specified tag name. If any matching children are found, we take the first matching child and immediately return its text content. We are therefore able to easily extract any piece of GBSeq data and immediately display it to the console.

In conclusion, Perl provides excellent support for XML. In this chapter, we have discussed the fundamental differences between tree-based and event-based XML parsers, and have illustrated these differences by exploring the SAX and DOM interfaces. Event-based parsers, such as SAX, are generally faster and require less memory than comparable tree-based parsers. However, tree-based parsers provide random access to any node or branch in the XML document and also provide facilities for modifying or creating new documents. If you have intense performance requirements, or are working with very large documents, you may have no choice but to use an event-based parser. However, for moderate-sized XML documents, you may find a tree-based interface easier to use. In either case, by sticking to well-defined public standards, such as SAX or DOM, you can more easily apply your XML knowledge to other programming languages, such as C++ or Java.

The Distributed Annotation System (DAS) 6

The genome sequence of an organism is an information resource unlike any that biologists have previously had access to. But the value of the genome is only as good as its annotation. It is the annotation that bridges the gap from the sequence to the biology of the organism. [81]

The Distributed Annotation System (DAS) [6; 82] is an XML-based protocol that facilitates the distribution and sharing of genome annotation data. Since its introduction, DAS has become one of the most widely used protocols for biological data exchange, and is now implemented at numerous laboratories, including UCSC, the Institute for Genomic Research (TIGR), and the European Bioinformatics Institute (EBI). The core of the DAS protocol is built around a small set of XML queries, and a corresponding set of XML Document Type Definitions (DTDs). It therefore provides an exciting window into the use of XML at the cutting edge of bioinformatics.

This chapter aims to provide you with a comprehensive introduction to DAS. The following topics are covered:

- introduction to genome annotation
- overview of available DAS clients
- DAS protocol overview
- comprehensive discussion of the main DAS queries
- working with DAS reference maps
- the future evolution of DAS

The next two chapters explore specific DAS queries further and use DAS examples to illustrate XML parsing in Java.

6.1 Genome Annotation

With the complete sequencing of numerous whole genomes, including the fruit fly (*Drosophila melanogaster*), the Japanese pufferfish (*Fugu rubripes*), *C. elegans*, and a working draft of the human genome, we now have gigabytes of raw DNA sequence data. Hidden within these raw sequences lie biological clues to the molecular machinery of life, the evolution of life on earth, and the origins of many human diseases. Much attention is therefore focused on the analysis and interpretation of raw genome sequences. This process is already well underway, and is likely to take decades and involve thousands of scientists across the globe.

A crucial step in genomic analysis and interpretation is genome annotation. At a very broad level, genome annotation is simply the process of analyzing regions of raw sequence data and adding notes, observations, and predictions. For example, annotation includes the identification of exons (protein-coding portions of genes) and introns (noncoding portions of genes), and the categorization of repeat-coding regions. Genome annotation may also include the linking of sequence data to already cataloged genes, making computerized predictions on the location of novel genes, and identifying sequence similarities across species. In short, annotation attempts to decipher and analyze raw sequence data and eventually connect it to biological function. If the genome represents the book of life, genome annotation represents our collective notes in the margins.

To make genome annotation more concrete, consider the Ensembl project [79], a joint collaboration between the European Bioinformatics Institute (EBI) and the Wellcome Trust Sanger Institute. The goal of Ensembl is to take eukaryotic genomes, including the human, mouse, and zebrafish, and provide *automated* genome annotation. In the case of Ensembl, automated annotation requires an astonishing array of computing power. This includes a complicated data pipeline process and a cluster of several hundred computers.

Ensembl currently provides several categories of annotations, but its main goal is to link sequence data to already known genes and to make computerized predictions regarding the location of novel genes. In fact, the gene analysis provided by Ensembl played a large part in the initial analysis of the human genome provided by the International Human Genome Sequencing Consortium (the public project).

Ensembl makes all of its software, genome annotations, and computerized predictions available freely though its web site. For example, Figure 6.1 shows a sample screenshot, showing a region of human chromosome 3. The display screen is divided into three main panels. The first panel

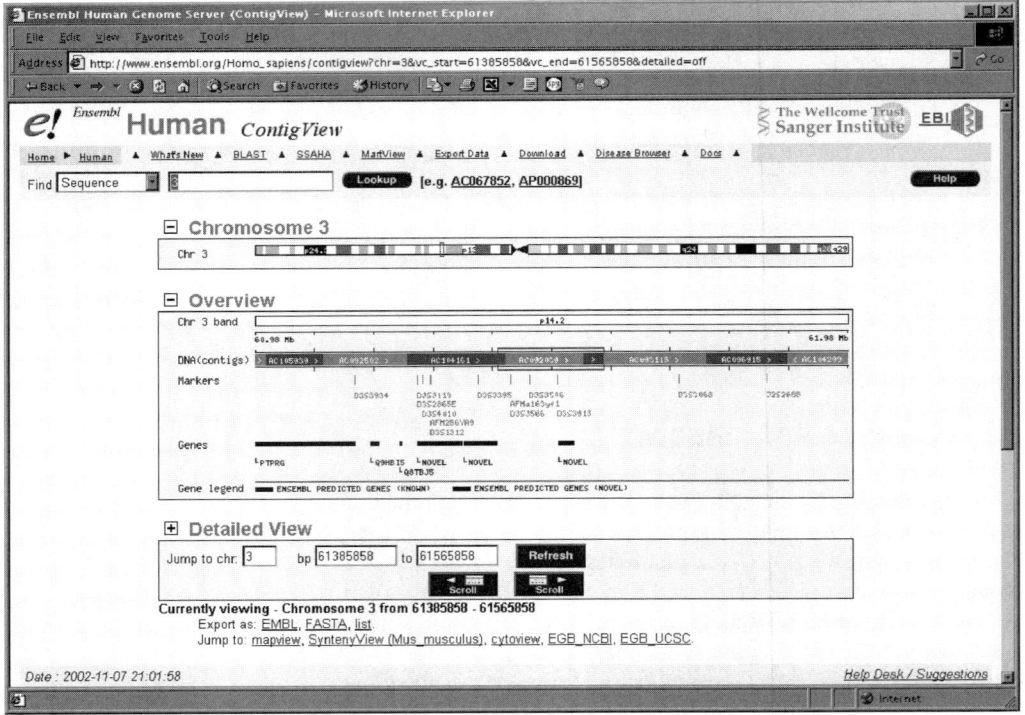

Figure 6.1 An example of genome annotation: a region of human chromosome 3, provided by the Ensembl web site. Ensembl provides automated annotation of the human genome. In the middle panel, labeled "Overview," Ensembl has annotated three known genes (left of panel), and three novel genes (middle of panel).

provides a bird's-eye view of the entire chromosome, the second panel provides an overview of the selected region, and the bottom panel provides a more detailed view of the selected region. Within the second panel under "genes," you can see that Ensembl has annotated three known genes (on the left), and predicted the location of three novel genes (middle of panel).

The computerized prediction of novel genes, such as that provided by Ensembl, is a particularly challenging endeavor. For an overview of the challenges and an introduction to available gene finding software, see "Gene Finders and Feature Detection in DNA" in Chapter 7 of *Developing Bioinformatics Computer Skills* (O'Reilly, 2001) [78].

Ensembl is but one example of genome annotation. In fact, Lincoln Stein (creator of the DAS protocol) currently divides annotation into three distinct layers: nucleotide-level, protein-level, and process-level (see Figure 6.2) [81]. Nucleotide-level annotation includes the mapping of genetic

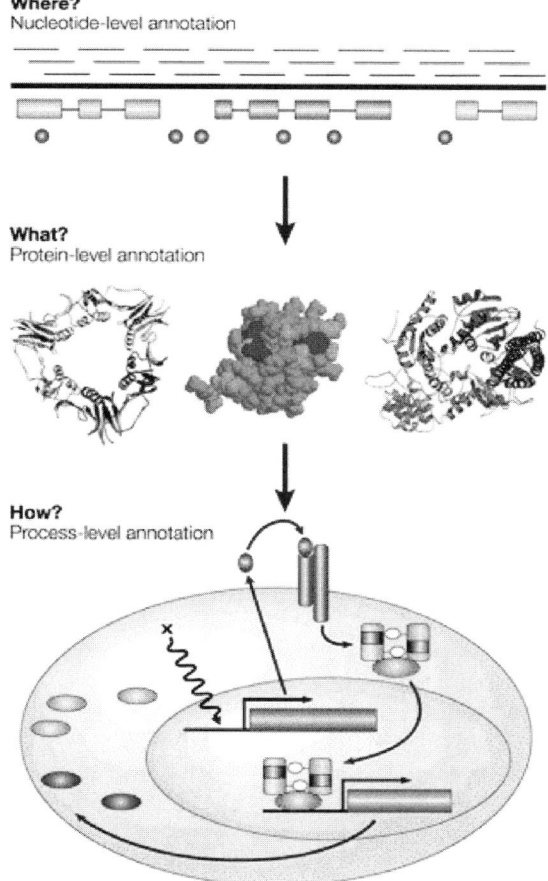

Where?
Nucleotide-level annotation

What?
Protein-level annotation

How?
Process-level annotation

Nature Reviews | Genetics

Figure 6.2 The three layers of genome annotation: where?, what?, and how? (Reprinted with permission, from Lincoln Stein [81].)

landmarks, the identification of genes, the categorization of repeat sequences, the mapping of sequence similarities across species, and the identification of sequence variability within individuals. Protein-level annotation attempts to connect sequences to their protein products and to create a comprehensive catalog of proteins. Process-level annotation attempts to determine protein function(s) within the organism and to determine biological interactions among genes, proteins, RNA, and other macromolecules. Each layer of annotation provides additional clues to biological function and ultimately leads to an overall view of the genome as a whole.

6.2   Introduction to DAS

Despite its enormous potential, genome annotation presents numerous technical challenges. First, annotation is highly decentralized and currently underway at dozens of laboratories throughout the world. Second, it is not likely that one organization will be able to coordinate and centralize all genomic annotations, and yet, we still need a mechanism to aggregate data from multiple laboratories. In response to these challenges, Lincoln Stein of Cold Spring Harbor Laboratory, along with Sean Eddy and LaDeana Hillier, both of Washington University at St. Louis, set out to build a distributed protocol for genome annotation.

DAS is formally specified by a client/server protocol and a set of XML Document Type Definitions (DTDs). Client applications connect to DAS servers, send queries, and receive XML encoded data back. For example, a client can request all genomic annotations within a specified region of human chromosome 11, or request only a subset of those annotations.

All DAS servers adhere to the same specification and encode annotation data in the same XML format. Client applications can therefore easily aggregate data from multiple servers. Without DAS, a user would need to manually surf through three different web sites to compare annotation data. With DAS, a user can open a single client application and simply specify three DAS servers. Behind the scenes, the client application connects to each DAS server, aggregates the response data, and presents a unified data view (see Figure 6.3).

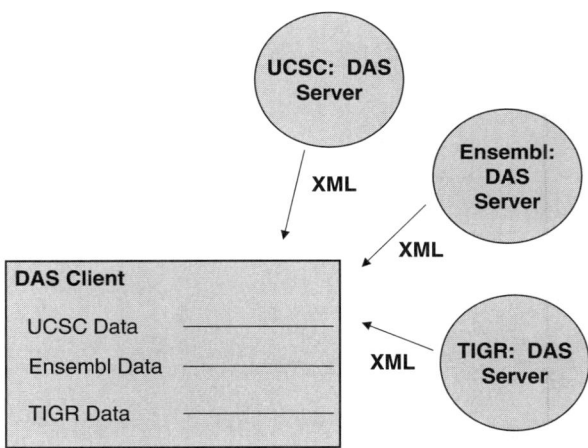

Figure 6.3 All DAS servers are required to create XML documents which adhere to common Document Type Definitions (DTDs). Clients can therefore easily aggregate annotation data from multiple DAS servers.

DAS specifies two types of servers: reference servers and annotation servers. Reference servers hold a reference map to the genome and store the raw genomic sequence data. Annotation servers hold the actual genomic annotations. The two server types are not necessarily mutually exclusive, and a single server can (and frequently does) act as both a reference and annotation server.

In order to work successfully, DAS requires that the community at large agree on a set of common genomic reference maps. For example, for the human genome, most DAS servers are using the public genome assembly, available from NCBI. Multiple versions of this assembly exist and new versions are continually published, as new data is generated and finalized. In order to accurately compare data from multiple DAS servers, each of the annotation servers must be using the same assembly, and the same version of the assembly. We will delve into specifics later in the chapter in "Working with Reference Maps" (Section 6.5).

Information in this chapter is based on DAS version 1.53. For a copy of the complete specification, go to: *http://www.biodas.org*.

A number of DAS clients are currently available, including web-based clients and stand-alone applications. Web-based clients perform data aggregation on the server side and therefore do not require local installation of any software. A summary of DAS clients is presented in Table 6.1. A summary of public DAS servers follows in Table 6.2.

6.2.1   The WormBase DAS Viewer

One of the first web-based DAS clients was created at WormBase.org [83], a site devoted to the study of *C. elegans*. The WormBase site includes a genome browser and an experimental DAS browser. Figure 6.4 shows a screenshot of the DAS browser. The bottom of the page indicates the chromosome region; in this case, we are viewing chromosome 4, 500–1000 base pairs. Directly above this is a list of selectable DAS data sources, including WormBase, the Knockout Consortium, and the Institute for Genomic Research (TIGR). If you select the "custom" option on any of the data sources, you can filter for specific types of annotation data.

Directly above the source list is a visualization of the annotation data. Each data source is represented as a separate horizontal track of data—in this case, we have chosen to display data from WormBase and TIGR. If you click on any of the individual annotation elements, such as the TIGR transcripts, you are directed to a TIGR page with annotation details.

The WormBase DAS browser also includes direct links to the WormBase DAS server. For example, if you click the "Ref Server Features" link, you will receive an XML document specifying the annotations within the selected region. This is a very useful feature for learning the DAS protocol, because it enables you to easily compare raw XML data with a graphical visualization.

6.3   DAS Protocol Overview

The DAS protocol is built on a simple pattern of requests and responses. DAS clients issue requests in the form of Internet URLs and servers issue responses encoded in XML (see Figure 6.5). Currently, there are only eight different DAS commands, and each command will trigger a different XML response from the server. For example, a client can request a list of data sources hosted by the

Table 6.1 A summary of DAS clients

DAS Client	Client Type	Description
Ensembl Genome Browser	Web Based	The Ensembl genome browser provides built-in support for DAS. Users can select from a preconfigured list of DAS servers or specify the URL for any arbitrary DAS server. Behind the scenes, the web site connects to all user-specified DAS servers and creates a single integrated view of the genomic data.
		URL: *http://www.ensembl.org*
WormBase Genome Browser	Web Based	The WormBase genome browser includes an experimental DAS viewer, named DASView. DASView enables users to select from a preconfigured list of DAS sources. Currently, this list includes: WormBase, the Knockout Consortium, and The Institute for Genomic Research (TIGR). Each DAS source is represented as a separate horizontal track of data. A sample screenshot is provided in Figure 6.4.
		URL: *http://www.wormbase.org*
TIGR DAS Viewer	Web Based	DAS Viewer hosted at The Institute for Genomic Research. Genome annotations are currently available for: Arabidopsis, Mosquito, Fugu, Human, Rice, Mouse, and C. elegans.
		URL: *http://www.tigr.org/tdb/DAS/DAS.shtml*
BioJava DAS Client	Stand alone	The BioJava DAS client is a stand-alone Java client application that relies extensively on the open source BioJava library. It is currently maintained by Matthew Pocock of the Sanger Institute. Users begin a client session by specifying the URL for a DAS server. The client automatically retrieves a list of data sources, and enables the user to zoom in on specific chromosomal regions. At any point, users can add additional DAS data sources.
		URL: *http://www.ensembl.org/das/client.html*
OmniGene	Stand alone	The OmniGene OmniView application includes experimental support for a SOAP-based version of DAS. In addition to connecting directly to DAS servers, the OmniView application can connect to the OmniGene server at MIT via a SOAP-based protocol. The OmniGene server then connects to the specified DAS server using the DAS 1.0 protocol, wraps these results in SOAP, and returns the results to the client. By using a "middleware" server, OmniView can retrieve both DAS data and non-DAS data. The user interface for OmniView is based on the Ensembl genome browser interface, and should be very familiar to Ensembl users. OmniGene is an open source project and is currently hosted on Source Forge.
		URL: *http://omnigene.sourceforge.net*
Geodesic	Stand alone	Geodesic is an experimental DAS client designed to work with an earlier version of the DAS specification. It does not work with the current version of DAS, and there are currently no plans to continue its development.
		URL: *http://biodas.org/geodesic*

Table 6.2 A partial list of DAS servers currently available online. A more complete list is maintained at: *http://www.tigr.org/tdb/DAS/das_server_list.html*

Name	URL
Ensembl/Sanger	*http://servlet.sanger.ac.uk:8080/das/dsn*
University of California, Santa Cruz	*http://genome.cse.ucsc.edu/cgi-bin/das/dsn*
WormBase.org	*http://www.wormbase.org/db/das/dsn*
The Institute for Genomic Research	*http://www.tigr.org/docs/tigr-scripts/tgi/das/dsn*
Max Planck Institute for Molecular Genetics	*http://tomcat.molgen.mpg.de:8080/das/dsn*

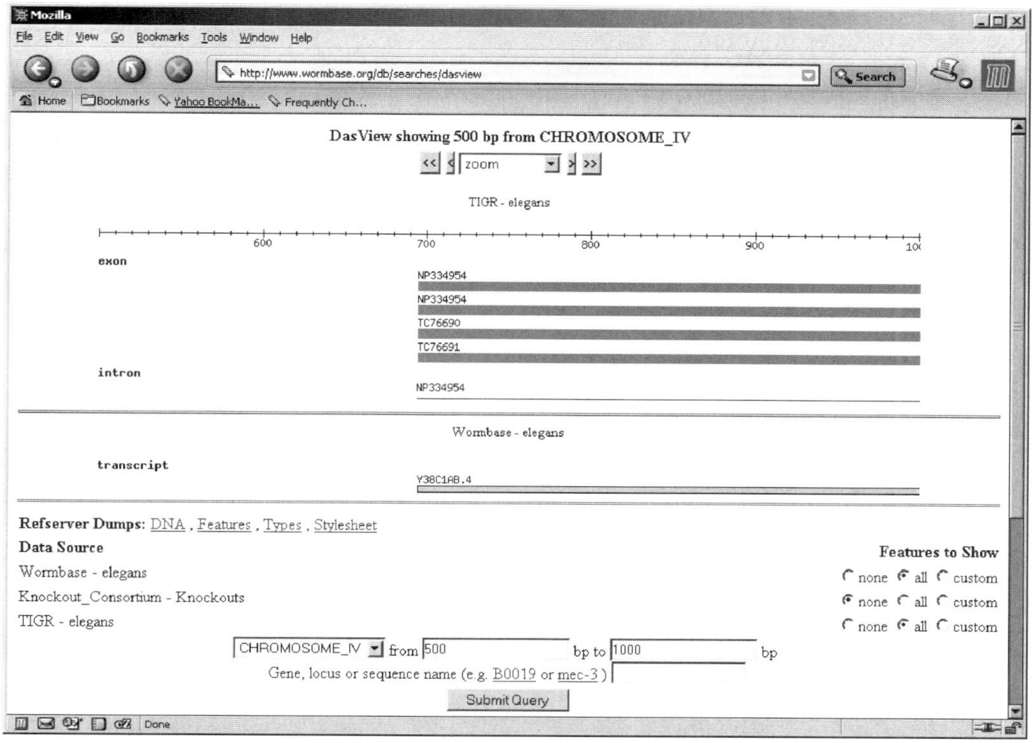

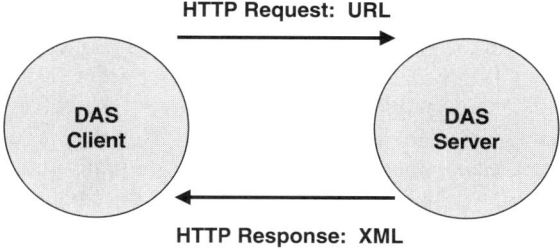

Figure 6.4 The WormBase DAS browser. Data from each DAS source is represented as a separate horizontal track of data. In this case, DAS data from WormBase and The Institute for Genomic Research (TIGR) is shown.

HTTP Request: URL

DAS
Client

DAS
Server

HTTP Response: XML

Figure 6.5 DAS clients issue requests in the form of Internet URLs and servers issue responses encoded in XML. Transportation is provided by HTTP.

server, retrieve annotations across a specific chromosomal region, or request raw DNA sequence data. An overview of the eight DAS commands is presented in Table 6.3.

The biodas.org web site maintains an active mailing list related to DAS issues. To subscribe or view the mail archives, go to: *http://biodas.org/mailman/listinfo/das.*

Table 6.3 Summary of DAS commands

Command	Description
dsn	Requests a list of data sources hosted by the DAS server. For example, a *dsn* request to the Ensembl site returns a complete list of its cataloged genomes, including: human, mosquito, mouse, and fruit fly
entry_points	Requests a list of entry points for accessing the genome. For most DAS servers, entry points refer to specific chromosomes within the genome. For example, an *entry_points* request to the Ensembl human genome returns a list of chromosomes 1–22, X, and Y
dna	Requests raw DNA sequence data
sequence	Requests raw sequence data (either DNA, RNA, or protein sequences). The *sequence* command is a newer, more general version of the *dna* command. For backward compatibility, the *dna* command is still maintained
types	Requests a summary of annotations across a specified genomic region
features	Requests full annotation records across a specific genomic region
link	Requests a web page associated with a specific annotation. The web page contains additional details regarding the annotation, but is formatted in HTML, not XML
stylesheet	Requests the DAS style sheet recommended by the server. The style sheet provides hints on how to visually render specific categories of annotations

Each of the DAS commands will be explored in detail in the next section. However, before we can understand the individual commands, we must first begin with a detailed overview of the protocol itself.

6.3.1   Getting Started

Transportation for the DAS protocol is handled by HTTP (HyperText Transfer Protocol), the same protocol used to connect web browsers and web servers. Because of this, you can easily use a regular web browser to experiment with and debug the DAS protocol. Therefore, before going any further, let us try a sample DAS query using Internet Explorer. Simply open up your browser and type the following URL:

http://servlet.sanger.ac.uk:8080/das/ensembl1533/dna?segment=1:100000,100100

As we will soon see, URLs associated with DAS queries have a very specific syntax. For now, note that we are connecting to the DAS server at Ensembl, and that the DAS portion of the request begins with a required /das prefix. This is followed by a data source (ensembl1533 refers to the Ensembl human genome, version 15.33), a DAS command (we are using the *dna* command to retrieve raw sequence data), and a segment parameter (indicating human chromosome 1, 100,000–100,100 base pairs). To summarize, the URL basically says: "Connect to Ensembl and give me the raw DNA sequence for human chromosome 1: 100,000–100,100 base pairs."

The *dna* command is arguably the simplest of the DAS queries and it is therefore a good place to begin. In response, the Ensembl DAS server returns the following XML document (see Figure 6.6).

```
<?xml version='1.0' standalone='no' ?>
<!DOCTYPE DASDNA SYSTEM 'dasdna.dtd' >
<DASDNA>
    <SEQUENCE id="1" version="15.33" start="100000" stop="100100">
    <DNA length="101">
            ttatgaattggtgttgagcttagtaagtcaccaaacaccttctgctcagcagcata
            aaggacatttccatgaaacctcccagggataatcttatttactct
    </DNA>
    </SEQUENCE>
</DASDNA>
```

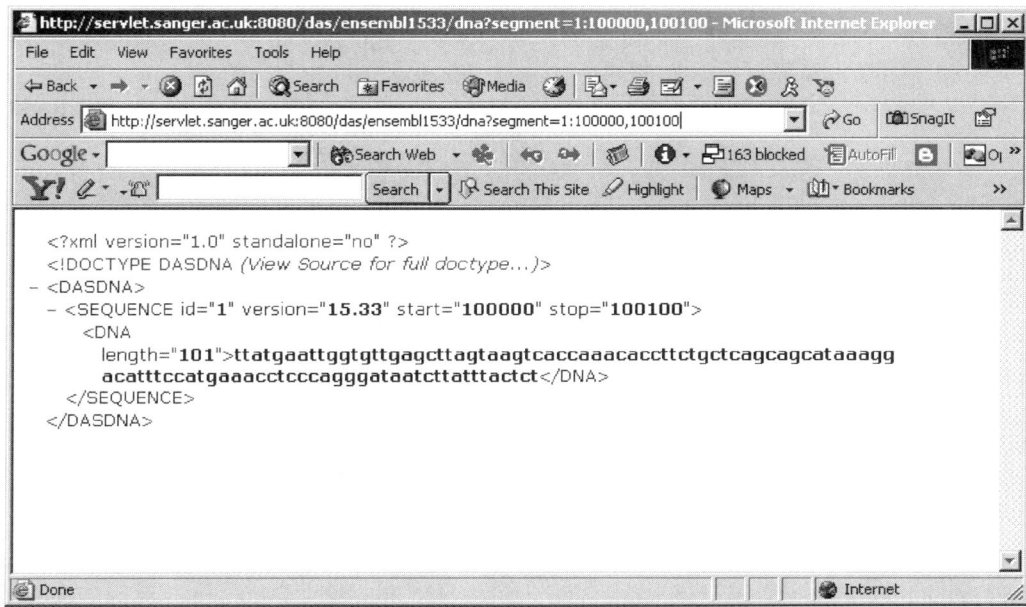

Figure 6.6 A sample DAS query, as seen through Internet Explorer. Since DAS uses existing web standards, including HTTP and XML, you can use a regular web browser to experiment with the protocol.

The DAS response is a valid XML document containing the raw sequence data. The prolog to the document specifies a Document Type Definition, named dasdna.dtd. The client can therefore validate the XML content against the DTD. The root of the document is specified by a DASDNA element, which contains an embedded SEQUENCE element. The SEQUENCE element specifies the chromosomal region and contains a DNA element containing the raw data.

Hopefully, this provides you with a taste of how the protocol works. Now, onto the details of requests and responses.

There are currently two open source DAS server software packages. The first is the Lightweight Distributed Annotation Server (LDAS). LDAS is written in Perl and runs on a MySQL database. Source code and installation instructions are available at: *http://www.biodas.org/servers*. The second is the BioJava Dazzle Server. Dazzle is written in Java and implemented as a Java servlet. Dazzle includes data source plug-ins for distributing EMBL (European Molecular Biology Laboratory) as well as GFF (General Feature Format) formatted flat files. Source code and installation instructions are available at: *http://www.biojava.org/dazzle*.

6.3.2   DAS Requests

Each DAS request is specified as an Internet URL. The URL is defined by five components, which appear in the following order:

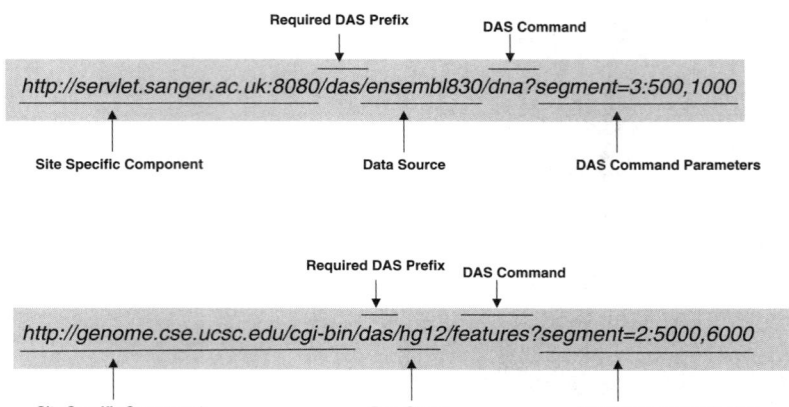

Figure 6.7 Two sample DAS requests. DAS requests are specified as Internet URLs, which must adhere to a specific syntax. Individual components of the URLs are noted.

- Site-specific component: this is the Internet domain name, followed by a path to the DAS server application. For example, the site-specific component of the WormBase DAS server is specified by *http://www.wormbase.org/db*, whereas the Ensemble DAS server is specified by *http://servlet.sanger.ac.uk:8080*.
- /das: a required prefix indicating the beginning of a DAS command.
- [Data source]: a data source element (required for most commands) indicating the data source of interest. Data sources vary by DAS server. For example, a data source of "mosquito1602" specifies the mosquito genome (version 16.2) at Ensembl.
- [DAS Command]: the actual DAS command, e.g., entry_points, dna, features, etc.
- [Command Parameters]: each DAS command can include specific parameters to refine the actual query. For example, a parameter of "segment=3:500,1000" refines the query to a specific region on chromosome 3, 500–1000 base pairs.

The first four components are specified as part of the URL. In contrast, command arguments are specified via HTTP name/value parameters. These parameters can be sent via HTTP GET (parameters are appended directly to the end of the URL), or via HTTP POST (parameters are embedded within the body of the HTTP request). Note that some servers restrict the length of URLs, and clients are therefore advised to send large sets of parameters via HTTP POST. Two sample DAS requests are illustrated in Figure 6.7.

6.3.3   DAS Responses

The DAS server response is embedded within the body of the HTTP response. For example, here is a request to the UCSC DAS server:

http://genome.cse.ucsc.edu/cgi-bin/das/hg12/features?segment=2:5000,6000

UCSC responds with the following:

```
HTTP/1.1 200 OK
Date: Thu, 07 Nov 2002 20:28:07 GMT
Server: Apache/1.3.26 (Unix)
```

```
X-DAS-Status: 200
X-DAS-Version: DAS/0.95
Keep-Alive: timeout=15, max=100
Connection: Keep-Alive
Transfer-Encoding: chunked
Content-Type: text/plain; charset=iso-8859-1
<?xml version="1.0" standalone="no"?>
<DASGFF>
    [... For brevity, the full XML document is not shown here]
</DASGFF>
```

As you can see, the DAS server specifies a number of regular HTTP headers, including: the current date, the server software, and the Content-Type.* However, the DAS server is also required to specify two additional headers. The first is: X-DAS-Version, indicating the DAS version implemented by the server. The second is: X-DAS-Status, indicating the status of the server response.

X-DAS-Status codes are divided into roughly the same categories as HTTP status codes. For example, a status code between 200–299 generally indicates that the response is OK, 400–499 indicates a bad request from the client, and 500–599 indicates some type of server error. A complete list of X-DAS-Status codes is presented in Table 6.4.

If you are curious to see a DAS error, try issuing the following invalid request to UCSC:

http://genome.cse.ucsc.edu/cgi-bin/das/human/features?segment=2:5000,5100

If you issue this request from a web browser, it looks as if UCSC simply returns a blank page. However, if you have some kind of network sniffer that is capable of intercepting HTTP packets (see side note below), you can view the full response from UCSC:

```
HTTP/1.1 200 OK
Date: Thu, 19 Sep 2002 21:55:20 GMT
Server: Apache/1.3.26 (Unix)
X-DAS-Status: 401
X-DAS-Version: DAS/0.95
Keep-Alive: timeout=15, max=100
Connection: Keep-Alive
Transfer-Encoding: chunked
Content-Type: text/plain; charset=iso-8859-1
```

The X-DAS-Status code of 401 indicates that we have specified an invalid data source. To fix the error, we have to change "human" to a valid UCSC data source, such as "hg12."

To view the full HTTP response from a DAS server, you will need a network "sniffer" that is capable of intercepting HTTP packets. A number of HTTP sniffer programs are available on the Internet. A particularly good program is Ethereal Network Analyzer, available at: *http://www.ethereal.com*. Ethereal is available for free, runs on Windows, Linux, and MacOS X, and is distributed under the GNU General Public License.

---

* The DAS Protocol does not formally specify a Content-Type. Some servers currently specify "text/plain," whereas others specify "text/xml."

Table 6.4 X-DAS-status codes

Status Code	Description
200 Success Codes	
200	**OK, request was successful.**
400 Client Error Codes	
400	**Invalid command specified by the client.**
	For example, if the client specifies an "annotations" command (the correct command is actually "features"), the DAS server will return a 400 status code
401	**Invalid data source specified by the client.**
	Specifically, the data source specified by the client is not hosted by the DAS server
402	**Invalid command arguments specified by the client.**
	For example, if the client issues a dna request to retrieve raw sequence data, but does not specify a chromosomal region, the server does not have enough information to process the request and returns a 402 status code
403	**Invalid reference ID specified by the client.**
	Reference IDs represent anchors for retrieving genomic data and frequently refer to chromosome number. For example, if the client incorrectly attempts to retrieve a portion of chromosome 30 from the human genome (humans only have 23 chromosomes), the DAS server will return a 403 status code
404	**Missing Stylesheet.**
	The DAS server does not have a stylesheet for the requested data source
405	**Invalid or out of bounds sequence coordinates specified by the client.**
	For example, a common error is to assume that sequences begin at 0, e.g., retrieve annotations for chromosome 2, 0–500 base pairs. This is actually out of bounds (biological sequences are numbered from 1), and will trigger a 405 error code. To fix the error, you must specify chromosome 2, 1–500 base pairs
500 Server Error Codes	
500	**Internal Server Error.**
	A 500 status code indicates that the client request was valid, but the server was still unable to fulfill the request. For example, the main database may be currently offline
501	**The server does not implement the requested feature.**
	For example, if the client connects to an annotation server and issues a command reserved for reference servers, such as a *sequence* command, the server will return a 501 status code

6.3.4   X-DAS-Capabilities Header

Starting with DAS version 1.5, servers can return an X-DAS-Capabilities header. As its name implies, the capabilities header summarizes the capabilities of the server and includes a complete list of all implemented functionality. The header is useful for two purposes. First, clients can check the capabilities header to determine which DAS commands and command arguments are supported, and thereby determine how to interface with the server. Second, DAS servers can implement new experimental functionality and easily inform the clients of the new feature.

Much like X-DAS-Status and X-DAS-Version, the X-DAS-Capabilities header is a regular HTTP header. The value of the header is a semicolon separated list of implemented functionality. For example, here is a response from Ensembl:

```
HTTP/1.1 200 OK
Server: Resin/2.0.5
Content-Encoding: gzip
X-DAS-Version: 1.5
X-DAS-Server: DazzleServer/0.97 (20020924; BioJava pre-1.3)
X-DAS-Capabilities: dsn/1.0; dna/1.0; types/1.0; stylesheet/1.0;
features/1.0; encoding-dasgff/1.0; encoding-xff/1.0; entry_points/1.0;
error-segment/1.0; unknown-segment/1.0; component/1.0
```

Table 6.5 The X-DAS-Capabilities header

Name	Description
component/1.0	Server returns map components via the *features* command. For details, refer to Section 6.5 of this chapter
dna/1.0	Server implements the DAS *dna* command. For details, refer to Section 6.4.3 of this chapter
dsn/1.0	Server implements the DAS *dsn* command. For details, refer to Section 6.4.1 of this chapter
entry_points/1.0	Server implements the DAS *entry_points* command. For details, refer to Section 6.4.2 of this chapter
error-segment/1.0	Server will respond to *feature* requests for invalid segments via an ERRORSEGMENT element. For details, refer to the error handling section of "Retrieving Annotations" (6.4.4) in this chapter
feature-by-id/1.0	Server supports the *feature_id* parameter in the DAS *features* command. For details, refer to the *features* command summary in Table 6.12
features/1.0	Server implements the DAS *features* command. For details, refer to Section 6.4.4 of this chapter
group-by-id/1.0	Server supports the *group_id* parameter in the DAS *features* command. For details, refer to the features command summary in Table 6.12
sequence/1.0	Server implements the DAS *sequence* command. For details, refer to Section 6.4.3 of this chapter
stylesheet/1.0	Server implements the DAS *stylesheet* command
Supercomponent/1.0	Server returns map super-components via the *features* command. For details, refer to Section 6.5 of this chapter
types/1.0	Server implements the DAS *types* command. For details, refer to Section 6.4.4 of this chapter
unknown-segment/1.0	Server will respond to *feature* requests for invalid segments via an UNKNOWNSEGMENT element. For details, refer to the error handling section of Section 6.4.4 of this chapter

```
X-DAS-Status: 200
Content-Type: text/xml
Transfer-Encoding: chunked
Date: Sat, 09 Nov 2002 16:29:21 GMT
```

A complete list of the possible X-DAS-Capabilities values is presented in Table 6.5.

6.4   DAS Command Reference

We are now ready to delve into the details of specific DAS commands. In general, DAS commands can be divided into four categories:

- retrieving data sources
- retrieving entry points
- retrieving sequence data
- retrieving annotation data

We explore each category below, in this order. For each DAS command, we have included a command reference table. The table includes a command description, the scope (reference or annotation server), request syntax, the response DTD, and real examples from public DAS servers.

6.4.1   Retrieving Data Sources

Once you have a URL for a DAS server, the first command to issue is usually the *dsn* command. The *dsn* command retrieves a list of all data sources hosted by the server. Note that both reference servers and annotation servers are required to respond to a *dsn* request. Reference servers will

Table 6.6 The DAS dsn command

Description: Requests a list of data sources hosted by the DAS server

Scope: Annotation and Reference Servers

Request Syntax:
PREFIX/das/dsn

Examples:
1. Retrieve data sources from the Max Planck Institute:
    http://tomcat.molgen.mpg.de:8080/das/dsn
2. Retrieve data sources from UCSC:
    http://genome.cse.ucsc.edu/cgi-bin/das/dsn

Response: *http://www.biodas.org/dtd/dasdsn.dtd*
```
<!ELEMENT DASDSN (DSN+)>
<!ELEMENT DSN (SOURCE, DESCRIPTION?, MAPMASTER)>
<!ELEMENT SOURCE (#PCDATA)>
<!ATTLIST SOURCE id CDATA #REQUIRED>
<!ATTLIST SOURCE version CDATA #IMPLIED>
<!ELEMENT DESCRIPTION (#PCDATA)>
<!ATTLIST DESCRIPTION href CDATA #IMPLIED>
<!ELEMENT MAPMASTER (#PCDATA)>
```

return a list of all sources for which it maintains genomic maps and raw sequence data. Annotation servers will return a list of all annotated data sources. Each annotated data source will include a URL to the correct reference server, and therefore enables clients to immediately tie the annotation server with the correct reference server.

The syntax for the *dsn* command is summarized in Table 6.6.

To get started, here is a *dsn* request to the Max Planck Institute DAS server:

http://tomcat.molgen.mpg.de:8080/das/dsn

The XML response is as follow:

```
<?xml version='1.0' standalone='no' ?>
<!DOCTYPE DASDSN SYSTEM 'dasdsn.dtd' >
<DASDSN>
    <DSN>
        <SOURCE id="CNBs" version="1.0">CNBs</SOURCE>
        <MAPMASTER>http://servlet.sanger.ac.uk:8080/das/ensembl729
          </MAPMASTER>
        <DESCRIPTION>Conserved Noncoding Blocks</DESCRIPTION>
    </DSN>
    <DSN>
        <SOURCE id="CNBs_Mouse" version="1.0">CNBs_Mouse</SOURCE>
        <MAPMASTER>http://servlet.sanger.ac.uk:8080/das/mouse53/
          </MAPMASTER>
        <DESCRIPTION>CNBs_Mouse</DESCRIPTION>
    </DSN>
    <DSN>
        <SOURCE id="EPD_Mouse" version="1.0">Mouse EPD Promoter
          </SOURCE>
```

```
        <MAPMASTER>http://servlet.sanger.ac.uk:8080/das/mouse53/
          </MAPMASTER>
        <DESCRIPTION>EPD_Mouse</DESCRIPTION>
      </DSN>
      <DSN>
        <SOURCE id="EPD_promoters" version="1.0">EPD_Promoters
          </SOURCE>
        <MAPMASTER>http://servlet.sanger.ac.uk:8080/das/ensembl729
          </MAPMASTER>
        <DESCRIPTION>Promoter positions</DESCRIPTION>
      </DSN>
  </DASDSN>*
```

The prolog of the document specifies the dasdsn.dtd document, and the root of the document is specified by the DASDSN element. Each data source is specified by a separate DSN element.

The SOURCE element specifies a data source id, which is to be used in subsequent DAS commands. The DESCRIPTION element is human readable text. The MAPMASTER element specifies the URL for the reference server. If this is a reference server, the URL simply points back to itself. As you can see in the example above, the Max Planck DAS server hosts four data sources, all of which use reference maps located at the Ensembl/Sanger DAS server.

6.4.2   Retrieving Entry Points

DAS reference servers maintain a detailed sequence map of the hosted genome. To navigate through a sequence map, clients must first retrieve a set of entry points. Entry points represent high-level or well-known elements on the reference sequence map. Frequently, entry points are simply chromosomes within the organism. For example, the entry points for the human genome are frequently listed as chromosomes 1–22, X, and Y. However, entry points can also refer to nonchromosomal entities, such as contigs or clones—the building blocks used to create the final genomic assembly.

To retrieve a list of entry points for a particular data source, the client issues a DAS *entry_points* command. The command syntax is summarized in Table 6.7.

For example, here is a sample query to the human genome assembly hosted at Ensembl:

http://servlet.sanger.ac.uk:8080/das/ensembl830/entry_points

Here is the XML response:

```
<?xml version='1.0' standalone='no' ?>
<!DOCTYPE DASEP SYSTEM 'dasep.dtd' >
<DASEP>
  <ENTRY_POINTS
    href="http://servlet.sanger.ac.uk:8080/das/ensembl830/entry_points"
    version="8.30">
```

---

* If you look carefully, you will note that this XML document is actually invalid. The DTD requires that the <DSN> element contain SOURCE, DESCRIPTION, and MAPMASTER in that specific order. However, the DAS 1.53 specification states that elements should appear in the order: SOURCE, MAPMASTER, DESCRIPTION, and many DAS servers have chosen to follow this convention. Unfortunately, this means there is a discrepancy between the DTD and the specification, and those servers which follow the specification are actually returning invalid DSN documents.

```
<SEGMENT id="Y" size="58368225" subparts="yes" />
<SEGMENT id="X" size="149249818" subparts="yes" />
<SEGMENT id="19" size="60013307" subparts="yes" />
<SEGMENT id="18" size="77516809" subparts="yes" />
<SEGMENT id="17" size="80052782" subparts="yes" />
<SEGMENT id="16" size="81671585" subparts="yes" />
<SEGMENT id="15" size="99217355" subparts="yes" />
<SEGMENT id="14" size="104324908" subparts="yes" />
[... For brevity, the full XML document is not shown here.]
  </ENTRY_POINTS>
</DASEP>
```

The prolog of the document specifies the dasep.dtd document, and the root of the document is specified by a DASEP element. The DASEP element contains a single ENTRY_POINTS element, which in turn contains 0 or more SEGMENT elements.

The SEGMENT element contains a number of important attributes. First up is the *id* attribute. This specifies a reference ID that can be used within subsequent DAS commands. The *size* attribute specifies the size, in base pairs, of the specific segment (alternatively, DAS servers can specify *start* and *stop* attributes).

The subparts attribute indicates whether the segment has an internal structure. If *subparts* is set to "yes," you can issue a features request to retrieve those subparts. Details are provided in Section 6.5 below.

Table 6.7 The DAS entry_points command

Description: Requests a list of entry points for accessing the genome. For most DAS servers, entry points correspond to specific chromosomes within the genome

Scope: Reference Servers Only

Request Syntax:
PREFIX /das/ *DSN* /entry_points

Examples:
1. Retrieve entry points for Ensembl human genome:
   http://servlet.sanger.ac.uk:8080/das/ensembl830/entry_points
2. Retrieve entry points for WormBase *C. elegans* genome:
   http://www.wormbase.org/db/das/elegans/entry_points

Response: *http://www.biodas.org/dtd/dasep.dtd*
```
<!ELEMENT DASEP (ENTRY_POINTS)>
<!ELEMENT ENTRY_POINTS (SEGMENT*)>
<!ATTLIST ENTRY_POINTS href CDATA #REQUIRED>
<!ATTLIST ENTRY_POINTS version CDATA #REQUIRED>
<!ATTLIST ENTRY_POINTS id CDATA #IMPLIED>
<!ELEMENT SEGMENT (#PCDATA)>
<!ATTLIST SEGMENT id CDATA #REQUIRED>
<!ATTLIST SEGMENT start CDATA #IMPLIED>
<!ATTLIST SEGMENT stop CDATA #IMPLIED>
<!ATTLIST SEGMENT orientation CDATA #IMPLIED>
<!ATTLIST SEGMENT subparts CDATA #IMPLIED>
<!ATTLIST SEGMENT size CDATA #IMPLIED>
<!ATTLIST SEGMENT class CDATA #IMPLIED>
```

If two DAS servers provide annotation for the same species, it is important that they share a common naming convention for entry points. For example, consider two DAS servers that provide data on the human genome. One uses chromosomes named "1," "2," etc., and the other uses chromosomes named "chr1," "chr2," etc. In order to aggregate data from both DAS servers, the client has to be smart enough to be able to map between two naming schemes. If the client does not contain the mapping (as is quite common for most current DAS clients), you cannot aggregate the data at all.

Unfortunately, this was a fairly common problem during early adoption of DAS, and the DAS community has therefore worked to informally standardize names within species. For example, human chromosomes are informally standardized as 1–22, X, Y, as opposed to "chr1," "CHR1," etc. In contrast, chromosomes for *C. elegans* are informally standardized as I, II, III, etc.

6.4.3   Retrieving Sequence Data

DAS provides two commands for retrieving sequence data. The first is the *dna* command, used exclusively for retrieving DNA sequences. The second is the *sequence* command (new to DAS version 1.5), used to retrieve any category of sequence data, including DNA, RNA, and protein. This section includes details on both commands.

The syntax of the DAS *dna* command is as follows:

PREFIX/das/ *DSN*/dna?segment=RANGE[;segment=RANGE]

In the syntax above, DSN refers to a data source (see Section 6.4.1). The segment refers to the specific genomic coordinates of the requested data. Note that you can specify multiple segments within one request.

For example, to request sequence data from human chromosome 13, you can issue the following request to UCSC:

http://genome.cse.ucsc.edu/cgi-bin/das/hg12/dna?segment=13:30875977,30876100

Alternatively, to retrieve two regions of data, you can issue this request:

http://genome.cse.ucsc.edu/cgi-bin/das/hg12/dna?segment=13:30875977,30876200; segment=13:30876005,30876240

Note that the range is specified by a reference ID, followed by the start and stop positions. If you do not specify start and stop positions, the server will return the complete sequence data for the specified segment.

In response to the second request, UCSC will return the following XML document:

```
<?xml version="1.0" standalone="no"?>
<DASDNA>
    <SEQUENCE id="13" start="30875977" stop="30876200" version="1.00">
    <DNA length="224">
            gtggcgcgagcttctgaaactaggcggcagaggcggagccgctgtggcac
            tgctgcgcctctgctgcgcctcgggtgtcttttgcggcggtgggtcgccg
            ccgggagaagcgtgaggggacagatttgtgaccggcgcggttttttgtcag
```

```
                  cttactccggccaaaaaagaactgcacctctggagcgggttagtggtggt
                  ggtagtgggttgggacgagcgcgt
        </DNA>
        </SEQUENCE>
        <SEQUENCE id="13" start="30876005" stop="30876240" version="1.00">
        <DNA length="236">
                  agaggcggagccgctgtggcactgctgcgcctctgctgcgcctcgggtgt
                  cttttgcggcggtgggtcgccgccgggagaagcgtgaggggacagatttg
                  tgaccggcgcggttttttgtcagcttactccggccaaaaaagaactgcacc
                  tctggagcgggttagtggtggtggtagtgggttgggacgagcgcgtcttc
                  cgcagtcccagtccagcgtggcgggggagcgcctca
        </DNA>
        </SEQUENCE>
</DASDNA>
```

The document must adhere to the dasdna.dtd Document Type Definition file, and the root of the document must be specified with a DASDNA element. For each segment in the request, the response will include a corresponding SEQUENCE element. The SEQUENCE element includes a single DNA element, which contains the actual sequence data. Sequence data must adhere to the standard IUPAC coding conventions (see Appendix).

The syntax of the *sequence* command is nearly identical to the *dna* command. The DAS response is nearly identical as well. The main differences are that the root element must be specified with a DASSEQUENCE element, instead of DASDNA , and that the SEQUENCE element must specify a *molecule* attribute. The *molecule* attribute specifies the type of sequence data. Valid options are: "DNA," "ssRNA" (single-stranded RNA), "dsRNA" (double-stranded RNA), or "Protein." All data must be specified using the standard IUPAC coding conventions (see Appendix).

Table 6.8 The DAS dna command

Description: Requests raw sequence data for a specific genomic region

Scope: Reference Servers Only

Request Syntax:
PREFIX/das/ *DSN*/dna?segment=RANGE[;segment=RANGE...]

Arguments:

- segment: indicates a specific region of sequence data. Segments are specified as: *referenceID* :*start* ,*stop* . For example, to request a portion of human chromosome 3, use: segment=3:50000,100000. If start and stop positions are omitted, the server will return the complete sequence for the specified reference ID. Requests can specify one or more segments, as needed.

Examples:

1. Retrieve single segment of DNA from UCSC:
   http://genome.cse.ucsc.edu/cgi-bin/das/hg12/dna?segment=13:30875977,30876100
2. Retrieve two segments of DNA from UCSC:
   http://genome.cse.ucsc.edu/cgi-bin/das/hg12/dna?segment=13:30875977,30876200;segment=13:30876005,30876240

Response: *http://www.biodas.org/dtd/dasdna.dtd*
```
<!ELEMENT DASDNA (SEQUENCE+)>
<!ELEMENT SEQUENCE (DNA)>
<!ATTLIST SEQUENCE id CDATA #REQUIRED>
<!ATTLIST SEQUENCE start CDATA #REQUIRED>
<!ATTLIST SEQUENCE stop CDATA #REQUIRED>
<!ATTLIST SEQUENCE version CDATA #REQUIRED>
<!ELEMENT DNA (#PCDATA)>
<!ATTLIST DNA length CDATA #REQUIRED>
```

Table 6.9 The DAS sequence command

Description: Requests raw sequence data (DNA, RNA, or protein sequence data). The sequence command is a newer, more general version of the *dna* command. For backward compatibility, the *dna* command is still maintained.

Scope: Reference Servers Only

Request Syntax:

PREFIX/das/ *DSN*/sequence?segment=RANGE[;segment=RANGE...]

Arguments:

• segment: indicates a specific region of sequence data. Segments are specified as: *referenceID: start, stop*. For example, to request a portion of human chromosome 3, use: segment=3:50000,100000. If start and stop positions are omitted, the server will return the complete sequence for the specified reference ID. Requests can specify one or more segments, as needed.

Examples:

1. Retrieve single segment of DNA from Ensembl:
   http://servlet.sanger.ac.uk:8080/das/ensembl830/sequence?segment=1:1000,1050

Response: *http://www.biodas.org/dtd/dassequence.dtd*

```
<!ELEMENT DASSEQUENCE (SEQUENCE+)>
<!ELEMENT SEQUENCE (#PCDATA)>
<!ATTLIST SEQUENCE id CDATA #REQUIRED>
<!ATTLIST SEQUENCE start CDATA #REQUIRED>
<!ATTLIST SEQUENCE stop CDATA #REQUIRED>
<!ATTLIST SEQUENCE version CDATA #REQUIRED>
<!ATTLIST SEQUENCE molecule CDATA #REQUIRED>
```

6.4.4   Retrieving Annotations

The DAS protocol provides three specific commands for retrieving annotation details. The *types* command retrieves a summary of available annotations. The *features* command retrieves a full list of annotation details. Finally, the *link* command retrieves a web page with detailed information regarding a specific annotation.

Prior to the invention of DAS, a number of other encoding formats were devised to annotate sequence data. Chief among these is the General Feature Format (GFF, formerly called the Gene Finding Format). GFF is a tab-delimited format used to encode biological features associated with sequences. DAS annotations are specified in XML, but the XML is based on the GFF specification. To understand annotations, we therefore begin with an overview of GFF.

Background: General Feature Format

The General Feature Format [77] is a text-based specification for annotating DNA, RNA, and protein sequences. GFF focuses on describing *features*, broadly defined as any property of a sequence that has biological significance. For example, given a long sequence of DNA, a GFF file might describe the predicted locations of genes, exons, and introns. Alternatively, a GFF file might mark certain regions of sequence data and indicate the sequence similarity between two or more species. GFF files are frequently bundled with one or more FASTA formatted files, which contain the actual sequence data.

GFF was originally created at the Wellcome Trust Sanger Institute. The full specification is available online at: *http://www.sanger.ac.uk/Software/formats/GFF/GFF_Spec.shtml*.

Below is a sample GFF file:

```
Sequence1    genscan      prediction    98689     98876     4.5100    +    .
Sequence1    genscan      prediction    112435    112605    7.3900    +    .
Sequence1    genscan      prediction    113507    113554    2.7300    +    .
Sequence1    EnsEMBL      exon          15135     15214     .         +    .
Sequence1    EnsEMBL      exon          90205     90389     .         +    .
Sequence1    EnsEMBL      exon          98689     98876     .         +    2
```

Each line indicates a unique feature, and each column indicates a different feature attribute. The current version of GFF (version 2) defines eight fields of data. These fields must be separated by tabs and must be specified in the following order:

```
<seqname> <source> <feature> <start> <end> <score> <strand> <frame>
```

A summary of the main fields is provided below:

- seqname: identifies the name of the sequence. Normally, the name is defined locally and usually references an accompanying FASTA file. Alternatively, the sequence name could reference the unique accession code of a record already stored in a public database, such as NCBI.
- source: identifies the source of the feature. Generally, this indicates the software package or institution that identified the feature. For example, in the sample above, the first three features were identified by the genscan gene finding software package.
- feature: identifies the type of feature. GFF does not formally enforce a predefined list of feature types. However, to ensure greater sharing of data, users are strongly encouraged to stick to the feature list already defined by NCBI/DDBJ/EMBL. The feature list is available online at: *http://www3.ebi.ac.uk/Services/WebFeat*. There are currently over sixty feature types defined, including: *gene*, *intron*, *exon*, *tRNA*, *repeat_region*, and *promoter*. The EBI web site includes detailed descriptions and examples of each feature type. Note that the DAS specification urges that its users stick to the same feature list.
- start, end: identifies the start and stop coordinates for the specified feature.
- score: if the feature was identified by a software package, the package can record a score, indicating the relative certainty that the feature actually exists. The actual score value can only be understood within the context of the software application, and you cannot necessarily compare scores generated by two different applications. If there is no score associated with the feature, you must specify a ".".
- strand: indicates the direction of transcription: + or −. Each base pair within a sequence has a unique position, starting at 1. If transcription is set to +, transcription occurs in the positive direction from min to max, or 5' to 3'. If transcription is set to −, transcription occurs in the negative direction from max to min, or 3' to 5'. If transcription direction is irrelevant to the feature, you must specify a ".".

Following the final field, you can specify a series of additional attributes. In an earlier version of the specification, the only attribute you could specify was a groupID. This enables users to specify that multiple features are part of the same group. For example, a series of exons and introns can be grouped together to form one gene. With GFF version 2, the attribute fields are much more open-ended. For details, refer to the specification online.

A number of GFF parsing tools are freely available on the web. In particular, the gff2ps tool takes GFF files and creates visualizations in PostScript format. For details, go to: *http://www1.imim.es/~jabril/GFFTOOLS/GFF2PS.html*.

The *types* command

The DAS *types* command retrieves a summary of annotations for a specific region of sequence data. The summary includes two categories of information:

- the type of annotation features available and
- the number of records for each feature type.

For example, a *types* command response may indicate that a specific region of data contains two types of features: exons and introns. Furthermore, it may indicate that the region contains a total of six exons and four introns. To retrieve details on each of these features, you must follow-up with a *features* command.

Unlike GFF, DAS annotation records are categorized via a two-level hierarchy. At the top level, DAS has annotation *categories*, which usually correspond to broad biological function. For example, the DAS specification defines nine annotation categories, including "transcription," "translation," and "similarity." Each category contains multiple types, which provide a finer grained description of the annotation record. For example, the transcription category contains types for "exons," "introns," and "mRNA."

Below is a sample *types* command issued to the UCSC DAS server:

http://genome.cse.ucsc.edu/cgi-bin/das/hg12/types?segment=3:50000,100000

The DAS response is provided below:

```
<?xml version="1.0" standalone="no"?>
<DASTYPES>
<GFF version="1.2" summary="yes" href="http://genome.cse.ucsc.edu/
cgi-bin/das/hg12/types">
<SEGMENT id="3" start="50000" stop="100000" version="1.00">
...
<TYPE id="all_sts_primer" category="other" >0</TYPE>
<TYPE id="all_sts_seq" category="other" >0</TYPE>
<TYPE id="bacEndPairs" category="other" >5</TYPE>
<TYPE id="chimpBac" category="other" >1</TYPE>
<TYPE id="chimpBlat" category="other" >0</TYPE>
<TYPE id="blastzBestMouse" category="transcription" method="BLAT" >31</TYPE>
<TYPE id="blastzMm2" category="other" >13</TYPE>
<TYPE id="blatFish" category="similarity" method="BLAT" >0</TYPE>
[For brevity, the full XML document is not shown here.]
<TYPE id="softberryGene" category="transcription" >0</TYPE>
<TYPE id="stsMap" category="other" >0</TYPE>
<TYPE id="twinscan" category="transcription" >0</TYPE>
...
</SEGMENT>
</GFF>
</DASTYPES>
```

In total, the UCSC server returns over 40 TYPE elements. Each TYPE element provides a summary of available annotations. Each element is also described via the two-level categorization described above. Specifically, each record has an annotation category for broad grouping of features and a type ID for finer grained descriptions. An optional *method* attribute identifies the method by which

Table 6.10 The DAS types command

Description: Requests a summary of annotations for a specific genomic region

Scope: Annotation and Reference Servers

Request Syntax:
PREFIX/das/*DSN*/types[?segment= *RANGE*]

        [;segment= *RANGE*]
        [;type= *TYPE*]
        [;type= *TYPE*]

Request Arguments:
- segment: indicates a specific region of sequence data. Segments are specified as: *referenceID: start, stop*. For example, to request a portion of human chromosome 3, use: segment=3:50000,100000. If no segment is specified, then all feature types for the specified data source are returned.
- type: filters the data set for records of the specified feature type. If multiple types are specified, the types are connected via a logical OR. For example, type=genscan;type=snpNih returns all records of type genscan or type snpNih.

Examples:
1. Retrieve UCSC feature types for human chromosome 3, base pairs: 50,000–100,000:
   http://genome.cse.ucsc.edu/cgi-bin/das/hg12/types?segment=3:50000,100000
2. Retrieve all UCSC feature types for data source: hg12:
   http://genome.cse.ucsc.edu/cgi-bin/das/hg12/types
3. Retrieve UCSC feature types for human chromosome 3, base pairs: 50,000–100,000; filter for feature types set to type=genscan or type=snpNih:
   http://genome.cse.ucsc.edu/cgi-bin/das/hg12/types?segment=3:50000,100000;type=genscan;type=snpNih

Response: *http://www.biodas.org/dtd/dastypes.dtd*
```
<!ELEMENT DASTYPES (GFF)>
<!ELEMENT GFF (SEGMENT)>
<!ATTLIST GFF version CDATA #REQUIRED>
<!ATTLIST GFF href CDATA #REQUIRED>
<!ELEMENT SEGMENT (TYPE+)>
<!ATTLIST SEGMENT id CDATA #REQUIRED>
<!ATTLIST SEGMENT start CDATA #REQUIRED>
<!ATTLIST SEGMENT stop CDATA #REQUIRED>
<!ATTLIST SEGMENT version CDATA #REQUIRED>
<!ATTLIST SEGMENT label CDATA #IMPLIED>
<!ELEMENT TYPE (#PCDATA)>
<!ATTLIST TYPE id CDATA #REQUIRED>
<!ATTLIST TYPE method CDATA #IMPLIED>
<!ATTLIST TYPE category CDATA #IMPLIED>
```

the annotation was identified. For example, the "blatFish" feature was identified via the UCSC BLAT sequence similarity tool.

As noted above, the DAS specification contains a list of nine predefined annotation categories. Each of these categories contains a list of recommended feature types. The list is not meant to be comprehensive, but does provide users with a starting reference point. For the most part, the DAS feature types are based on the same NCBI/DDBJ/EMBL features list used by GFF (see above). The main difference is that DAS attempts to group these features into broad categories. See Table 6.11 for details.

Just like GFF, categories and types are not enforced by the specification, and annotation servers are free to return whatever they like. This can lead to some confusion for users. For example, in the UCSC response above, one transcription type is specified as "softberryGene." The softberryGene is not part of the predefined list of DAS types, nor is it part of the NCBI feature list. To find out what it actually means, you can issue a *link* command to retrieve additional details (defined below). However, if the link command is not implemented (as in the case of UCSC), or

Table 6.11 DAS categories and types: The DAS specification includes a list of nine predefined annotation categories and their associated feature types. The list is not meant to be comprehensive, and annotation servers are free to return whatever categories or types they choose

Category Name	Description	Example Types
Component	Any annotation that identifies genomic map components. For details, refer to Section 6.5 of this chapter	• chromosome • super-contig • contig • bac • read
Experimental	A miscellaneous annotation category for new experimental results	
Repeat	Any annotation that identifies repetitive sequence content	• microsatellite • inverted • tandem • transposable_element • LINE • SINE • misc_repeat
Similarity	Any annotation that identifies sequence similarity between two sequences, including cross-species sequence similarity	• NN (nucleotide-to-nucleotide) • NP (nucleotide-to-protein) • PN (protein-to-nucleotide) • PP (protein-to-protein) • misc_homology
Structural	Any annotation related to mapping, sequencing, and assembly of the genome	• clone • primer_left • primer_right • oligo • assembly_tag • misc_structural
Supercomponent	Any annotation that identifies genomic map super-components. For details, refer to Section 6.5 of this chapter	• chromosome • super-contig • contig • bac • read
Transcription	Any annotation related to the transcription of DNA to RNA	• exon • intron • tRNA • mRNA • 5'Cap • PolyA • Splice5 • Splice3 • misc_transcribed
Translation	Any annotation related to the translation of RNA to protein	• stop • ATG • CDS • 5'UTR • 3'UTR • misc_translated
Variation	Any annotation that identifies sequence variation or polymorphism	• insertion • deletion • substitution • misc_variation • SNP (single nucleotide polymorphism)

Table 6.12 The DAS features command

Description: Requests full annotation records across a specific genomic region

Scope: Annotation and Reference Servers

Request Syntax:

```
PREFIX/das/ DSN /features?[segment= RANGE]
      [;segment= RANGE...]
      [;type= TYPE]
      [;type= TYPE]
      [;category= CATEGORY]
      [;category= CATEGORY]
      [;categorize= yes| no]
      [;feature_id= ID]
      [;group_id= ID]
```

Request Arguments:

- segment: indicates a specific region of sequence data. Segments are specified as: *referenceID: start, stop*. For example, to request a portion of human chromosome 3, use: segment=3:50000,100000. Users are free to specify multiple segments, as necessary.
- type: filters the data set for records of the specified feature type. If multiple types are specified, the types are connected via a logical OR. For example, type=genscan;type=snpNih returns all records of type genscan or type snpNih.
- category: filters the data set for records of the specified category. If multiple categories are specified, the categories are connected via a logical OR.
- categorize: indicates whether the returned feature records include a functional category. If set to "yes," features must include category information.
- feature_id: instead of specifying a segment, users can opt to retrieve a specific feature by specifying the feature_id argument. This argument is new to DAS version 1.5, and not yet widely implemented. Check the X-DAS-Capabilities header "feature-by-id/1.0" to determine if the server implements this functionality.
- group_id: instead of specifying a segment, users can opt to retrieve a specific group of features by specifying the group_id argument. This argument is new to DAS version 1.5, and not yet widely implemented. Check the X-DAS-Capabilities header "group-by-id/1.0" to determine if the server implements this functionality.

Examples:

1. Retrieve all UCSC annotations for human chromosome 3, 50000–100000:
   http://genome.cse.ucsc.edu/cgi-bin/das/hg12/features?segment=3:50000,100000
2. Retrieve all UCSC annotations for human chromosome 3, 50000–100000; filter for features types set to blastzBestMouse:
   http://genome.cse.ucsc.edu/cgi-bin/das/hg12/features?segment=3:50000,100000;type=blastzBestMouse
3. Retrieve all UCSC annotations for human chromosome 3, 50000–100000; filter for features related to transcription:
   http://genome.cse.ucsc.edu/cgi-bin/das/hg12/features?segment=3:50000,100000;category=transcription
4. Retrieve WormBase feature by feature_id:
   http://www.wormbase.org/db/das/elegans/features?feature_id=Sequence:cb25.fpc0039/3015360

Response: *http://www.biodas.org/dtd/dasgff.dtd*

```
<!ELEMENT DASGFF (GFF)>
<!ELEMENT GFF (SEGMENT | ERRORSEGMENT | UNKNOWNSEGMENT)+>
<!ATTLIST GFF version CDATA #REQUIRED>
<!ATTLIST GFF href CDATA #REQUIRED>
<!ELEMENT SEGMENT (FEATURE+)>
<!ATTLIST SEGMENT id CDATA #REQUIRED>
<!ATTLIST SEGMENT start CDATA #REQUIRED>
<!ATTLIST SEGMENT stop CDATA #REQUIRED>
<!ATTLIST SEGMENT version CDATA #REQUIRED>
<!ATTLIST SEGMENT label CDATA #IMPLIED>
<!ELEMENT FEATURE (TYPE, METHOD, START, END, SCORE, ORIENTATION,
  PHASE, GROUP?, LINK?, NOTE?, TARGET?)>
<!ATTLIST FEATURE id CDATA #REQUIRED>
```

Table 6.12 (cont.)

```
<!ATTLIST FEATURE label CDATA #IMPLIED>
<!ATTLIST FEATURE version CDATA #IMPLIED>
<!ELEMENT TYPE (#PCDATA)>
<!ATTLIST TYPE id CDATA #IMPLIED>
<!ATTLIST TYPE category CDATA #IMPLIED>
<!ATTLIST TYPE reference CDATA "no">
<!ATTLIST TYPE subparts CDATA "no">
<!ELEMENT METHOD (#PCDATA)>
<!ATTLIST METHOD id CDATA #IMPLIED>
<!ELEMENT START (#PCDATA)>
<!ELEMENT END (#PCDATA)>
<!ELEMENT SCORE (#PCDATA)>
<!ELEMENT ORIENTATION (#PCDATA)>
<!ELEMENT PHASE (#PCDATA)>
<!ELEMENT GROUP (NOTE*, LINK*, TARGET*)>
<!ATTLIST GROUP id CDATA #REQUIRED>
<!ELEMENT NOTE (#PCDATA)>
<!ELEMENT LINK (#PCDATA)>
<!ATTLIST LINK href CDATA #REQUIRED>
<!ELEMENT TARGET (#PCDATA)>
<!ATTLIST TARGET id CDATA #REQUIRED>
<!ATTLIST TARGET start CDATA #REQUIRED>
<!ATTLIST TARGET stop CDATA #REQUIRED>
<!ELEMENT ERRORSEGMENT EMPTY>
<!ATTLIST ERRORSEGMENT id CDATA #REQUIRED>
<!ATTLIST ERRORSEGMENT start CDATA #IMPLIED>
<!ATTLIST ERRORSEGMENT stop CDATA #IMPLIED>
<!ELEMENT UNKNOWNSEGMENT EMPTY>
<!ATTLIST UNKNOWNSEGMENT id CDATA #REQUIRED>
<!ATTLIST UNKNOWNSEGMENT start CDATA #IMPLIED>
<!ATTLIST UNKNOWNSEGMENT stop CDATA #IMPLIED>
```

the DAS server does not have information about this specific type, you only have two options left—search the web site associated with the DAS server or contact the maintainers of the server directly via email. In the specific case of the softberryGene type, a quick search of Google will quickly lead you to the UCSC Genome Browser User Guide, where a detailed explanation is provided.

The Sequence Ontology project aims to create a controlled vocabulary for describing sequence features, and may be adopted in a future version of DAS. With a controlled vocabulary, DAS annotation servers would be constrained to use a predefined list of terms to describe all sequence features. According to Lincoln Stein, "the existence of a shared ontology allows an integrator to merge two databases with some guarantee that a term used in one database corresponds to the same term used in the other" [11]. This would add some overhead to the existing protocol, but would also significantly improve the ability of DAS to aggregate data from multiple servers. Complete details regarding the Sequence Ontology project are available online at: *http:// song.sourceforge.net*.

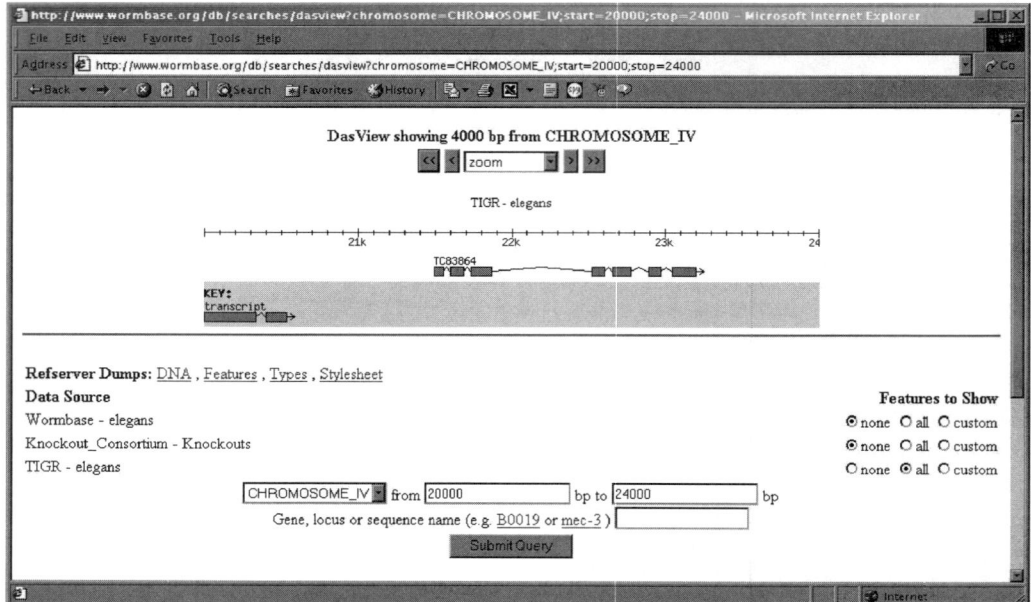

Figure 6.8 A screenshot of TIGR annotation data viewed via the WormBase DAS viewer. Exons are represented as blue rectangles and the introns are represented as single black lines.

The *features* command

The DAS *features* command returns the actual genomic annotation records stored within a DAS annotation server. The request syntax and the resulting XML are more complicated than other DAS commands, and this section therefore includes a more detailed example.

One of the best ways to learn the *features* command is to view genomic annotations through a DAS viewer and compare the visualizations provided with the actual XML data. For an introductory example, we consider annotation data for the *C. elegans* genome, provided by the Institute for Genomic Research (TIGR). We first examine a small sample of the data via the WormBase DAS viewer. We then examine the same data in its native XML format.

To get started, open a web browser and go to the WormBase DAS viewer:

http://www.wormbase.org/db/searches/dasview?chromosome=CHROMOSOME_IV;
start=20000;stop=24000

At the bottom of the page, you will see multiple data sources, including "WormBase-elegans" and "TIGR-elegans." Under the panel named "Features to show," select "none" for WormBase and "all" for TIGR. You should now see a screen that looks similar to Figure 6.8. The visualization provided by WormBase shows a single coding sequence for the *C. elegans* Y38C1AB gene. Exons are represented as blue rectangles and the introns are represented as single black lines. All told, the visualization shows seven exons and seven introns.

Next, issue a *features* request to the TIGR DAS server and request the same set of data:

http://www.tigr.org/docs/tigr-scripts/tgi/das/elegans/features?segment=CHROMOSOME_IV: 20000,24000

The TIGR response is displayed in Listing 6.1. For brevity, only a portion of the XML response is shown. However, you can see that several of the FEATURE elements correspond to the visualization provided by WormBase. For example, the first two features correspond to exons starting at location 23046 and 21606, and the third feature corresponds to an intron starting at location 23199.

The features response must reference the dasgff.dtd document, and must contain a root GFF element. To help understand the DTD, a visual representation is presented in Figure 6.9.

The FEATURE element represents the actual genome annotation record, and therefore deserves a close look. Fortunately, most of the elements associated with features correspond to the GFF format described earlier. Below is a summary of each of these elements:

- TYPE : indicates the feature type. This includes a type *id* and *category*. For example, in the TIGR response, the first feature specifies an exon , which is part of the transcription category.
- METHOD : indicates the method by which the feature was identified. For example, this might specify a gene finding software package.
- START : indicates the start position of the feature.
- END : indicates the stop position of the feature.
- SCORE : if the feature was identified by a software package, the package can record a score, indicating the relative certainty that the feature actually exists. The actual score value can only be understood within the context of the software application and you cannot necessarily compare scores generated by two different applications. This field corresponds to the "score" column in GFF.
- ORIENTATION : indicates the direction of transcription: + or −. If transcription is set to +, transcription occurs in the positive direction. If transcription is set to −, transcription occurs in the negative direction. If transcription is set to 0, transcription direction is irrelevant to this feature. This field corresponds to the "strand" column in GFF.
- PHASE : specifies the number of base pairs to shift before reading codons. A codon is a sequence of three base pairs that codes for a specific amino acid. For example, a phase value of 1 indicates that the sequence should be shifted be one, and that codons should be read starting at position 2. A value of "−" indicates that the phase is irrelevant to the feature.
- GROUP : indicates that the specified feature is part of a group of features. For example, in the TIGR example, all the exons and introns correspond to a single gene, and are therefore grouped together with a common group id, TC83864.
- LINK : a URL that provides further information about the feature.
- NOTE : human readable documentation about the feature.
- TARGET : indicates the target sequence for a sequence similarity match. The element requires three attributes: an *id* reference, and *start* and *stop* values.

Starting in DAS version 1.5, DAS servers can indicate invalid segment requests by returning either an ERRORSEGMENT or UNKNOWNSEGMENT element. Reference servers are required to return an ERRORSEGMENT element. In contrast, annotation servers are required to return an UNKNOWNSEG-MENT element. The difference arises because annotation servers do not store a complete genomic map, and they are therefore unable to determine if the requested segment is actually invalid or just outside the set of data it contains.

Listing 6.1 DAS response from TIGR. Note that the complete DAS response includes a total of 13 FEATURE elements. For brevity, only the first three FEATURE elements are displayed here.

```xml
<?xml version="1.0" standalone="yes"?>
<!DOCTYPE DASGFF SYSTEM  "http://biodas.org/dtd/dasgff.dtd">
<DASGFF>
<GFF version="0.995" href= "http://www.tigr.org/docs/tigr-
scripts/tgi/das/elegans/features">
<SEGMENT id="Sequence:CHROMOSOME_IV" start="20000" stop="24000"
  version="1.0">
  <FEATURE id="TC83864"  label="TC83864">
    <TYPE id="exon" category="transcription">transcript</TYPE>
    <METHOD id="curated">curated</METHOD>
    <START>23046</START>
    <END>23198</END>
    <SCORE>-</SCORE>
    <ORIENTATION>+</ORIENTATION>
    <PHASE>0</PHASE>
    <GROUP id="TC83864">
       <LINK href="http://www.tigr.org/docs/tigr-
       scripts/tgi/tc_report.pl?species=elegans;tc=TC83864">TC83864
       </LINK>
    </GROUP>
  </FEATURE>
  <FEATURE id="TC83864" label="TC83864">
    <TYPE id="exon" category="transcription">transcript</TYPE>
    <METHOD id="curated">curated</METHOD>
    <START>21606</START>
    <END>21689</END>
    <SCORE>-</SCORE>
    <ORIENTATION>+</ORIENTATION>
    <PHASE>0</PHASE>
    <GROUP id="TC83864">
       <LINK href="http://www.tigr.org/docs/tigr-
       scripts/tgi/tc_report.pl?species=elegans;tc=TC83864">TC83864
       </LINK>
    </GROUP>
  </FEATURE>
  <FEATURE id="TC83864" label="TC83864">
    <TYPE id="intron" category="transcription">transcript</TYPE>
    <METHOD id="curated">curated</METHOD>
    <START>23199</START>
    <END>21605</END>
    <SCORE></SCORE>
    <ORIENTATION>+</ORIENTATION>
    <PHASE>0</PHASE>
    <LINK href="http://www.tigr.org/docs/tigr-
    scripts/tgi/tc_report.pl?species=elegans;tc=TC83864">""</LINK>
    <GROUP id="TC83864">
    </GROUP>
  </FEATURE>
  ...
</SEGMENT>
</GFF>
</DASGFF>
```

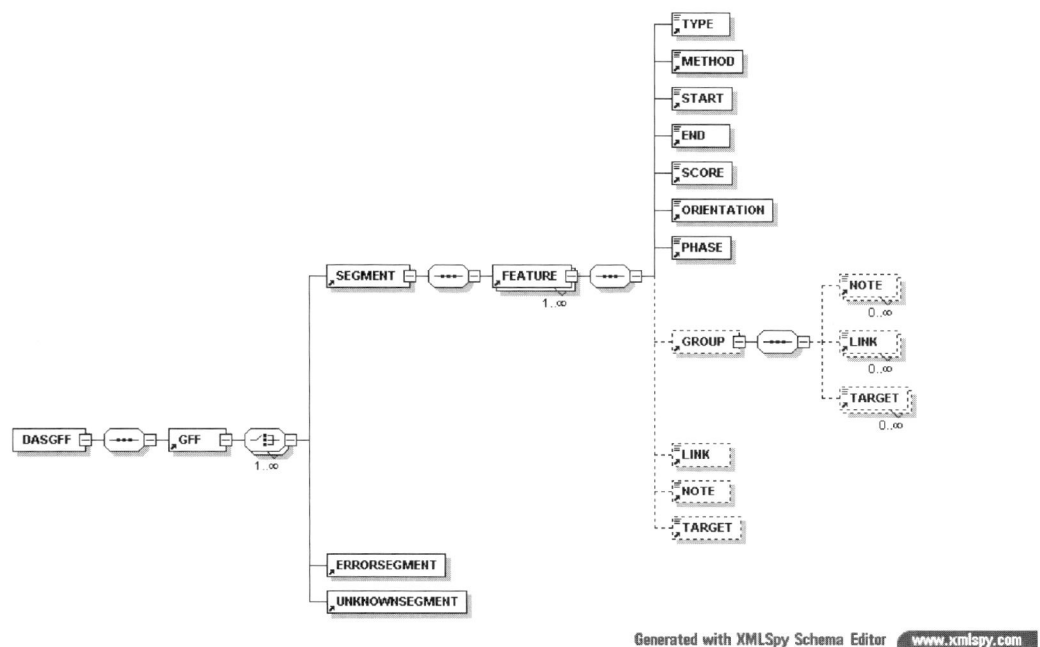

Generated with XMLSpy Schema Editor   www.xmlspy.com

Figure 6.9 A visual representation of the dasgff DTD. The diagram was created with XMLSpy®.

The following is an invalid request to the Ensembl DAS reference server:

http://servlet.sanger.ac.uk:8080/das/ensembl830/features?segment=30:100000,200000

DAS style sheets

When creating a visualization of annotations, DAS clients can base the visualization on a style sheet retrieved from the annotation server. Annotation servers are free to maintain separate style sheets for each data source, or simply maintain one master style sheet. Furthermore, clients are not required to actually follow the style sheets—in practice, many of the current DAS viewers simply ignore the style sheet or implement a limited set of functionality.

It is important to note the DAS style sheets bear no resemblance to other style sheet technologies, such as Cascading Style Sheets (CSS) or eXtensible Stylesheet Language (XSL). DAS style sheets are completely homegrown and have their own unique syntax and set of conventions.

A complete discussion of DAS style sheets is beyond the scope of this chapter. However, a quick example will provide you with a taste of how it works. To request a style sheet, you simply invoke the *stylesheet* command. For example, to retrieve the style sheet for the human genome provided by Ensembl, use this command:

http://servlet.sanger.ac.uk:8080/das/ensembl830/stylesheet

The style sheet document associates feature categories and types with objects known as glyphs. Glyphs represent graphical objects, such as lines, boxes, triangles, and arrows. Each glyph can be customized with attributes, such as height, color, outline color, and labeling conventions. For example, here is an excerpt from the Ensembl style sheet:

```
<TYPE id="similarity">
  <GLYPH>
    <BOX>
      <HEIGHT>15</HEIGHT>
      <COLOR>black</COLOR>
      <OUTLINECOLOR>green</OUTLINECOLOR>
      <LINEWIDTH>1</LINEWIDTH>
    </BOX>
  </GLYPH>
</TYPE>
```

This snippet of XML basically recommends that similarity features be rendered with a black box, 15 pixels high and outlined with a green border. The remainder of the Ensembl style sheet contains similar recommendations for other feature types.

There is no human chromosome 30 and Ensembl will therefore return the following response:

```
<?xml version='1.0'  standalone='no' ?>
<!DOCTYPE DASGFF SYSTEM  'dasgff.dtd' >
<DASGFF>
  <GFF version="1.0"
    href="http://servlet.sanger.ac.uk:8080/das/ensembl830/features?
    segment= 30:1000000,2000000">
    <ERRORSEGMENT id="30" start="100000" stop="200000" />
  </GFF>
</DASGFF>
```

If Ensembl was acting purely as an annotation server, it would return an UNKNOWNSEGMENT element instead. To determine if a server supports either type of error messages, check the X-DAS Capabilities header for "error-segment/1.0" or "unknown-segment/1.0."

The *link* command

The final DAS command to explore is the *link* command. The link command retrieves a web page with additional information regarding an individual annotation. The response is a regular HTML page, and unlike all the other DAS commands, the page does not contain any structured XML content. The purpose of the *link* command is to provide end users with additional human readable information about individual annotations.

It is important to note that the *link* command is not yet widely implemented. Instead of implementing the link command, most DAS servers have elected to return URL links within the LINK element provided by a *features* command request (see the *features* command above). By using the LINK element, each feature comes with a URL link automatically, and the clients have no need to issue a separate *link* command.

Complete request/response syntax for the *link* command is provided in Table 6.13.

Table 6.13 The DAS link command

Description: Retrieves a web page with additional information regarding an individual feature

Scope: Annotation and Reference Servers

Request Syntax:

PREFIX/das/ *DSN*/link?field= *FIELD*;id=*ID*

Request Arguments:
- field: indicates the specific type of data to retrieve. Possible values are:
  - feature: retrieves information about the feature itself
  - type: retrieves type information about the specified feature
  - method: retrieves method information about the specified feature
  - category: retrieves category information about the specified feature
  - target: retrieves target information about the specified feature. Usually used to retrieve sequence similarity information
- id: indicates the feature ID

Response:
The response is an HTML web page

The Bio::Das Perl Module

In the next two chapters, we will discuss the specifics of parsing DAS data using Java. However, if you are eager to programmatically interface with DAS and do not want to delve into the specifics of XML parsing, you may want to consider using one of the open source Bio* libraries. For example, both BioJava and BioPerl now provide DAS support. To use DAS in Perl, you will need to download the Bio::Das Perl module, created by Lincoln Stein. This module is available via CPAN and requires the installation of XML::Parser and LWP (The World-Wide Web library for Perl). Once properly installed, you can easily make programmatic calls to any DAS server. For example, the code below connects to Ensembl and issues a DSN request:

```perl
#!/usr/local/bin/perl
use Bio::Das;

print "Running Bio::Das DSN Request\ n";

# Create New BioDas Object, with timeout of 5 seconds.
$das = Bio::Das->new(5);

# Issue DAS DSN Request
# Use the Sanger/Ensembl DAS Server
$response = $das->dsn
   ('http://servlet.sanger.ac.uk:8080/das');

# Display the URL Request
print "DAS Request URL: ", $response->url,"\ n";

# Check if Request was successful.
# If successful, print all DSN results;
# Otherwise print the DAS error message.
```

```
if ($response->is_success) {
    @dsns = $response->results;
    foreach $dsn (@dsns) {
            print "Data Source ID: ", $dsn->id;
            print "\ n\tName: ",$dsn->name;
            print "\ n\tDescription: ",$dsn->description;
            print "\ n\tMapmaster: ",$dsn->master;
            print "\ n";
    }
} else {
    print "Error: ",$response->error,"\ n";
}
```

As you can see, you issue DAS requests by first instantiating a `Das` object and subsequently use this object to issue queries. For example, in the code above, we are using the dsn query and passing the URL to the Sanger/Ensembl DAS server. You can then check for possible error codes and display the complete contents of the DAS server response. Complete documentation, along with more detailed examples, is available within the module.

6.5   Working with Reference Maps

With a complete understanding of the full suite of DAS commands, we can now return to a critical issue in DAS—working with reference maps. We already know that users can issue an *entry_points* command to receive a list of starting points to the genomic map. Most reference servers will return a set of chromosome numbers. For example, the Ensembl reference server returns chromosomes 1–22, X, and Y. These chromosome numbers are used as reference IDs within subsequent DAS commands. For example, you can retrieve annotations for human chromosome 1 with this *features* command:

> *http://servlet.sanger.ac.uk:8080/das/ensembl830/features?segment=1:100000,200000*

So far, all of the examples in this chapter have used chromosome reference IDs, but you are not restricted to using just chromosomes. In fact, as we will soon see, chromosome coordinates may shift between each new version of the genomic assembly, and DAS therefore allows you to use other reference IDs. DAS even provides the functionality for drilling down within a genomic map.

The DAS drill-down functionality is based on the genomic map maintained by the reference server. For example, many genomic assemblies are created via the "clone-contig" approach—individual clones are sequenced and the clones are then tiled together to form "contigs." To make this more concrete, consider Figure 6.10. The figure shows a portion of human chromosome 3 (1–1,000,000 base pairs) via the UCSC genome browser. The map for human chromosome 3 is divided into hundreds of contigs, and the screenshot shows a portion of just one of these contigs, NT_005927. The screenshot also shows the clones, which make up the contig, and you can clearly see the tiling path of overlapping clones. For example, the first clone is AC066595 and its end overlaps with the beginning of clone AC026187.

DAS enables you to drill down the genomic map, from chromosomes to contigs to clones. To do so, you must issue a cascading set of DAS commands. A summary of these steps is presented in Figure 6.11. A detailed example is presented below.

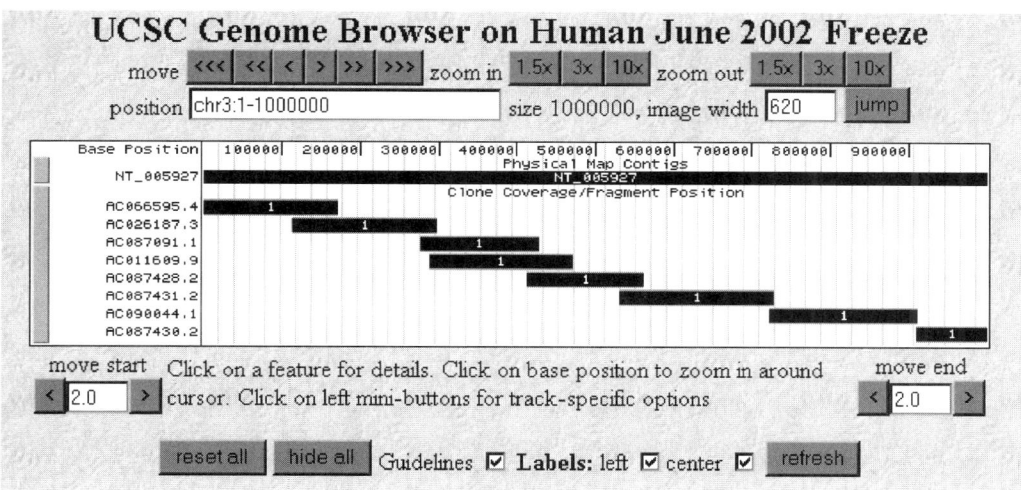

Figure 6.10 Screenshot of the USCS genome browser. A portion of chromosome 3 is shown. The browser has been configured to show clone coverage and map contigs.

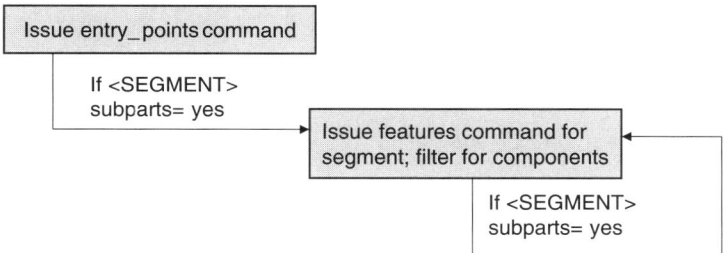

Figure 6.11 To drill down within a genomic map, you must issue a cascade of DAS commands. You begin with an entry_points command and follow-up with a series of feature commands. At each point, you can check the subparts attribute to determine if there is further substructure to explore.

6.5.1   Traversing the Ensembl Reference Map

The first step in traversing a genomic map is to retrieve a set of entry points. Here is an excerpt from the Ensembl *entry_points* response:

```
<?xml version='1.0' standalone='no' ?>
<!DOCTYPE DASEP SYSTEM 'dasep.dtd' >
<DASEP>
  <ENTRY_POINTS href="http://servlet.sanger.ac.uk:8080/das/
    ensembl830/entry_points" version="8.30">
    <SEGMENT id="Y" size="58368225" subparts="yes" />
    <SEGMENT id="X" size="149249818" subparts="yes" />
    <SEGMENT id="19" size="60013307" subparts="yes" />
    <SEGMENT id="18" size="77516809" subparts="yes" />
    . . .
```

If the *subparts* attribute is set to "yes," then the segment has some internal structure. To determine just what the internal structure is, you must issue a *features* command and restrict the result set to "components" only. For example, the following command queries Ensembl for all components within a specific portion of chromosome 3:

http://servlet.sanger.ac.uk:8080/das/ensembl830/features?segment=3:1,20000000; category=component

Here is an excerpt from the Ensembl response:

```
<?xml version='1.0' standalone='no' ?>
<!DOCTYPE DASGFF SYSTEM 'dasgff.dtd' >
<DASGFF>
  <GFF version="1.0"
    href="http://servlet.sanger.ac.uk:8080/das/ensembl830/features
    ?segment=3:1,20000000;category=component">
    <SEGMENT id="3" version="8.30" start="1" stop="20000000">
      <FEATURE id="components/NT_005927">
        <TYPE id="static_golden_path_contig" reference="yes"
          subparts="yes">static_golden_path_contig</TYPE>
        <METHOD id="ensembl">ensembl</METHOD>
        <START>1</START>
        <END>17431026</END>
        <SCORE>-</SCORE>
        <ORIENTATION>+</ORIENTATION>
        <PHASE>-</PHASE>
        <TARGET id="NT_005927" start="1" stop="17431026" />
      </FEATURE>
      ...
    </SEGMENT>
  </GFF>
</DASGFF>
```

All told, the response includes four features, each of which corresponds to a single contig within the genomic map. If you use the UCSC genome browser, you can see the same four contigs (see Figure 6.12). The first contig within the UCSC map and within the XML response is NT_005927. If you look carefully at the FEATURE element, you can see that the *reference* attribute is set to "yes." This means that NT_005927 can be used as a reference ID in subsequent DAS commands.

Also note that the FEATURE *subparts* attribute is set to "yes." This indicates that the contig itself has even more internal structure. You can therefore query for the components of the contig:

http://servlet.sanger.ac.uk:8080/das/ensembl830/features?segment=NT_005927:1,1000000; category=component

Here is an excerpt from the Ensembl response:

```
<?xml version='1.0' standalone='no' ?>
<!DOCTYPE DASGFF SYSTEM 'dasgff.dtd' >
<DASGFF>
  <GFF version="1.0"
    href="http://servlet.sanger.ac.uk:8080/das/ensembl830/features?
    segment=NT_005927:1,1000000;category=component">
    <SEGMENT id="NT_005927" version="8.30" start="1" stop="1000000">
    ...
```

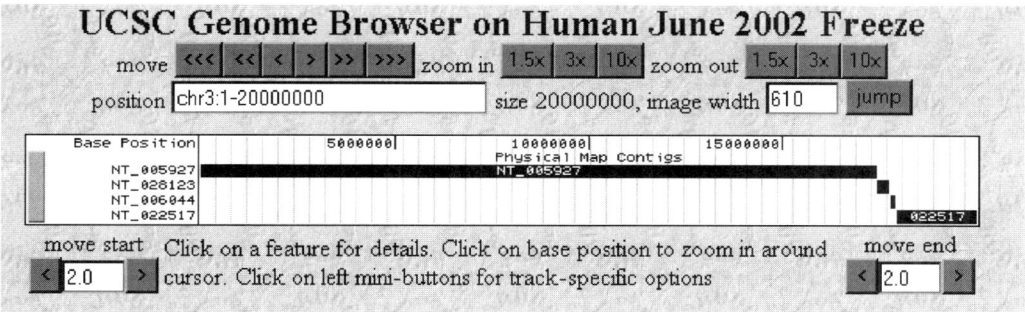

Figure 6.12 Screenshot of the UCSC genome browser. A portion of chromosome 3 is shown. The browser has been configured to show map contigs only.

```
<FEATURE id="components/AC066595">
  <TYPE id="static_golden_path_clone" reference="yes"
     subparts="no">static_golden_path_clone</TYPE>
  <METHOD id="ensembl">ensembl</METHOD>
  <START>1</START>
  <END>172113</END>
  <SCORE>-</SCORE>
  <ORIENTATION>-</ORIENTATION>
  <PHASE>-</PHASE>
  <TARGET id="AC066595" start="5" stop="172117" />
  </FEATURE>
</GFF>
</DASGFF>
```

We have now drilled all the way down to the clone level. All told, the response document references a total of eight clones (each of the clones appears within Figure 6.10). In the one clone shown above, you can see that its *subparts* attribute is set to "no," meaning that it does not contain any further substructure. Its *reference* attribute is, however, set to "yes." This means that you can use the clone as a reference ID. For example, this command retrieves all features for the clone:

http://servlet.sanger.ac.uk:8080/das/ensembl830/features?segment=AC066595

Some DAS servers also support the ability to query for "supercomponents." When querying for components, you are basically drilling down the genomic map; for example, you can drill down from chromosomes to contigs to clones. By querying for supercomponents, you essentially work in reverse—you can begin with a clone, discover its contig, and then move up to its chromosome. At the time of this writing, the Ensembl DAS server does not support supercomponent functionality.

6.5.2   Working with Evolving Reference Maps

Now that you understand reference maps, contigs, and clones, we are ready to tackle one remaining issue: working with evolving reference maps. To highlight the specific issues involved, consider a simple scenario: how to retrieve all genomic annotations related to the ADAM2 gene in human from two different annotation servers.

The first step is to determine that the two annotation servers share a common reference map. In the ideal case, both annotation servers reference the same DAS reference server, and you can therefore be certain that they share a common genomic map. However, this is not always the case. For example, both UCSC and Ensembl are currently set up as annotation servers and reference servers. In other words, if you issue a DSN query to UCSC, you can see that its annotations reference UCSC maps. Annotations from Ensembl reference Ensembl maps. Therefore, based on information from DAS, it is impossible to tell if UCSC and Ensembl are actually referencing the same genomic map. In fact, the only way to determine this type of information is usually to check the web site associated with the DAS server. For example, you can check the UCSC Release FAQ, and determine that the July 2003 release (hg16) uses NCBI Assembly Build 34. Similar information can be found from the Ensembl web site. If the two data sources use identical NCBI Assembly builds, they will share a common genomic map, and you can proceed with the second step.

DAS primarily deals with retrieval of annotations based on genomic coordinates (it is also possible to retrieve annotations based on feature id, but this is not yet widely implemented). Hence, the second step is to determine the genomic coordinates of the ADAM2 gene. DAS does not provide such a lookup facility, and you will therefore need to manually obtain this information from one of the main genome browsers, such as NCBI, UCSC, or Ensembl. For example, you can connect to UCSC, enter the term "ADAM2," and find a matching gene location. If you select the July 2003 assembly, you can determine that ADAM2 is located at chr8:39618624-39713096. You can also determine that the gene is located in contig NT_008251 and clone AP005902.

As the third and final step, you can issue a *feature* request for each annotation server and specify the chromosomal coordinates for the ADAM2 gene. However, you may also retrieve annotation data via contig or clone coordinates. Which option is best? To answer this question, consider that each time a new genomic assembly is created, sequences are added, deleted, and refined, and chromosomal (or absolute) coordinates inevitably shift. For example, a new assembly may insert 100 nucleotides at the beginning of chromosome 8 and this will shift all features by 100. ADAM2 is therefore not guaranteed to always be located at the same absolute coordinates. In fact, if you select the April 2003 assembly from UCSC, you can see that ADAM2 is now located at chr8:39342200-39436675—it has shifted approximately 276,000 base pairs. However, you can also determine that ADAM2 is still located in contig NT_008251 and clone AP005902.

In general, chromosomal coordinates tend to change with each new version of the assembly, but contig/clone coordinates tend to stay the same. Therefore, if you want to reliably compare features across multiple versions of an assembly, you are advised to use contig or clone coordinates.

6.6 The Future of DAS

DAS is supported by an active community of developers, and work on DAS 2.0 is already well under way. The specification process is open to the entire DAS community, and the biodas web site maintains a Request for Comment (RFC) section (see *http://www.biodas.org/RFCs/index.html*). Anyone with a good idea can simply write a new RFC and submit it directly to the web site. Users can then comment on specific RFCs via the biodas mailing list.

Over a dozen RFCs have already been submitted for DAS version 2.0. Collectively, they provide some hints regarding the future evolution of DAS. Some RFCs are focused on architectural issues.

For example, two RFCs propose that DAS adopt a formal SOAP specification; another RFC proposes the creation of a DAS registry service that will enable clients to automatically discover new DAS servers. Other RFCs are focused on creating new functionality for the future. For example, two RFCs propose a new coordinate mapping service for mapping between different genomic assemblies; another RFC proposes the creation of a DAS visualization service. Check the biodas.org web site for details on each RFC.

Parsing DAS Data with SAX 7

The Simple API for XML (SAX) is a standard event-based interface for parsing XML documents. It is particularly well suited for bioinformatics applications because it is very fast, takes up little memory, and is capable of parsing very large documents.

This chapter provides a comprehensive overview of the Java SAX API (the Perl SAX API is covered in Chapter 5). The chapter includes a detailed description of the main SAX interfaces and detailed examples to illustrate core concepts. We begin with an introduction to the two most important SAX interfaces: the `XMLReader` interface implemented by the XML parser; and the `ContentHandler` interface, used to receive SAX events as they occur. We then continue with additional topics, including: document validation against DTDs and XML Schemas, XML namespace issues new to SAX 2.0, and options for converting SAX events into custom data structures. The chapter concludes with an in-depth example that parses Distributed Annotation System (DAS) feature data via SAX and displays those features via the open source BioJava toolkit. Throughout the chapter, the sample code uses two SAX 2.0 compliant XML parsers: the lightweight nonvalidating Piccolo XML parser, and the fully featured Xerces2 XML parser from the Apache group.

7.1 Introduction to SAX

The Simple API for XML (SAX) [71] is a standard event-based interface for parsing XML documents. Unlike XML itself or the Document Object Model (DOM), SAX is not an official standard of any organization, such as the World Wide Web Consortium (W3C). Rather, SAX is a de facto standard, developed by a group of volunteers, freely available to the public, and widely implemented by dozens of XML parsers. SAX was originally designed for Java, but SAX and SAX-inspired implementations are now available for other languages, including Perl, Python, C++, Visual Basic, and Pascal.

For a complete introduction to SAX, and a description of tree-based vs. event-based interfaces, refer to Chapter 5.

7.1.1 A First Example

Let us dive right in with our first SAX example. Listing 7.1 shows a sample XML document, retrieved from the Ensembl DAS server. We will use this sample document in subsequent examples. Listing 7.2 provides the source code for `BasicSAX.java`. The goal of the program is to capture

Listing 7.1 ensembl.dna.xml

```
<?xml version="1.0" standalone="no" ?>
<!DOCTYPE DASDNA SYSTEM
"http://servlet.sanger.ac.uk:8080/das/dasdna.dtd">
<DASDNA>
  <SEQUENCE id="1" version="8.30" start="1000" stop="1050">
    <DNA length="51">
      taatttctcccattttgtaggttatcacttcactctgttgactttcttttg
    </DNA>
  </SEQUENCE>
  <SEQUENCE id="2" version="8.30" start="1000" stop="1050">
    <DNA length="51">
      taatgcaactaaatccaggcgaagcatttcagcttaacccccgagacttttg
    </DNA>
  </SEQUENCE>
</DASDNA>
```

Listing 7.2 SAXBasic.java

```java
package org.xmlbio.sax;

import org.xml.sax.Attributes;
import org.xml.sax.ContentHandler;
import org.xml.sax.Locator;
import org.xml.sax.SAXException;
import org.xml.sax.XMLReader;
import org.xml.sax.helpers.XMLReaderFactory;

import java.io.IOException;

/**
 * Basic SAX Example.
 * Illustrates basic implementation of the SAX Content Handler.
 */
public class SAXBasic implements ContentHandler {

    public void startDocument() throws SAXException{
        System.out.println("Start Document");
    }

    public void characters(char[] ch, int start, int length)
        throws SAXException {
        String str = new String(ch, start, length);
        System.out.println("Characters: " + str.trim());
    }

    public void endDocument() throws SAXException {
        System.out.println("End Document");
    }
```

Listing 7.2 *(cont.)*

```java
public void endElement(String namespaceURI, String localName,
   String qName) throws SAXException {
   System.out.println("End Element: " + localName);
}

public void endPrefixMapping(String prefix) throws SAXException {
    // No-op
}

public void ignorableWhitespace(char[] ch, int start, int length)
   throws SAXException {
   // No-op
}

public void processingInstruction(String target, String data)
     throws SAXException {
   // No-op
}
public void setDocumentLocator(Locator locator) {
   // No-op

}
public void skippedEntity(String name) throws SAXException {
   // No-op
}

public void startElement(String namespaceURI, String localName,
   String qName, Attributes atts) throws SAXException {
   System.out.println("Start Element: " + localName);
}

public void startPrefixMapping(String prefix, String uri)
   throws SAXException {
   // No-op
}

/**
 * Prints Command Line Usage
 */
private static void printUsage() {
  System.out.println ("usage: SAXBasic xml-file");
  System.exit(0);
}

/**
 * Main Method
 * Options for instantiating XMLReader Implementation:
 * 1) XMLReader parser = XMLReaderFactory.createXMLReader();
 * 2) XMLReader parser = XMLReaderFactory.createXMLReader
```

Listing 7.2 (*cont.*)

```
 *          ("org.apache.xerces.parsers.SAXParser");
 * 3) XMLReader parser = new org.apache.xerces.parsers.SAXParser();
 */
public static void main(String[] args) {
  if (args.length != 1) {
    printUsage();
  }
  try {
      SAXBasic saxHandler = new SAXBasic();
      XMLReader parser = XMLReaderFactory.createXMLReader
          ("org.apache.xerces.parsers.SAXParser");
      parser.setContentHandler(saxHandler);
      parser.parse(args[0]);
  } catch (SAXException e) {
      System.out.println ("SAXException: "+e.getMessage());
  } catch (IOException e) {
      System.out.println ("IOException: "+e.getMessage());
  }
 }
}
```

the major SAX events as they occur and output them to the console. For example, if we run this program on our DNA example from Listing 7.1, we receive the following output:

```
Start Document
Start Element: DASDNA
Start Element: SEQUENCE
Start Element: DNA
Characters: taatttctcccattttgtaggttatcacttcactctgttgactttcttttg
Characters:
End Element: DNA
End Element: SEQUENCE
Start Element: SEQUENCE
Start Element: DNA
Characters: taatgcaactaaatccaggcgaagcatttcagcttaaccccgagacttttg
Characters:
End Element: DNA
End Element: SEQUENCE
End Element: DASDNA
End Document
```

As we continue this section, we will flesh out the full details of SAXBasic.java. For now, consider a bird's-eye view. First, the main() method is responsible for creating an XMLReader class. The XMLReader interface provides a simple API for interacting with any SAX compliant XML parser. For example, the XMLReader.parse() method directs the parser to retrieve the specified XML document and immediately start parsing.

To receive call-backs from the SAX parser, SAXBasic.java provides an implementation of the SAX ContentHandler interface. For each major parsing event, the parser calls the registered content handler and invokes the appropriate method. For example, when the parser encounters a start tag, it invokes the startElement() method. In the SAXBasic class, some methods are

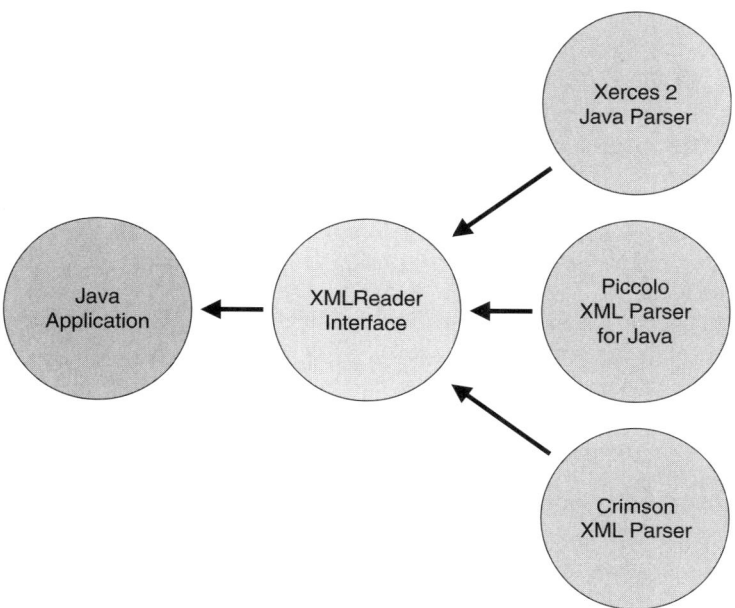

Figure 7.1 The `XMLReader` interface provides a common API for interfacing with any SAX 2 compliant XML parser. You can therefore easily swap XML parsers in and out.

implemented to simply output the event to `System.out`. Other methods, such as `startPrefixMapping()`, have empty implementations and are therefore silently ignored. In the event of a fatal error, the parser will throw a `SAXException` and the program will display a full stack trace.

7.1.2   The `XMLReader` Interface

As noted above, the `XMLReader` interface provides a simple API for interacting with any SAX compliant XML parser. For example, the `XMLReader` interface provides methods for setting parser properties or features. From the developer perspective, the `XMLReader` interface provides a common set of methods, but the implementation details are conveniently hidden from view. SAX therefore provides a simple plug and play facility for swapping XML parsers. For example, your application can easily swap between the Xerces2 XML parser, the Crimson XML parser, and the Piccolo XML parser. See Figure 7.1.

There are several options for retrieving an instance of an `XMLReader`. The first option is to specify your XML parser via a system property and then call the SAX `XMLReaderFactory.createXMLReader()`. For example, you can modify the `main()` method as follows:

```
try {
    SAXBasic saxHandler = new  SAXBasic();
    XMLReader parser = XMLReaderFactory.createXMLReader ();
    parser.setContentHandler(saxHandler);
    parser.parse(args[0]);
} catch (SAXException e) {
```

```
      e.printStackTrace();
   } catch (IOException e) {
      e.printStackTrace();
   }
```

To specify an XML parser at runtime, you must provide a system property for the SAX driver: `org.xml.sax.driver`. The value of the property must point to the `XMLReader` implementation provided by the XML parser. This varies by parser. For example, for Xerces2, the system property must be set to `org.apache.xerces.parser.SAXParser` ; for Piccolo, the property must be set to `com.bluecast.xml.Piccolo`. If you are using another XML parser, check its documentation for the exact class name.

You normally specify system properties via the -D option to the `java` command line program. For example, the following command line invokes our newly modified SAXBasic program with the Xerces2 XML parser:

```
java -Dorg.xml.sax.driver=org.apache.xerces.parsers.SAXParser
org.xmlbio.sax.SAXBasic ensembl_dna.xml
```

This command line invokes SAXBasic with the Piccolo XML parser:

```
java -Dorg.xml.sax.driver=com.bluecast.xml.Piccolo
org.xmlbio.sax.SAXBasic ensembl_dna.xml
```

The advantage of using this approach is that you can easily swap parsers without modifying or recompiling any code. Note that if the `XMLReaderFactory` is unable to determine any valid system defaults, it will throw a `SAXException` , with a specific message: "System property org.xml.sax.driver not specified."

If the `XMLReaderFactory` is unable to locate the SAX driver specified by `org.xml.sax.driver`, it will next check all the JAR files within your CLASSPATH. JAR files can contain a META-INF/services directory for registering implementations of well-known interfaces. If an implementation of `XMLReader` is registered, the `XMLReaderFactory` will use this one.

There are two caveats to using this approach: first, not all XML parsers actually register their services in this manner; for example, the Xerces2 JAR file specifically does not register any services. Second, the `XMLReaderFactory` goes through each JAR file in your CLASSPATH and finds the first registered implementation. If you have multiple XML parsers in your CLASSPATH (quite common these days, because many software distributions automatically include XML parsers), and each one implements the `XMLReader` interface, the first matching parser is chosen. Adding a new JAR file may therefore have a side effect of changing the chosen parser, and this new parser may not necessarily implement the features your code requires. To avoid this type of problem, it is best to explicitly set the driver via system properties.

If you have read the warning above, you will note that it is sometimes tricky to determine which XML parser you are actually using. To determine the current parser, simply print out the full class name of the `XMLReader` implementation object. The following code illustrates the basic idea:

```
XMLReader parser = XMLReaderFactory.createXMLReader();
System.out.println ("XMLReader: "+parser.getClass().getName());
```

For example, based on the following sample output:

```
XMLReader: com.bluecast.xml.Piccolo
```

you can determine that the Piccolo XML parser was chosen.

The second option is to call the `XMLReaderFactory.createXMLReader()` method and explicitly pass the full path to the `XMLReader` implementation. For example, the following code loads the Xerces2 SAX driver:

```
XMLReader parser = XMLReaderFactory.createXMLReader
    ("org.apache.xerces.parsers.SAXParser");
```

The third option is to bypass the `XMLReaderFactory` altogether and instantiate the `XMLReader` implementation directly. For example, this code also loads the Xerces2 SAX driver:

```
XMLReader parser = new org.apache.xerces.parsers.SAXParser();
```

The downside to this approach is that you must recompile your code if you decide to swap in a different parser.

As a fourth option, you can use Sun's JAXP API (Java API for XML Parsing [85]) to instantiate an XML parser object. JAXP 1.1 is now included in Java JDK 1.4 and the following code illustrates the basic functionality:

```
SAXParserFactory factory = SAXParserFactory.newInstance();
SAXParser parser = factory.newSAXParser();
parser.parse(args[0], handler);
```

This code will instantiate a `SAXParser` object via the JAXP `SAXParserFactory`. System properties determine which parser gets instantiated. If you are using JDK 1.4 and have no other parsers within your CLASSPATH, the factory will default to the built-in Crimson XML parser. Once you have a `SAXParser` object, it essentially serves as a wrapper to the SAX `XMLReader` interface and enables you to set properties/features and initiate XML parsing.

Most of the power of JAXP comes from providing a vendor neutral API for interfacing with DOM parsers and XSL transformers. When it comes to SAX, however, JAXP does not provide much new functionality, and it is often simpler to stick with the regular SAX `XMLReaderFactory` class. For full details regarding JAXP, go to: *http://java.sun.com/xml/jaxp.*

About Xerces2 and Piccolo

The examples in this chapter will work with any SAX 2.0 compliant parser. However, to illustrate the range of available XML parsers, the examples alternate between using the Xerces2 Java parser and the Piccolo XML parser for Java.

Table 7.1 The SAX `XMLReader` interface. The `XMLReader` interface provides a common API for interfacing with any SAX 2 compliant XML parser. Copied from the official SAX 2.0 JavaDoc API [71]

	Method Summary
`ContentHandler`	`getContentHandler()` Return the current content handler
`DTDHandler`	`getDTDHandler()` Return the current DTD handler
`EntityResolver`	`getEntityResolver()` Return the current entity resolver
`ErrorHandler`	`getErrorHandler()` Return the current error handler
`boolean`	`getFeature(String name)` Look up the value of a feature flag
`Object`	`getProperty(String name)` Look up the value of a property
`void`	`parse(InputSource input)` Parse an XML document
`void`	`parse(String systemId)` Parse an XML document from a system identifier (URI)
`void`	`setContentHandler(ContentHandler handler)` Allow an application to register a content event handler
`void`	`setDTDHandler(DTDHandler handler)` Allow an application to register a DTD event handler
`void`	`setEntityResolver(EntityResolver resolver)` Allow an application to register an entity resolver
`void`	`setErrorHandler(ErrorHandler handler)` Allow an application to register an error event handler
`void`	`setFeature(String name, boolean value)` Set the value of a feature flag
`void`	`setProperty(String name, Object value)` Set the value of a property

Xerces2 [87] is available from the Apache group and represents a complete rewrite of the very popular Xerces 1 parser. Xerces2 is extremely full-featured, includes support for SAX and DOM, and is capable of validating against both DTDs and XML Schemas. Full details and downloads are available at: *http://xml.apache.org/xerces2-j/index.html.* Note that the current Xerces2 release includes two JAR files. The first, xmlParserAPIs.jar, includes all the standard XML APIs, including DOM and SAX. The second, xercesImpl.jar, includes the Xerces implementation of these APIs. To use Xerces, you must include both JAR files within your CLASSPATH.

The Piccolo XML parser [86] is a fast, nonvalidating XML parser, hosted on SourceForge. You can download the Piccolo distribution at: *http://piccolo.sourceforge.net.*

7.1.3 The `ContentHandler` Interface

The SAX `ContentHandler` interface represents the core of any SAX application. As your XML parser encounters significant parsing events, it calls the appropriate methods in the `ContentHandler`. For example, when the parser encounters a start tag, it calls the `startElement()` method. Your implementation of the interface determines if and how the event is recorded.

In total, the `ContentHandler` interface defines 11 call-back methods. Each of these methods is summarized in Table 7.2. For now consider the five most important methods:

Table 7.2 The SAX `ContentHandler` Interface. Copied from the official SAX 2.0 JavaDoc API [71]

Method Summary	
void	`characters(char[] ch, int start, int length)` Receive notification of character data
void	`endDocument()` Receive notification of the end of a document
void	`endElement(String uri, String localName, String qName)` Receive notification of the end of an element
void	`endPrefixMapping(String prefix)` End the scope of a prefix-URI mapping
void	`ignorableWhitespace(char[] ch, int start, int length)` Receive notification of ignorable whitespace in element content
void	`processingInstruction(String target, String data)` Receive notification of a processing instruction
void	`setDocumentLocator(Locator locator)` Receive an object for locating the origin of SAX document events
void	`skippedEntity(String name)` Receive notification of a skipped entity
void	`startDocument()` Receive notification of the beginning of a document
void	`startElement(String uri, String localName, String qName, Attributes atts)` Receive notification of the beginning of an element
void	`startPrefixMapping(String prefix, String uri)` Begin the scope of a prefix-URI Namespace mapping

- `startDocument()`: indicates that the parser has encountered the beginning of an XML document. This is usually a good place to execute any initialization procedures.
- `startElement()`: indicates that the parser has encountered a start XML tag. The method receives a total of four parameters. The first three parameters provide information about the element name and its associated XML namespace. The fourth parameter is an `Attributes` object, which encapsulates all attribute data associated with this element. We will explore namespace issues and attributes in detail in the next section.
- `endElement()`: indicates that the parser has encountered an end XML tag. Much like the `startElement()` method, `endElement()` also receives three parameters, which provide element name and XML namespace data.
- `characters()`: indicates that the parser has encountered character data. For performance optimization, the method receives a character array, instead of a `String` or `StringBuffer` object. To extract the correct character data, you must use the start and length parameters. For example, this code extracts the current character data into a regular `String` object:

```
public void characters(char[] ch, int start, int length)
      throws SAXException {
      String str = new String(ch, start, length);
   ...
 }
```

Parsers are free to call the `characters()` method in any way they see fit. For example, consider the following snippet of XML:

```
<DNA>taatttctcccattttgtaggttatc</DNA>
```

One XML parser might choose to call `characters()` once with all the text. Another parser might choose to break the text into two chunks of data and call `characters()` twice. Yet another parser might choose to call `characters()` for each single character encountered. You have no way of knowing which strategy your parser will take, and your application therefore needs some method of buffering character data. For example, the following implementation of `characters()` appends to a `StringBuffer` object:

```
private StringBuffer currentText;

...

public void characters(char[] ch, int start, int length)
    throws SAXException {
    String str = new String(ch, start, length);
    currentText.append(str);
}
```

- `endDocument()`: indicates that the parser has reached the end of the XML document. This is usually a good place to execute any finalization procedures or free any resources.

7.1.4 Extending the `DefaultHandler`

When you implement the SAX `ContentHandler`, you must implement a total of 11 methods. Most of the time, however, your application really only needs a few of these methods. To help out in this common situation, SAX provides a helper class, named `DefaultHandler`. The `Default-Handler` provides an empty, no-operation implementation of the `ContentHandler` interface. (Actually, the `DefaultHandler` provides empty implementations for several SAX interfaces, including the SAX `ErrorHandler`—more on this shortly.)

To create a SAX `ContentHandler`, you simply extend the `DefaultHandler`, and override only those methods that you need. For example, Listing 7.3 provides a rewrite of our first example.

Listing 7.3 SAXDefaultHandler.java

```
package org.xmlbio.sax;

import org.xml.sax.helpers.DefaultHandler;
import org.xml.sax.helpers.XMLReaderFactory;
import org.xml.sax.SAXException;
import org.xml.sax.Attributes;
import org.xml.sax.XMLReader;

import java.io.IOException;
/**
 * Basic SAX Example.
 * Illustrates extending of DefaultHandler
 */
public class SAXDefaultHandler extends DefaultHandler {

    public void startDocument() throws SAXException {
        System.out.println("Start Document");
    }
```

Listing 7.3 *(cont.)*

```java
public void characters(char[] ch, int start, int length)
    throws SAXException {
    String str = new String(ch, start, length);
    System.out.println("Characters: " + str.trim());
}


public void endDocument() throws SAXException {
    System.out.println("End Document");
}


public void endElement(String namespaceURI, String localName,
    String qName) throws SAXException {
    System.out.println("End Element: " + localName);
}


public void startElement(String namespaceURI, String localName,
    String qName, Attributes atts) throws SAXException {
    System.out.println("Start Element: " + localName);
}


/**
 * Prints Command Line Usage
 */
private static void printUsage() {
    System.out.println ("usage: SAXDefaultHandler xml-file");
    System.exit(0);
}


/**
 * Main Method
 */
public static void main(String[] args) {
    if (args.length != 1) {
        printUsage();
    }
    try {
        SAXDefaultHandler saxHandler = new SAXDefaultHandler();
        XMLReader parser = XMLReaderFactory.createXMLReader
            ("com.bluecast.xml.Piccolo");
        parser.setContentHandler(saxHandler);
        parser.parse(args[0]);
    } catch (SAXException e) {
        System.out.println ("SAXException: "+e.getMessage());
    } catch (IOException e) {
        System.out.println ("IOException: "+e.getMessage());
    }
}
}
```

The class extends the `DefaultHandler` and overrides a total of five elements. Note that the code is much more compact than our original example. Note also that for variety, we have now switched to the Piccolo XML parser.

In SAX 1.0, the helper class was named `HandlerBase`. The class is now deprecated.

7.1.5   Using `InputSource` Objects

So far, our examples have passed a URI String to the `XMLReader.parse()` method. For example, we currently have the following code:

```
XMLReader parser = XMLReaderFactory.createXMLReader
     ("com.bluecast.xml.Piccolo");
parser.setContentHandler(saxHandler);
parser.parse(args[0]);
```

The URI argument can represent a path to a local file or an absolute URL to an external file. For example, this command line parses a file in the current working directory:

```
java org.xmlbio.sax.SAXDefaultHandler ensembl_dna.xml
```

In contrast, this command line retrieves the same file directly from the Ensembl DAS server:

```
java org.xmlbio.sax.SAXDefaultHandler\
http://servlet.sanger.ac.uk:8030/das/ensembl830/dna?\
segment=1:1000,1050;segment=2:1000,1050
```

The parser automatically downloads the specified file and handles all the networking details for you.

For full flexibility, SAX also enables you to specify XML documents via the SAX `InputSource` class. (See Table 7.3 for the full API.) The advantage of using `InputSource` objects is that you can process XML files via URIs, character streams, or byte streams.

`InputSource` objects can be useful in numerous situations. For example, you can parse Strings, which contain XML documents. You can even upload an XML document to a Java servlet, and parse the document on the server side. The following code excerpt illustrates the basic concepts for server-side parsing:

```
public class SAXServlet extends HttpServlet {

    /**
     * Processes HTTP Post Requests
     */

    public void doPost(HttpServletRequest request,
            HttpServletResponse response)
            throws IOException, ServletException {

        ...
        parseXML(request, response);
        ...
    }
    /**
     * Parses XML document specified by "xml" parameter
     */
    private void parseXML(HttpServletRequest request,
```

```
        HttpServletResponse response) {

    PrintWriter out = null;
    try {
        out = response.getWriter();
        SAXHTMLHandler saxHandler = new SAXHTMLHandler(out);
        XMLReader parser = XMLReaderFactory.createXMLReader
                ("org.apache.xerces.parsers.SAXParser");
        parser.setContentHandler(saxHandler);
        String xml = request.getParameter("xml");
        StringReader reader = new StringReader(xml);
        InputSource inputSource = new InputSource(reader);
        parser.parse(inputSource);
    } catch (SAXException e) {
                e.printStackTrace(out);
    } catch (IOException e) {
                e.printStackTrace(out);
    }
  }
}
```

Table 7.3 The SAX `InputSource` class. Copied from the official SAX 2.0 JavaDoc API [71]

Constructor Summary
`InputSource()`   Zero-argument default constructor
`InputSource(InputStream byteStream)`   Create a new input source with a byte stream
`InputSource(Reader characterStream)`   Create a new input source with a character stream
`InputSource(String systemId)`   Create a new input source with a system identifier

Method Summary	
`InputStream`	`getByteStream()` Get the byte stream for this input source
`Reader`	`getCharacterStream()` Get the character stream for this input source
`String`	`getEncoding()` Get the character encoding for a byte stream or URI
`String`	`getPublicId()` Get the public identifier for this input source
`String`	`getSystemId()` Get the system identifier for this input source
`void`	`setByteStream(InputStream byteStream)` Set the byte stream for this input source
`void`	`setCharacterStream(Reader characterStream)` Set the character stream for this input source
`void`	`setEncoding(String encoding)` Set the character encoding, if known
`void`	`setPublicId(String publicId)` Set the public identifier for this input source
`void`	`setSystemId(String systemId)` Set the system identifier for this input source

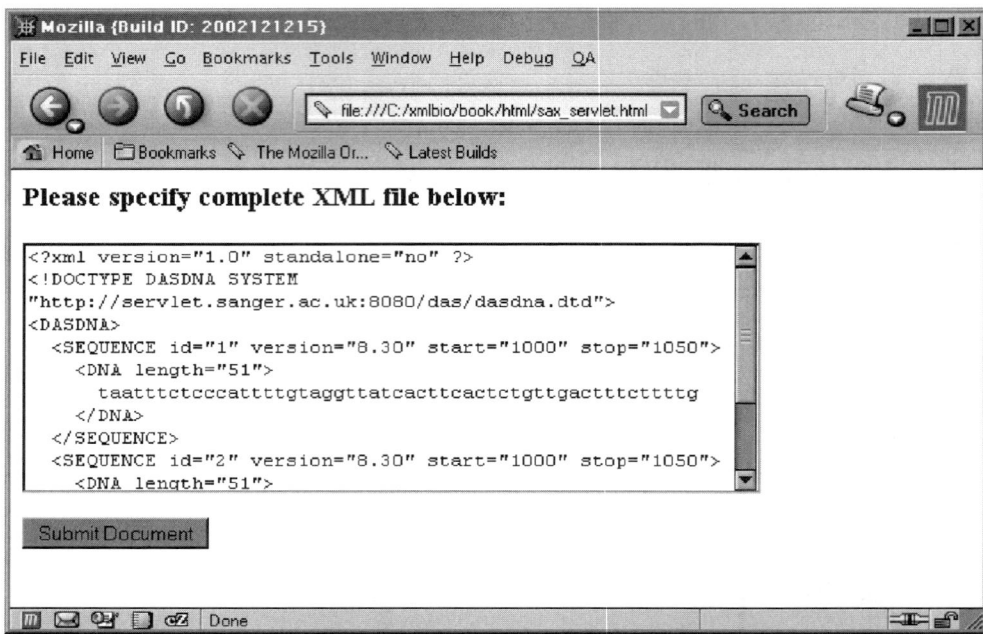

Figure 7.2 Sample screenshot of HTML form. The form enables you to cut and paste XML documents and submit them for server-side processing. See Figure 7.3 for sample output.

In the Servlet API, you receive HTML form parameters via the request `getParameter()` method. In this case, we are retrieving an "xml" parameter, wrapping this in a `StringReader` object, and then wrapping this in an `InputSource` object. Sample screenshots of the servlet in action are provided in Figures 7.2 and 7.3. You can download the full servlet code from the web site that accompanies this book.

The `InputSource` API enables you to specify a byte stream, a character stream, or a URI. Because it is possible to specify more than one of these options (and the options may not even be related), the class follows a strict order of evaluation. Character streams have highest priority, followed by byte streams, and then by URIs. In other words, if you create an InputSource object and specify both a character stream and a URI, the parser will always retrieve the document from the character stream.

7.2   Validating XML Documents

7.2.1   Checking for Well-Formedness

By default, XML parsers will automatically check for well-formedness. To review, a document is said to be well-formed if every start tag has a corresponding end tag, all tags are properly nested, and all attributes are enclosed in quotes. If any of these constraints are not met, the XML parser will immediately stop parsing and throw a `SAXException`.

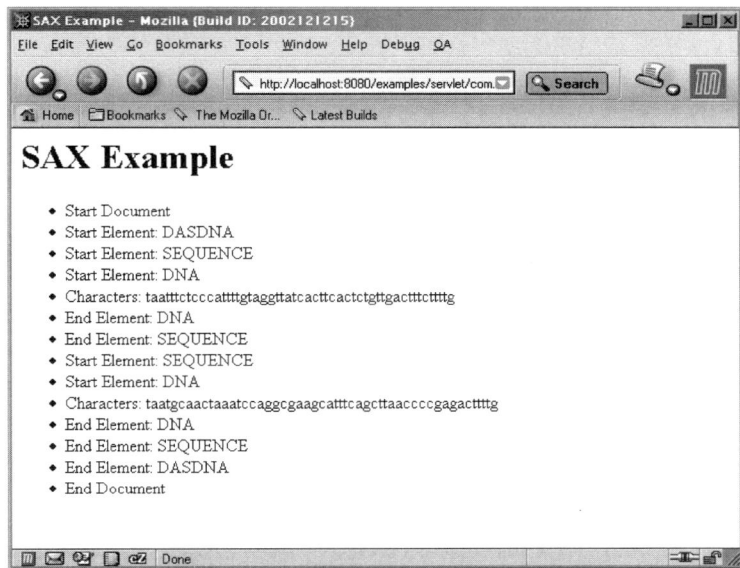

Figure 7.3 Sample screenshot of server-side XML processing. See Figure 7.2 for sample input.

To capture exceptions, our first two examples surround `parse()` with a try/catch block:

```
try {
    ...
    parser.parse(args[0]);
} catch (SAXException e) {
    System.out.println ("SAXException: "+e.getMessage());
} catch (IOException e) {
    System.out.println ("IOException: "+e.getMessage());
}
```

To view a well-formedness error, let us modify our working DNA example by removing one of the end `</DNA>` tags. (See Listing 7.4.) When we run the example through our `SAXBasic` program, we get the following output:

```
Start Document
Start Element: DASDNA
Start Element: SEQUENCE
Start Element: DNA
Characters: taatttctcccattttgtaggttatcacttcactctgttgactttcttttg
Characters:
End Element: DNA
End Element: SEQUENCE
Start Element: SEQUENCE
Start Element: DNA
Characters: taatgcaactaaatccaggcgaagcatttcagcttaaccccgagacttttg
Characters:
SAXException: The element type "DNA" must be terminated by
the matching end-tag "</DNA>".
```

Listing 7.4 ensembl_dna_error.xml. The second end </DNA> tag has been removed, and the document is therefore no longer well-formed

```
<?xml version='1.0' standalone='no' ?>
<!DOCTYPE DASDNA SYSTEM
"http://servlet.sanger.ac.uk:8080/das/dasdna.dtd">
<DASDNA>
    <SEQUENCE id="1" version="8.30" start="1000" stop="1050">
      <DNA length="51">
            taatttctcccattttgtaggttatcacttcactctgttgactttcttttg
      </DNA>
    </SEQUENCE>
    <SEQUENCE id="2" version="8.30" start="1000" stop="1050">
      <DNA length="51">
            taatgcaactaaatccaggcgaagcatttcagcttaaccccgagacttttg
    </SEQUENCE>
</DASDNA>
```

As you can see, a `SAXException` is thrown, and the exception includes specific details about the missing end tag. The specific error message is created by the XML parser, and therefore varies from parser to parser. For example, if we switch from Xerces to Piccolo and parse the same document, we now get the following slightly different error message: "SAXException: </SEQUENCE> does not close tag <DNA>."

It is important to note that the XML parser reports all events prior to the error. For example, in the output above, you can see that the parser reports several events before reporting the actual error. Depending on your application, your code may need to take special precautions to handle this situation. For example, imagine that your application takes each DNA element and stores it to a relational database. After the first DNA element, you store one string of DNA to the database. While processing the second element, you encounter the error, and no data is stored. You may now have a data integrity issue because you have only stored half the DNA data. One potential solution is to create a database transaction. If you complete parsing without any exceptions, you can commit the transaction. If, however, a `SAXException` occurs, you can choose to roll back the entire transaction.

7.2.2 Validating XML Documents: Overview

Checking for well-formedness is easy and automatic. Checking for validity requires more work. Below is an overview of the four-step process:

- First, you need to pick a validating XML parser. For example, Xerces is validating, but Piccolo is not.
- Second, you must explicitly turn XML validation on. You do this via the `XMLReader.setFeature()` method.
- Third, you must provide an implementation of the SAX `ErrorHandler` interface.
- Fourth, you need to explicitly register your `ErrorHandler`.

In the sections that follow, we provide implementation details for each of the steps.

7.2.3 Activating the SAX Validation Feature

Once you have selected a validating parser, the first step is to turn the SAX validation feature on:

```
try {
    parser.setFeature
    ("http://xml.org/sax/features/validation", true);
} catch (SAXNotRecognizedException e) {
    System.out.println ("SAX Not Recognized: "+e.getMessage());
} catch (SAXNotSupportedException e) {
    System.out.println ("SAX Not Supported: "+e.getMessage());
}
```

The first parameter to `setFeature()` takes a String argument, indicating the feature that you want to activate or deactivate. Generally, features are divided into two categories: standard SAX features, common to all parsers; and vendor-specific features, which are specific to a single XML parser. Standard features begin with the prefix "http://xml.org/sax/features/." For example, "http://xml.org/sax/features/validation" turns validation on. Vendor-specific features usually begin with the associated domain. For example, Xerces-specific features begin with the prefix "http://apache.org/xml/features/."

Even though SAX features are specified as URLs, these URLs do not point to anything meaningful. For example, if you type "http://xml.org/sax/features/validation" into a web browser, you actually get a 404 Not Found Error. It is therefore best to think of the features as identifiers only. A full list of the standard SAX features is provided in the JavaDoc API at *http://www.saxproject.org*.

The `setFeature()` method can throw two possible exceptions. The first, `SAXNotRecognizedException`, indicates that the specified feature is not recognized. For example, your feature string may have a typo or you may have just switched parsers, and the vendor-specific feature is no longer recognized. The second, `SAXNotSupportedException`, indicates that the feature is recognized, but not supported. For example, if you attempt to activate validation for Piccolo, a nonvalidating parser, it will throw a `SAXNotSupportedException`.

7.2.4 The `ErrorHandler` Interface

Once you have activated the validation feature, the parser will check all validity constraints, but validity errors will be silently ignored. To receive notification of validation errors, you must provide an implementation of the SAX `ErrorHandler` interface and explicitly register your `ErrorHandler` with the parser.

The `ErrorHandler` interface defines three error methods, corresponding to the three levels of errors defined in the XML 1.0 specification:

- Fatal Errors: Primarily refer to errors in well-formedness.
- Errors: Primarily refer to errors in validity.
- Warnings: Catch-all category for reporting low-level warnings.

Table 7.4 The SAX `ErrorHandler` interface. Copied from the official SAX 2.0 JavaDoc API [71]

Method Summary	
void	`error(SAXParseException exception)`
	Receive notification of a recoverable error
void	`fatalError(SAXParseException exception)`
	Receive notification of a nonrecoverable error
void	`warning(SAXParseException exception)`
	Receive notification of a warning

Table 7.5 The `SAXException` API. Copied from the official SAX 2.0 JavaDoc API [71]

public class SAXException extends Exception

Method Summary	
Exception	`getException()`
	Return the embedded exception, if any
String	`getMessage()`
	Return a detail message for this exception
String	`toString()`
	Override toString to pick up any embedded exception

public class SAXParseException extends SAXException

Method Summary	
int	`getColumnNumber()`
	The column number of the end of the text where the exception occurred
int	`getLineNumber()`
	The line number of the end of the text where the exception occurred
String	`getPublicId()`
	Get the public identifier of the entity where the exception occurred
String	`getSystemId()`
	Get the system identifier of the entity where the exception occurred

Each of the SAX error methods receives a `SAXParseException` parameter and declares that it can throw a `SAXException`. The `SAXParseException` encapsulates information about the error, including the specific error message and its location within the XML document.

Upon receiving an error notification, the error handler has two main options. The first option is to simply record the error and choose not to throw the exception. For example, the implementation can output the error to `System.out` or record it to a log file. Because the method does not throw an exception, the parser continues normal processing.

The second option is to perform some type of logging and then explicitly throw the exception. By throwing the exception, all normal XML processing is stopped. For example, the following implementation stops all parsing when a validity error is encountered:

```
public void error(SAXParseException exception) throws SAXException {
    reportError(exception);
    throw exception;
}
```

By definition, *fatal errors* are nonrecoverable, and will always trigger a SAXParseException. Even if your error handler does not explicitly throw any exceptions, the parser will still stop normal processing and throw its own SAXParseException.

A complete example of XML validation is provided in Listing 7.5. This program extends the SAX DefaultHandler helper class. As you may recall, the DefaultHandler provides

Listing 7.5 SAXValidator.java

```java
package org.xmlbio.sax;

import org.xml.sax.*;
import org.xml.sax.helpers.DefaultHandler;
import org.xml.sax.helpers.XMLReaderFactory;

import java.io.IOException;
/**
 * SAX Validator.
 * Illustrates Basic Error Handling.
 */
public class SAXValidator extends DefaultHandler {
    private boolean isValid = true;

    /**
     * Receives notification of a recoverable error.
     * Validation Errors are reported here.
     */
    public void error(SAXParseException exception) throws SAXException {
        isValid = false;
        reportError("Error", exception);
    }

    /**
     * Receives notification of a warning.
     */
    public void warning(SAXParseException exception) throws SAXException {
        reportError("Warning", exception);
    }
    /**
     * Reports SAXParseException Information
     */
    private void reportError(String errorType, SAXParseException exception) {
        System.out.println(errorType + ": " + exception.getMessage());
        System.out.println(" Line: " + exception.getLineNumber());
        System.out.println(" Column: " + exception.getColumnNumber());
    }
    /**
```

Listing 7.5 (*cont.*)

```
 * Returns isValid boolean flag
 */
public boolean isValid() {
  return isValid;
}

/**
 * Prints Command Line Usage
 */
private static void printUsage() {
  System.out.println("usage: SAXValidator xml-file");
  System.exit(0);
}

/**
 * Main Method
 */
public static void main(String[] args) {
    if (args.length != 1) {
        printUsage();
    }
    try {
        SAXValidator errorHandler = new SAXValidator();
        XMLReader parser = XMLReaderFactory.createXMLReader
                ("org.apache.xerces.parsers.SAXParser");

        // Turn Validation On and Set Error Handler
        turnValidationOn(parser);
        parser.setErrorHandler(errorHandler);
        parser.parse(args[0]);

        // If SAXException has not been thrown,
        // document must be well-formed
        System.out.println("The Document is well-formed.");
        if (errorHandler.isValid()) {
           System.out.println("The Document is valid.");
        }
    } catch (SAXException e) {
        System.out.println("SAXException: " + e.getMessage());
    } catch (IOException e) {
        System.out.println("IOException: " + e.getMessage());
    }
}

/**
 * Turns Validation On
 * Includes specific exception handling for
 * SAXNotRecognizedException and SAXNotSupportedException.
 */
private static void turnValidationOn(XMLReader parser) {
```

Listing 7.5 (*cont.*)

```
    try {
        parser.setFeature
           ("http://xml.org/sax/features/validation", true);
    } catch (SAXNotRecognizedException e) {
        System.out.println("SAX Not Recognized: " + e.getMessage());
    } catch (SAXNotSupportedException e) {
        System.out.println("SAX Not Supported: " + e.getMessage());
    }
  }
}
```

Listing 7.6 ensembl_dna_invalid.xml. The <SEQUENCE> *id* attributes have been removed and the document is therefore no longer valid

```
<?xml version='1.0' standalone='no' ?>
<!DOCTYPE DASDNA SYSTEM
"http://servlet.sanger.ac.uk:8080/das/dasdna.dtd">
<DASDNA>
    <SEQUENCE version="8.30" start="1000" stop="1050">
        <DNA length="51">
            taatttctcccattttgtaggttatcacttcactctgttgactttcttttg
        </DNA>
    </SEQUENCE>
    <SEQUENCE version="8.30" start="1000" stop="1050">
        <DNA length="51">
            taatgcaactaaatccaggcgaagcatttcagcttaaccccgagacttttg
        </DNA>
    </SEQUENCE>
</DASDNA>
```

implementations of several interfaces, including `ContentHandler` and `ErrorHandler`. In this case, we choose to override the `error()` and `warning()` methods—in each case, we simply record the error to `System.out`. If no errors are encountered, the program declares that the document is both well-formed and valid.

Note that by default, the `DefaultHandler fatalError()` method will throw a `SAXException` and terminate normal processing. Note also that you must explicitly register your error handler via the `setErrorHandler()` method:

```
parser.setErrorHandler(errorHandler);
```

Consider the program output for an invalid XML file. For example, consider the revised example in Listing 7.6. This program is still well-formed, but the required <SEQUENCE> *id* attributes have been deleted. When run through the SAX validator, we get a listing of all the validation errors:

```
Error: Attribute "id" is required and must be specified for
element type "SEQUENCE".
    Line: 5
    Column: 53
```

```
Error: Attribute "id" is required and must be specified for
element type "SEQUENCE".
    Line: 10
    Column: 53
The Document is well-formed.
```

If you are using Xerces and turn validation on, your document must specify a DTD or XML Schema. Otherwise, you receive the following error: "Document is invalid. No grammar found." To get around this, you can activate the Xerces dynamic validation feature *http://apache.org/xml/features/validation/dynamic*. When activated, documents with grammars are validated; documents without grammars are not validated.

7.2.5   Validating against XML Schemas

Not all validating parsers are capable of validating against an XML Schema. The Xerces2 XML parser provides this functionality, but requires that you explicitly activate the regular validation feature *and* a schema-specific validation feature. The schema validation feature is specified by the URL *http://apache.org/xml/features/validation/schema*.

To redo our SAXValidator example and add explicit support for schema validation, we can rewrite the turnValidationOn() method as follows:

```
private static void turnValidationOn(XMLReader parser) {
   try {
       parser.setFeature
          ("http://apache.org/xml/features/validation/schema", true);
       parser.setFeature
          ("http://xml.org/sax/features/validation", true);
   } catch (SAXNotRecognizedException e) {
       System.out.println ("SAX Feature Not Recognized:"+e.getMessage());
   } catch (SAXNotSupportedException e) {
       System.out.println ("SAX Feature Not Supported:"+e.getMessage());
   }
}
```

Errors in schema validity are reported to the error handler, just like DTD validity errors.

To determine if your parser supports schema validation, check the parser's original documentation.

In addition to the ContentHandler and the ErrorHandler, SAX 2.0 provides three additional handlers: EntityResolver, DTDHandler, and LexicalHandler. For details on these handlers, an excellent description is provided in Elliotte Rusty Harold, *Processing XML with Java* (Addison-Wesley Professional; November 5, 2002) [57]. The complete book is available online at: *http://cafeconleche.org/books/xmljava*.

7.3   Elements, Attributes, and Namespaces

Now that you have a solid understanding of the basic SAX interfaces, let us dig a little deeper into elements, attributes, and namespaces. Support for XML Namespaces is the biggest addition to SAX 2.0, and deserves to be explored in detail. To make all the concepts concrete, we will examine a new SAX program, and its output for two sample XML documents. The first sample document is the DAS DNA example, already introduced earlier in the chapter. As you may have already noticed, DAS documents do not use XML Namespaces. The second sample document is an excerpt of TrEMBL data in its SPTr-XML format. Unlike DAS, the SPTr-XML format does use XML Namespaces. By comparing the output of the two documents, we can explore element and namespace issues in detail. Following this, we will delve into the SAX attributes API.

7.3.1   Working with Elements and Namespaces

To get started, let us examine a very simple XHML document that utilizes XML Namespaces:

```
<?xml version="1.0"?>
<xhtml:html xmlns:xhtml="http://www.w3.org/TR/REC-html40">
    <xhtml:head>
        <xhtml:title>XML for Bioinformatics</xhtml:title>
    </xhtml:head>
    <xhtml:body>
        <xhtml:p>Welcome!</xhtml:p>
    </xhtml:body>
</xhtml:html>
```

The document has one declared namespace, specified by the prefix "xhtml." Each of the elements is specifically declared within the xhtml namespace via the use of *qualified names*. As a quick review, qualified names are specified by a namespace prefix, followed by a colon, and the local name. For example, "xhtml:body" specifies that the body element is associated with the xhtml namespace.

Next, let us examine our new SAX program. The source code is provided in Listing 7.7.

Listing 7.7 SAXElementNamespace.java

```
package org.xmlbio.sax;

import org.xml.sax.Attributes;
import org.xml.sax.Locator;
import org.xml.sax.SAXException;
import org.xml.sax.XMLReader;
import org.xml.sax.helpers.DefaultHandler;
import org.xml.sax.helpers.XMLReaderFactory;

import java.io.IOException;
/**
 * SAXElementNamespace.
 * Illustrates Element and Namespace Functionality.
 * Also illustrates use of Document Locator object.
 */
```

Listing 7.7 (cont.)

```java
public class SAXElementNamespace extends DefaultHandler {
    private Locator _locator;

    /**
     * Prints out all three name/namespace parameters.
     */
    public void startElement(String namespaceURI, String localName,
                String qName, Attributes atts) throws SAXException {
            namespaceURI = checkEmptyString(namespaceURI);
            System.out.println("Start Element: ");
            System.out.println("... Line: " + _locator.getLineNumber());
            System.out.println ("... Column: " + _locator.getColumnNumber());
            System.out.println("... Namespace URI: " + namespaceURI);
            System.out.println("... Local Name:    " + localName);
            System.out.println("... qName:         " + qName);
    }

    /**
     * Signals Start Prefix Mapping for XML Namespaces
     */
    public void startPrefixMapping(String prefix, String uri)
                throws SAXException {
            prefix = checkEmptyString(prefix);
            uri = checkEmptyString(uri);
            System.out.println("Start Prefix Mapping: ");
            System.out.println("... Prefix: " + prefix);
            System.out.println("... URI: "    + uri);
    }

    /**
     * Signal End Prefix Mapping for XML Namespaces
     */
    public void endPrefixMapping(String prefix) throws SAXException {
            System.out.println("End Prefix Mapping: " + prefix);
    }

    /**
     * Stores Document Locator
     */
    public void setDocumentLocator(Locator locator) {
            this._locator = locator;
    }

    /**
     * Checks for Empty Strings
     */
    private String checkEmptyString(String str) {
            if (str.equals(""))
                    return new String("[Empty String]");
            else
```

Listing 7.7 *(cont.)*

```
                return str;
        }

        /**
         * Prints Command Line Usage
         */
        private static void printUsage() {
                System.out.println("usage: SAXElementNamespace xml-file");
                System.exit(0);
        }

        /**
         * Main Method
         */
        public static void main(String[] args) {
                if (args.length != 1) {
                        printUsage();
                }
                try {
                        SAXElementNamespace saxHandler = new SAXElementNamespace();
                        XMLReader parser = XMLReaderFactory.createXMLReader
                                        ("org.apache.xerces.parsers.SAXParser");
                        parser.setContentHandler(saxHandler);
                        parser.parse(args[0]);
                } catch (SAXException e) {
                        e.printStackTrace();
                } catch (IOException e) {
                        e.printStackTrace();
                }
        }
}
```

There are a number of important items to note about the example. First, the `startElement()` method implementation is now much more detailed. Specifically, it now prints information about each of the following method parameters:

- `namespaceURI`: indicates the namespace URI for the specified element. For example, in the document above, the <xhtml:body> element will have its namespace URI set to *http://www.w3.org/TR/REC-html40*. If the element is not associated with any namespace, the parameter is specified by an empty string.
- `localName`: indicates the local element name. The local name is the element name, minus any namespace information. For example, the <xhtml:body> element will have its local name set to "body." If the element is not associated with any namespace, the local name is simply the regular element name. If you do not want to worry about any namespace issues, and just want the regular element name, this is the parameter you want.
- `qualifiedName`: indicates the element qualified name. For example, the <xhtml:body> element will have its qualified name set to "xhtml:body." If the element is not associated with a namespace or your document has declared a default namespace, the qualified name will be identical to the local name and can frequently be ignored.

Determining Namespace Scope

Namespaces are declared for a specific document scope. For example, in the XHTML document above, the xhtml namespace is defined for the root element and all of its descendants. To determine namespace scoping, SAX provides the `startPrefixMapping()` and `endPrefixMapping()` methods. When the XML parser finds an XML namespace declaration, it signals a start prefix mapping event, and passes the namespace prefix and URI. It then calls the `startElement()` method associated with the namespace declaration. For example, when parsing the html root element above:

```
<xhtml:html xmlns:xhtml="http://www.w3.org/TR/REC-html40">
```

the parser will first call the `startPrefixMapping()` method, followed by the `startElement()` method. When the namespace goes out of scope, the parser signals an end prefix mapping event and passes the same prefix and URI.

By default, all SAX 2.0 parsers support XML Namespaces. If you want to disable namespace support, set the SAX namespace feature to false:

```
parser.setFeature ("http://xml.org/sax/features/namespaces", false);
```

With namespace support disabled, the parser will not signal any start/end prefix mapping events. Additionally, namespace URIs and local names are simply ignored and passed as empty strings.

The Document `Locator`

As a final item, note that the new example also provides an implementation for the `setDocumentLocator()` method. This method is called at the very beginning of parsing, and receives a SAX `Locator` object. If you save a local reference to the locator object, you can use it to determine the specific location of parsing events. This can be extremely useful for debugging purposes or error handling.

For example, our `startElement()` method now prints out the line number and column number for each start tag:

Table 7.6 The SAX `Locator` interface. Copied from the official SAX 2.0 JavaDoc API [71]

Method Summary	
int	`getColumnNumber()`
	Return the column number where the current document event ends
int	`getLineNumber()`
	Return the line number where the current document event ends
String	`getPublicId()`
	Return the public identifier for the current document event
String	`getSystemId()`
	Return the system identifier for the current document event

```
public void startElement(String namespaceURI, String localName,
   String qName, Attributes atts) throws SAXException {
   System.out.println ("Start Element: ");
   System.out.println ("... Line: " + _locator.getLineNumber());
   System.out.println ("... Column: " + _locator.getColumnNumber());
   ...
}
```

Sample Output

As promised, let us now run two sample documents through our new program. First, we process the DAS DNA example from Listing 7.1. An excerpt of the program output is provided below:

```
Start Element:
... Line: 4
... Column: 9
... Namespace URI:   [Empty String]
... Local Name:      DASDNA
... qName:           DASDNA
Start Element:
... Line: 5
... Column: 60
... Namespace URI:   [Empty String]
... Local Name:      SEQUENCE
... qName:           SEQUENCE
Start Element:
... Line: 6
... Column: 22
... Namespace URI:   [Empty String]
... Local Name:      DNA
... qName:           DNA
[Output continues...]
```

As you can see, we can now determine the location of each start element event. Since the DAS document does not use XML Namespaces, the Namespace URI parameters are specified as empty strings, and local names are identical to qualified names.

Next, let us examine the sample SPTr-XML in Listing 7.8. This sample document declares a default namespace in the root <sptr> element. This default namespace is set to *urn:uk:ac:ebi:spml*, and applies to all element descendants. When run through our new program, we get the following output:

```
Start Prefix Mapping:
... Prefix:  [Empty String]
... URI:     urn:uk:ac:ebi:spml
Start Prefix Mapping:
... Prefix:  xsi
... URI:     http://www.w3.org/2001/XMLSchema-instance
Start Element:
... Line: 5
... Column: 2
... Namespace URI:  urn:uk:ac:ebi:spml
```

```
... Local Name:      sptr
... qName:           sptr
Start Element:
... Line: 6
... Column: 118
... Namespace URI: urn:uk:ac:ebi:spml
... Local Name:      entry
... qName:           entry
[Output continues...]
```

As you can see, the elements <sptr> and <entry> are both defined within the defined spml namespace. Because the document uses a default namespace, qualified names are not required, and local names are identical to qualified names in the program output.

7.3.2   Working with Attributes

When the startElement() method is invoked, the parser will pass all attribute information in a SAX Attributes object. The Attributes interface provides easy access to all attribute information, including attribute names, values, types, and associated namespaces. You can easily retrieve attribute

Listing 7.8 Sample SPTr-XML file. For brevity, we have only included a small excerpt of the file

```
<?xml version="1.0" encoding="UTF-8"?>
<sptr xmlns="urn:uk:ac:ebi:spml"
 xmlns:xsi="http://www.w3.org/2001/XMLSchema-instance"
 xsi:schemaLocation="urn:uk:ac:ebi:spml http://www.ebi.ac.uk/krunte/
sp-ml/src/xsd/SP-ML.xsd"
>
<entry accession="Q99ME3" database="TrEMBL" name="Q99ME3"
firstPublic="2001-06-01" lastAnnotationUpdate="2002-06-01">
   <protein>
      <name>Synphilin-1</name>
   </protein>
   <geneList>
     <gene>
       <name>
         Sncaip
         <evList>
            <ev ref="EI3" />
         </evList>
       </name>
     </gene>
   </geneList>
   <organismList>
      <organism iRefID="1">
         <name type="scientific name">Mus musculus</name>
         <name type="common name">Mouse</name>
         <dbReferenceList>
            <dbReference db="NCBI Taxonomy" id="10090" iRefID="2">
 ...
```

information by index value or by name. For example, the following method implementation loops through all attributes by index value:

```
public void startElement(String namespaceURI, String localName,
    String qName, Attributes atts) throws SAXException {
    System.out.println ("Start Element: ");
    System.out.println ("... Local Name:    "+localName);
    System.out.println ("... qName:         "+qName);
    for (int i=0; i< atts.getLength(); i++) {
        System.out.println ("--> Attribute: ");
        System.out.println (" ... URI:          "+atts.getURI(i));
        System.out.println (" ... Local Name: "+atts.getLocalName(i));
        System.out.println (" ... QName:        "+atts.getQName(i));
        System.out.println (" ... Type:         "+atts.getType(i));
        System.out.println (" ... Value:        "+atts.getValue(i));
    }
}
```

Just like elements, attributes can be specified as qualified names and can be associated with specific XML Namespaces.

If an element has more than one attribute, the parser may not maintain the order of those attributes. In fact, the precise order of attributes is implementation dependent.

Table 7.7 The SAX `Attributes` interface. Copied from the official SAX 2.0 JavaDoc API [71]

Method Summary	
int	`getIndex(String qName)` Look up the index of an attribute by XML 1.0 qualified name
int	`getIndex(String uri, String localName)` Look up the index of an attribute by Namespace name
int	`getLength()` Return the number of attributes in the list
String	`getLocalName(int index)` Look up an attribute's local name by index
String	`getQName(int index)` Look up an attribute's XML 1.0 qualified name by index
String	`getType(int index)` Look up an attribute's type by index
String	`getType(java.lang.String qName)` Look up an attribute's type by XML 1.0 qualified name
String	`getType(String uri, String localName)` Look up an attribute's type by Namespace name
String	`getURI(int index)` Look up an attribute's Namespace URI by index
String	`getValue(int index)` Look up an attribute's value by index
String	`getValue(String qName)` Look up an attribute's value by XML 1.0 qualified name
String	`getValue(String uri, String localName)` Look up an attribute's value by Namespace name

7.4   Building Custom Data Structures with SAX

For our final SAX topic, we explore the mechanics of building custom data structures. SAX parsing is a series of linear events, but we frequently need to convert these events into objects or collections of objects. For example, you can easily transform SAX events into a hierarchical tree of node objects. You can also transform events into custom objects better suited for your specific application. For example, you can transform DAS XML documents into discrete sets of sequence or features objects.

For our final example, we will parse DAS documents into feature objects and then display the features using the open source BioJava toolkit [84]. To make the example more manageable, we have broken the code into two parts. The first part handles the SAX processing and converts the SAX events into our own `Feature` objects. The second part uses the BioJava API to render the features. The second part is not necessary for understanding SAX, but makes the example much more compelling.

7.4.1   Parsing DAS Feature Data

Our goal is to parse XML results from a DAS *features* request. As a quick refresher, here is an excerpt from a DAS response:

```
<?xml version='1.0' standalone='no' ?>
<!DOCTYPE DASGFF SYSTEM 'dasgff.dtd' >
<DASGFF>
    <GFF version="1.0"
href="http://servlet.sanger.ac.uk:8080/das/ensembl930/features?
segment=NT_008045; type=transcript">
    <SEGMENT id="NT_008045" version="9.30" start="1" stop="1433313">
      <FEATURE id="ENST00000079954">
        <TYPE id="transcript">transcript</TYPE>
        <METHOD id="ensembl">ensembl</METHOD>
        <START>118771</START>
        <END>147304</END>
        <SCORE>-</SCORE>
        <ORIENTATION>+</ORIENTATION>
        <PHASE>-</PHASE>
      </FEATURE>
  . . .
```

Features contain multiple fields of data, but for now we will focus solely on the feature ID, start and end location. Our first task is therefore to build a `Feature` class that encapsulates this data. The code for the `Feature` class is presented in Listing 7.9. As you can see, the class has `get()` methods for each of the main properties.

> The BioJava API includes its own `Feature` interface. We are using our own `Feature` class to keep the example simpler, especially for those readers not yet familiar with BioJava.

Our next task is to create a `ContentHandler` capable of parsing a DAS response document and extracting the feature data. This code is presented in Listing 7.10. There are a few important elements

Listing 7.9 Feature.java

```java
package org.xmlbio.sax;


/**
 * Encapsulates Basic DAS Feature Information.
 */
public class Feature {
    private int _start;
    private int _end;
    private String _id;

    /**
     * Constructor.
     * @param id Feature ID
     * @param start Start Base Pair Location
     * @param end Stop Base Pair Location
     */
    public Feature (String id, int start, int end) {
        this._id = id;
        this._start = start;
        this._end = end;
    }


    /**
     * Gets Feature ID.
     * @return Feature ID
     */
    public String getID () {
        return this._id;
    }


    /**
     * Gets Start Base Pair Location.
     * @return Base Pair Location
     */
    public int getStart () {
        return this._start;
    }


    /**
     * Gets End Base Pair Location.
     * @return Base Pair Location
     */
    public int getEnd () {
        return this._end;
    }
}
```

Listing 7.10 DASHandler.java

```java
package org.xmlbio.sax;

import org.xml.sax.Attributes;
import org.xml.sax.SAXException;
import org.xml.sax.helpers.DefaultHandler;

import java.util.ArrayList;

/**
 * SAX DAS Processor.
 * Processes results from DAS Feature Commands.
 */
public class DASHandler extends DefaultHandler {
    private final static String FEATURE_ELEMENT = "FEATURE";
    private final static String START_ELEMENT = "START";
    private final static String END_ELEMENT = "END";
    private final static String ID_ATTRIBUTE = "id";

    private ArrayList _features;
    private int _startLocation;
    private int _endLocation;
    private String _featureID;
    private StringBuffer _currentText;

    /**
     * Constructor.
     */
    public DASHandler () {
        this._features = new ArrayList ();
    }

    /**
     * Gets ArrayList of Fetures.
     */
    public ArrayList getFeatures () {
        return this._features;
    }

    /**
     * Processes Start Element Events.
     */
    public void startElement (String namespaceURI, String localName,
                String qName, Attributes atts) throws SAXException {
        if (localName.equals (FEATURE_ELEMENT)) {
                _featureID = atts.getValue (ID_ATTRIBUTE);
        }
        _currentText = new StringBuffer ();
    }

    /**
     * Processes Character Events.
     */
```

Listing 7.10 (*cont.*)

```
public void characters (char[] ch, int start, int length)
                throws SAXException {
        String str = new String (ch, start, length);
        _currentText.append (str);
}

/**
 * Processes End Element Events.
 */
public void endElement (String namespaceURI, String localName,
        String qName) throws SAXException {
    if (localName.equals (START_ELEMENT)) {
        _startLocation = Integer.parseInt (_currentText.toString ());
    } else if (localName.equals (END_ELEMENT)) {
        _endLocation = Integer.parseInt (_currentText.toString ());
    } else if (localName.equals (FEATURE_ELEMENT)) {
        Feature feature = new Feature
                (_featureID, _startLocation, _endLocation);
        _features.add (feature);
    }
}

/**
 * Prints Feature Objects.
 * Primarily Used for Debugging Purposes.
 */
public void printFeatures () {
    System.out.println ("DAS Features");
    for (int i = 0; i < _features.size (); i++) {
        Feature feature = (Feature) _features.get (i);
        System.out.println ("Feature:");
        System.out.println ("... ID: " + feature.getID ());
        System.out.println ("... Start: " + feature.getStart ());
        System.out.println ("... End: " + feature.getEnd ());
    }
}
}
```

to note. First, the handler contains an `ArrayList` of `Feature` objects. When the handler encounters an end `</FEATURE>` tag, a new `Feature` object is created and added to the `ArrayList`. Second, note that we explicitly extract the Feature *id* attribute via the `Attributes.getValue()` method. Third, note that the code handles character "chunking" by appending to a `StringBuffer` object. The handler also includes a handy `printFeatures()` method used to print out all the feature objects—this is particularly helpful for debugging the application.

The BioJava API also includes several classes for automatically parsing DAS data. We could have used these as well, but then we could not illustrate the mechanics of SAX!

7.4.2 Integrating with BioJava

As our next step, we can use the open source BioJava package to render the DAS features. The BioJava API includes a `SequencePanel` class capable of rendering sequences and their associated features. The full rendering code is shown in Listings 7.11 and 7.12.

Listing 7.11 FeatureViewer.java

```java
package org.xmlbio.sax;

import org.biojava.bio.Annotation;
import org.biojava.bio.BioException;
import org.biojava.bio.gui.sequence.*;
import org.biojava.bio.seq.DNATools;
import org.biojava.bio.seq.Sequence;
import org.biojava.bio.seq.StrandedFeature;
import org.biojava.bio.seq.impl.SimpleSequence;
import org.biojava.bio.symbol.DummySymbolList;
import org.biojava.bio.symbol.RangeLocation;
import org.biojava.bio.symbol.SymbolList;
import org.biojava.utils.ChangeVetoException;

import javax.swing.*;
import java.awt.*;
import java.awt.event.WindowAdapter;
import java.awt.event.WindowEvent;
import java.util.ArrayList;

/**
 * Simple Feature Browser.
 * Utilizes the open source BioJava Libaray.
 */
public class FeatureViewer extends JFrame {
    private final static int SEQUENCE_LENGTH = 1000000000;
    private Sequence _sequence;
    private ArrayList _features;
    private int _rangeStart;
    private int _rangeEnd;

    /**
     * Constructor.
     */
    public FeatureViewer (ArrayList features)
                throws BioException, ChangeVetoException {
        this._features = features;
        this._sequence = createDummySequence ();
        determineWindowRange ();
        addFeatures (_sequence);
        createGUI ();
        setExit ();
        show ();
    }
```

Listing 7.11 (*cont.*)

```
/**
 * Creates a Dummy Sequence.
 * We create a Dummy Sequence, because we do not have the actual
 * sequence data.
 */
private Sequence createDummySequence () {
        SymbolList dummyList = new DummySymbolList (DNATools.getDNA (),
                        SEQUENCE_LENGTH);
        Sequence sequence = new SimpleSequence (dummyList, "ensembl",
                        "ensembl", Annotation.EMPTY_ANNOTATION);
        return sequence;
}


/**
 * Adds Features to Sequence.
 */
private void addFeatures (Sequence sequence)
                throws BioException, ChangeVetoException {
        for (int i = 0; i < _features.size (); i++) {
                Feature feature = (Feature)_features.get (i);
                StrandedFeature.Template bioFeature =
                                new StrandedFeature.Template ();
                bioFeature.location = new RangeLocation (feature.getStart (),
                                feature.getEnd ());
                bioFeature.annotation = Annotation.EMPTY_ANNOTATION;
                sequence.createFeature (bioFeature);
        }
}


/**
 * Determines Window Range for Sequence Panel.
 */
private void determineWindowRange () {
        if (_features.size () > 0) {
                Feature feature = (Feature)_features.get (0);
                this._rangeStart = feature.getStart ();
                this._rangeEnd = feature.getEnd ();
                for (int i = 1; i <_features.size (); i++) {
                        feature = (Feature)_features.get (i);
                        int start = feature.getStart ();
                        int end = feature.getEnd ();
                        if (start <_rangeStart)
                                _rangeStart = start;
                        if (end >_rangeEnd)
                                _rangeEnd = end;
                }
        }
}
```

Listing 7.11 (*cont.*)

```java
/**
 * Create the User Interface.
 */
private void createGUI () throws ChangeVetoException {
        // Create Master Panel
        JPanel panel = new JPanel ();
        panel.setLayout (new BorderLayout ());
        panel.setBorder (BorderFactory.createTitledBorder
                        ("SAX Feature Viewer"));

        // Add Sequence Panel inside Scroll Pane
        SequencePanel sequencePanel = createSequencePanel ();
        JScrollPane scrollPane = new JScrollPane (sequencePanel);
        panel.add (scrollPane, BorderLayout.CENTER);

        // Add Master Panel to Frame
        Container contentPane = this.getContentPane ();
        contentPane.add (panel);

        // Set Frame Title and Size
        setTitle ("SAX Feature Viewer");
        setSize (600, 120);
}

/**
 * Creates the Sequence Panel for Displaying Features.
 */
private SequencePanel createSequencePanel ()
                throws ChangeVetoException {
        // Create Sequence Panel
        SequencePanel sequencePanel = new SequencePanel ();

        // Create Ruler to display Base Pair Locations
        RulerRenderer ruler = new RulerRenderer ();

        // Create Feature Renderer to Display Features
        FeatureBlockSequenceRenderer features =
                        new FeatureBlockSequenceRenderer ();
        RectangularBeadRenderer featr = new RectangularBeadRenderer();
        features.setFeatureRenderer (featr);

        // Add Renderers to the MultiLineRenderer.
        // Enables us to display features along with a ruler.
        MultiLineRenderer mlr = new MultiLineRenderer ();
        mlr.addRenderer (features);
        mlr.addRenderer (ruler);
        sequencePanel.setRenderer (mlr);

        // Set the Sequence to Render, Range, and Scale
        sequencePanel.setSequence (_sequence);
```

Listing 7.11 *(cont.)*

```java
            sequencePanel.setRange (new RangeLocation (_rangeStart, _rangeEnd));
            sequencePanel.setScale (.001);
            return sequencePanel;
    }
    /**
     * Sets Window Exit Event Handling.
     */
    private void setExit () {
            this.addWindowListener (
                            new WindowAdapter () {
                                    public void windowClosing (WindowEvent e) {
                                            System.exit (0);
                                    }
                            });
    }
}
```

Listing 7.12 DASProcessor.java

```java
package org.xmlbio.sax;

import org.biojava.bio.BioException;
import org.biojava.utils.ChangeVetoException;
import org.xml.sax.SAXException;
import org.xml.sax.XMLReader;
import org.xml.sax.helpers.XMLReaderFactory;

import java.io.IOException;
import java.util.ArrayList;

/**
 * DAS Processor.
 * Retrieves features via DAS Protocol and displays them via
 * the BioJava Sequence Panel.
 */
public class DASProcessor {
    private ArrayList_features;
    private String_urn;

    /**
     * Constructor.
     */
    public DASProcessor (String urn) throws BioException,
                ChangeVetoException, IOException, SAXException {
            this._urn = urn;
            loadFeatures ();
            showViewer ();
    }
```

Listing 7.12 *(cont.)*

```java
/**
 * Loads Features from Specified Data Source.
 */
private void loadFeatures () throws SAXException, IOException {
        DASHandler dasHandler = new DASHandler ();
        XMLReader parser = XMLReaderFactory.createXMLReader
                        ("org.apache.xerces.parsers.SAXParser");
        parser.setContentHandler (dasHandler);
        parser.parse (_urn);
        dasHandler.printFeatures ();
        _features = dasHandler.getFeatures ();
}

/**
 * Show Feature Viewer.
 */
private void showViewer () throws BioException, ChangeVetoException {
        FeatureViewer featureViewer = new FeatureViewer (_features);
}

/**
 * Prints Command Line Usage.
 */
private static void printUsage () {
        System.out.println ("usage: DASProcessor URL | xml-file");
        System.exit (0);
}

/**
 * Main Method.
 */
public static void main (String[] args) throws Exception {
        if (args.length != 1) {
                printUsage ();
        }
        DASProcessor processor = new DASProcessor (args[0]);
}
}
```

To fully understand the code in Listing 7.11, you will need some passing familiarity with the Java Swing API and the BioJava API. For now, note that the code transforms our custom `Feature` objects into BioJava `Feature` objects. The code also makes use of the BioJava `MultiLineRenderer` class:

```java
MultiLineRenderer mlr = new MultiLineRenderer ();
mlr.addRenderer (features);
mlr.addRenderer (ruler);
```

This enables us to display features directly above a ruler that displays the base pair locations.

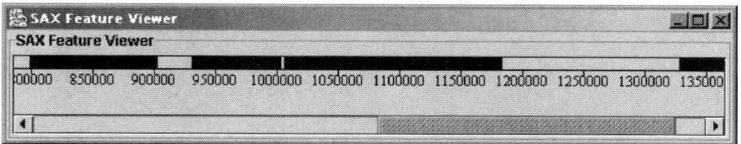

Figure 7.4 The SAX DAS processor program in action. The program receives DAS feature data over the Internet, parses the data via SAX, and renders the features via the BioJava API.

To run the program, you must specify a fully qualified DAS features request on the command line. For example, the following invocation requests transcript features from Ensembl for contig NT_008045:

```
java org.xmlbio.sax.DASProcessor
"http://servlet.sanger.ac.uk:8080/das/ensembl930/features?
segment=NT_008045;type=transcript"
```

A sample screenshot of the program in action is provided in Figure 7.4. Each feature is represented as a black rectangle, and you can use the scroll-bar to scroll through the complete set of features.

Parsing DAS Data with JDOM   8

JDOM is a popular open source library for reading, writing, and modifying XML documents. Unlike the Document Object Model (DOM), JDOM was specifically created for Java and provides a very elegant, intuitive API.

JDOM provides easy facilities for reading and validating XML documents, traversing through document contents, and working with XML Namespaces. It also provides built-in functionality for creating new XML documents from scratch, moving elements within a document, and outputting the final result using a number of formatting options. Furthermore, JDOM includes built-in support for XSL transformations, and a new JDOM extension for XPath functionality.

This chapter takes you through the basics of JDOM. Along the way, each new JDOM concept is illustrated with an example specifically tailored to the Distributed Annotation System (DAS). For example, when discussing JDOM traversal functionality, we provide several examples that parse and extract DAS XML response data. We also explore options for validating DAS documents, creating new DAS documents, and modifying existing documents.

For a larger case study, the chapter concludes with a discussion of JDAS, an open source DAS client library that makes extensive use of the JDOM API.

8.1   JDOM Basics

JDOM [91] is an open source library for reading, writing, and modifying XML documents. It is currently released under an Apache-style open source license and available for download from *http://www.jdom.org*. The original release of JDOM was created by Brett McLaughlin and Jason Hunter. Since its original release, JDOM has been officially accepted into the Sun Java Community Process (JCP) [88], and may eventually find its way into the core Java distribution. For latest details on JDOM's progress within the JCP, check out the main JCP web site at: *http://www.jcp.org*.

8.1.1   JDOM Package Overview

JDOM consists of five main packages:

- **org.jdom:** Contains the core classes for representing XML components. For example, this package contains XML `Document`, `Element`, and `Attribute` classes. Each of these classes provides accessor methods for extracting pieces of data, such as element text, element children, or element namespace information.

- **org.jdom.input:** Contains classes for reading in XML documents from a variety of sources. For example, the SAXBuilder class builds documents by utilizing the Simple API for XML (SAX).
- **org.jdom.output:** Contains classes for outputting XML documents to a variety of sources. For example, the XMLOutputter class provides several options for "pretty printing" XML documents.
- **org.jdom.filter:** Contains a Filter interface and several prebuilt filters for filtering the contents of an XML document.
- **org.jdom.transform:** Contains several classes for performing XSL transformations and interfacing with the new Java Transformation API for XML (TRaX).

8.1.2 Parsing XML Documents with JDOM

JDOM is not actually an XML parser. Rather, JDOM is a set of classes for representing XML documents and easily interfacing with existing XML parsers. You can therefore use just about any XML parser with JDOM. The JDOM distribution does, however, include the Apache Xerces XML parser and JDOM therefore works "out of the box."

When parsing XML documents via JDOM, you have two main options. The first option is to use the SAXBuilder class. This class builds a JDOM document by intercepting SAX events (for details on SAX events, see the previous chapter). The second option is to use the DOMBuilder class. The DOMBuilder class parses the entire XML document and builds a complete DOM tree; it then traverses the entire tree and creates a second JDOM document tree with the identical information.

Note that the DOMBuilder requires that you build two complete trees, and this takes up more time and memory than the SAXBuilder . It is therefore generally recommended that you use the SAXBuilder over the DOMBuilder .

SAXBuilder Basics

To parse a document via the SAXBuilder class, you instantiate a SAXBuilder object and call its build() method. For example:

```
SAXBuilder builder = new SAXBuilder();
File file = new File("test_data/ensembl_dna.xml");
Document document = builder.build (file);
```

The build() method is overloaded and includes options for building documents from files, input streams, character streams, and URLs. Once invoked, the build() method will locate your specified file, parse it, and return a fully formed JDOM Document object. The Document object will contain the complete contents of the XML file and represents the entry point for tree traversal.

By default, XML validation will be turned *off*. By passing true to the SAXBuilder() constructor, you turn XML validation on. For example:

```
SAXBuilder builder = new SAXBuilder(true);
Document document = builder.build (file);
```

If any errors occur during the parsing process, JDOM will throw a JDOMException. For example, errors in well-formedness and validity are reported as JDOMExceptions .

Behind the scenes, JDOM uses the Java API for XML (JAXP) to locate your XML parser. If you are using JDK 1.4, JAXP defaults to the built-in Crimson XML parser. If JDOM is unable to determine your JAXP settings, it is hard coded to use the Xerces XML parser included in the JDOM distribution.

If you want to override the defaults and use a specific XML parser, you have two main options. The first option is to specify a JAXP system property, usually from the command line. The JAXP property is "javax.xml.parsers.SAXParserFactory," and it must point to the SAXParserFactory implementation provided by your XML parser. For example, the following command line specifies the Xerces XML parser:

```
java - Djavax.xml.parsers.SAXParserFactory=org.apache.xerces.jaxp.
   SAXParserFactoryImpl org.xmlbio.jdom.JDOMBasic ensembl_dna.xml
```

The second option is to pass the XML parser information to the SAXBuilder() or DOMBuilder() constructors. For this to work, you must pass a fully qualified reference to the XMLReader implementation provided by your XML parser. For example, the following code also specifies the Xerces XML parser:

```
SAXBuilder builder = new SAXBuilder
   ("org.apache.xerces.parsers.SAXParser", true);
```

If you are using a parser other than Xerces, check the parser documentation directly for information on the correct class settings.

With validation turned on, JDOM will only validate against DTDs. To validate against an XML Schema, you need to take a few additional steps. First, make sure that your XML parser is capable of validating against XML Schemas. For example, both Apache Xerces 1 and Xerces 2 provide Schema support (Xerces 1 is currently included in the JDOM distribution). Second, make sure that JDOM is using your Schema capable parser. You can set this via a JAXP system property or via the SAXBuilder() constructor (see above).

Third, you need to explicitly turn on Schema validation. In Xerces, you must activate the "validation/schema" feature. Your code would therefore look like this:

```
SAXBuilder builder = new SAXBuilder
   ("org.apache.xerces.parsers.SAXParser", true);
builder.setFeature
   ("http://apache.org/xml/features/validation/schema", true);
```

Once everything is set up, Xerces will automatically validate your document against the XML Schema referenced by your document. If any validation errors are encountered, JDOM will throw a JDOMException.

XMLOutputter Basics

To output the contents of your XML document, use the JDOM XMLOutputter class. For example, the following code outputs a JDOM Document to System.out:

```
XMLOutputter out = new XMLOutputter();
out.output (document, System.out);
```

By default, `XMLOutputter` will output your document exactly as it is stored in memory. However, the `XMLOutputter` class also includes several configuration options. For example, you can force new line characters after the end of each element. You can also specify the exact size of indentation. For example, the following code outputs your XML document in "pretty print" format:

```
XMLOutputter out = new XMLOutputter();
out.setNewlines(true);
out.setIndentSize(4);
out.output(document, System.out);
```

You can also use `XMLOutputter` to write out the document contents to a file. For example:

```
XMLOutputter out = new XMLOutputter();
File file = new File ("document.xml");
FileWriter writer = new FileWriter (file);
out.output(doc, writer);
```

Basic JDOM Examples

We are now ready for some actual code. Our first example parses any XML file, checks it for well-formedness, and outputs its contents to the screen. See Listing 8.1 for the full code.

Listing 8.1 JDOMBasic.java

```
package org.xmlbio.jdom;

import org.jdom.Document;
import org.jdom.JDOMException;
import org.jdom.input.SAXBuilder;
import org.jdom.output.XMLOutputter;

import java.io.IOException;

/**
 * Basic JDOM Example.
 * Illustrates the JDOM SAXBuilder and XMLOutputter.
 */
public class JDOMBasic {

    /**
     * Build and Output the Specified XML Document
     */
    public void process(String systemID) {
        try {
            SAXBuilder builder = new SAXBuilder();
            Document doc = builder.build(systemID);
            XMLOutputter out = new XMLOutputter();
            out.output(doc, System.out);
```

Listing 8.1 (*cont.*)

```
            // If we get here without a JDOMException,
            // document is well-formed.
            System.out.println("Document is well-formed.");
        } catch (JDOMException e) {
            System.out.println("JDOM Exception: " + e.getMessage());
        } catch (IOException e) {
            System.out.println("IOException: " + e.getMessage());
        }
    }

    /**
     * Prints Command Line Usage
     */
    private static void printUsage() {
        System.out.println("usage: JDOMBasic xml-file");
        System.exit(0);
    }

    /**
     * Main Method
     */
    public static void main(String[] args) {
        if (args.length != 1) {
            printUsage();
        }
        System.out.println("JDOMBasic Example");
        System.out.println("Parsing file: " + args[0]);
        JDOMBasic app = new JDOMBasic();
        app.process(args[0]);
    }
}
```

There are a number of important elements to note about Listing 8.1. First, we are using the SAXBuilder object to build a JDOM Document object. We are using the no-argument SAXBuilder() constructor, and XML validation is, by default, turned off. If JDOM detects any errors in well-formedness, it will immediately throw a JDOMException. Otherwise, the program exits with the message "Document is well-formed." The program also outputs the document to the screen via the XMLOutputter class.

Our second example performs basic XML validation. See Listing 8.2 for the complete code. This time, we pass true to the SAXBuilder() constructor. If JDOM detects any errors in

Listing 8.2 JDOMValidator.java

```
package org.xmlbio.jdom;

import org.jdom.Document;
import org.jdom.JDOMException;
import org.jdom.input.SAXBuilder;
```

Listing 8.2 (*cont.*)

```java
/**
 * DAS Validator.
 * Illustrates XML Validation.
 */
public class JDOMValidator {

    /**
     * Build and Validate the Specified XML Document
     */
    public void process(String systemID) {
        try {
            // By passing true to SAXBuilder Constructor, we
            // turn XML validation on
            SAXBuilder builder = new SAXBuilder(true);
            Document doc = builder.build(systemID);
            // If we get here without a JDOMException,
            // document is well-formed and valid.
            System.out.println("Document is well-formed.");
            System.out.println("Document is valid.");
        } catch (JDOMException e) {
            // Indicates error in well-formedness or validity
            System.out.println("JDOM Exception: " + e.getMessage());
        }
    }

    /**
     * Prints Command Line Usage
     */
    private static void printUsage() {
        System.out.println("usage: JDOMValidator xml-file");
        System.exit(0);
    }

    /**
     * Main Method
     */
    public static void main(String[] args) {
        if (args.length != 1) {
            printUsage();
        }
        JDOMValidator app = new JDOMValidator();
        app.process(args[0]);
    }
}
```

well-formedness or validity, it will immediately throw a JDOMException. Otherwise, the program exits with the message "Document is well-formed. Document is valid."

The second example is a handy program to keep around. You can use it to validate any XML document and immediately determine the source of any validity errors.

8.2 Parsing DAS Documents with JDOM

8.2.1 Introduction to the JDOM Element API

Once you have processed your XML document via the `SAXBuilder` or `DOMBuilder` classes, your application will receive a JDOM `Document` object. The `Document` object contains the complete contents of your XML document. It also serves as the entry point to the root element of the document and the starting point for complete tree traversal.

To obtain the root element of your document, simply call the `getRootElement()` method. For example:

```
SAXBuilder builder = new SAXBuilder();
Document doc = builder.build (systemID);
Element root = doc. getRootElement();
```

The JDOM `Element` class contains numerous accessor methods. For example, you can determine the element's name, extract its attributes, get a list of all its child elements, add new content, or remove existing content. A list of these methods is provided in Table 8.1. Take a moment now to skim the table. We will not be covering every single one of these methods here, but it is helpful to understand the range of functionality provided.

Table 8.1 Methods of the JDOM `Element` class. Copied directly from the official JDOM JavaDoc API [91], with the permission of Jason Hunter. For the complete JDOM API go to *http://www.jdom.org/docs/apidocs/index.html*

Method Summary	
Element	`addContent(CDATA cdata)`
	This adds a CDATA section as content to this element
Element	`addContent(Comment comment)`
	This adds a comment as content to this element
Element	`addContent(Element element)`
	This adds element content to this element
Element	`addContent(EntityRef entity)`
	This adds entity content to this element
Element	`addContent(String str)`
	This adds text content to this element
Element	`addContent(Text text)`
	This adds text content to this element
void	`addNamespaceDeclaration(Namespace additional)`
	This will add a namespace declaration to this element
java.lang.Object	`clone()`
	This returns a deep clone of this element
Element	`detach()`
	This detaches the element from its parent, or does nothing if the element has no parent
boolean	`equals(java.lang.Object ob)`
	This tests for equality of this Element to the supplied Object, explicitly using the == operator.
java.util.List	`getAdditionalNamespaces()`
	This will return any namespace declarations on this element that exist, excluding the namespace of the element itself, which can be obtained through `getNamespace()`
Attribute	`getAttribute(String name)`
	This returns the attribute for this element with the given name and within no namespace, or null if no such attribute exists

Table 8.1 (cont.)

	Method Summary
Attribute	`getAttribute(String name, Namespace ns)` This returns the attribute for this element with the given name and within the given namespace, or null if no such attribute exists
java.util.List	`getAttributes()` This returns the complete set of attributes for this element, as a List of Attribute objects in no particular order, or an empty list if there are none
String	`getAttributeValue(String name)` This returns the attribute value for the attribute with the given name and within no namespace, null if there is no such attribute, and the empty string if the attribute value is empty
String	`getAttributeValue(String name, Namespace ns)` This returns the attribute value for the attribute with the given name and within the given namespace, null if there is no such attribute, and the empty string if the attribute value is empty
String	`getAttributeValue(String name, Namespace ns, String def)` This returns the attribute value for the attribute with the given name and within the given namespace, or the passed-in default if there is no such attribute
String	`getAttributeValue(String name, String def)` This returns the attribute value for the attribute with the given name and within no namespace, or the passed-in default if there is no such attribute
Element	`getChild(String name)` This returns the first child element within this element with the given local name and belonging to no namespace
Element	`getChild(String name, Namespace ns)` This returns the first child element within this element with the given local name and belonging to the given namespace
java.util.List	`getChildren()` This returns a List of all the child elements nested directly (one level deep) within this element, as Element objects
java.util.List	`getChildren(String name)` This returns a List of all the child elements nested directly (one level deep) within this element with the given local name and belonging to no namespace, returned as Element objects
java.util.List	`getChildren(String name, Namespace ns)` This returns a List of all the child elements nested directly (one level deep) within this element with the given local name and belonging to the given namespace, returned as Element objects
String	`getChildText(String name)` This convenience method returns the textual content of the named child element, or returns an empty String ("") if the child has no textual content
String	`getChildText(String name, Namespace ns)` This convenience method returns the textual content of the named child element, or returns null if there is no such child
String	`getChildTextNormalize(String name)` This convenience method returns the normalized textual content of the named child element, or returns null if there is no such child
String	`getChildTextNormalize(String name, Namespace ns)` This convenience method returns the normalized textual content of the named child element, or returns null if there is no such child
String	`getChildTextTrim(String name)` This convenience method returns the trimmed textual content of the named child element, or returns null if there is no such child
String	`getChildTextTrim(String name, Namespace ns)` This convenience method returns the trimmed textual content of the named child element, or returns null if there is no such child

Table 8.1 (*cont.*)

`java.util.List`	`getContent()`
	This returns the full content of the element as a List that may contain objects of type `Text`, `Element`, `Comment`, `ProcessingInstruction`, `CDATA`, and `EntityRef`
`java.util.List`	`getContent(Filter filter)`
	Return a filter view of this Element's content
`Document`	`getDocument()`
	This retrieves the owning Document for this Element, or null if not currently a member of a Document
`String`	`getName()`
	This returns the (local) name of the Element, without any namespace prefix, if one exists
`Namespace`	`getNamespace()`
	This will return this Element's Namespace
`Namespace`	`getNamespace(String prefix)`
	This returns the namespace in scope on this element for the given prefix (this involves searching up the tree, so the results depend on the current location of the element)
`String`	`getNamespacePrefix()`
	This returns the namespace prefix of the Element, if one exists
`String`	`getNamespaceURI()`
	This returns the URI mapped to this Element's prefix (or the default namespace if no prefix)
`Element`	`getParent()`
	This will return the parent of this Element
`String`	`getQualifiedName()`
	This returns the full name of the Element, in the form [namespacePrefix]:[localName]
`String`	`getText()`
	This returns the textual content directly held under this element
`String`	`getTextNormalize()`
	This returns the textual content of this element with all surrounding whitespace removed and internal whitespace normalized to a single space
`String`	`getTextTrim()`
	This returns the textual content of this element with all surrounding whitespace removed
`boolean`	`hasChildren()`
	Test whether this element has a child element
`int`	`hashCode()`
	This returns the hash code for this element
`boolean`	`isAncestor(Element element)`
	Determines if this element is the ancestor of another element
`boolean`	`isRootElement()`
	This returns a boolean value indicating whether this Element is a root Element for a JDOM Document
`boolean`	`removeAttribute(Attribute attribute)`
	This removes the supplied Attribute should it exist
`boolean`	`removeAttribute(String name)`
	This removes the attribute with the given name and within no namespace
`boolean`	`removeAttribute(String name, Namespace ns)`
	This removes the attribute with the given name and within the given namespace
`boolean`	`removeChild(String name)`
	This removes the first child element (one level deep) with the given local name and belonging to no namespace
`boolean`	`removeChild(String name, Namespace ns)`
	This removes the first child element (one level deep) with the given local name and belonging to the given namespace
`boolean`	`removeChildren()`
	This removes all child elements
`boolean`	`removeChildren(String name)`
	This removes all child elements (one level deep) with the given local name and belonging to no namespace

Table 8.1 *(cont.)*

	Method Summary
boolean	`removeChildren(String name, Namespace ns)` This removes all child elements (one level deep) with the given local name and belonging to the given namespace
boolean	`removeContent(CDATA cdata)` This removes the specified CDATA
boolean	`removeContent(Comment comment)` This removes the specified Comment
boolean	`removeContent(Element element)` This removes the specified Element
boolean	`removeContent(EntityRef entity)` This removes the specified EntityRef
boolean	`removeContent(ProcessingInstruction pi)` This removes the specified ProcessingInstruction
boolean	`removeContent(Text text)` This removes the specified Text
void	`removeNamespaceDeclaration(Namespace additionalNamespace)` This will remove a namespace declaration from this element
Element	`setAttribute(Attribute attribute)` This sets an attribute value for this element
Element	`setAttribute(String name, String value)` This sets an attribute value for this element
Element	`setAttribute(String name, String value, Namespace ns)` This sets an attribute value for this element
Element	`setAttributes(java.util.List newAttributes)` This sets the attributes of the element
Element	`setChildren(java.util.List children)` This sets the content of the element the same as setContent(java.util.List), except only Element objects are allowed in the supplied list
Element	`setContent(java.util.List newContent)` This sets the content of the element
protected Element	`setDocument(Document document)` This sets the Document parent of this element and makes it the root element
Element	`setName(String name)` This sets the (local) name of the Element
Element	`setNamespace(Namespace namespace)` This sets this Element's Namespace
protected Element	`setParent(Element parent)` This will set the parent of this Element
Element	`setText(String text)` This sets the content of the element to be the text given
String	`toString()` This returns a String representation of the Element, suitable for debugging

8.2.2   Traversing DAS Documents

To illustrate the core methods of the JDOM `Element` class, let us try out another example. List-ing 8.3 provides code for a `DASWalker` program. The program accepts a single URL argument

Listing 8.3 DASWalker.java

```java
package org.xmlbio.jdom;

import org.jdom.Attribute;
import org.jdom.Document;
import org.jdom.Element;
import org.jdom.JDOMException;
import org.jdom.input.SAXBuilder;

import java.util.List;

/**
 * DAS Walker.
 * Illustrates the basics of JDOM Traversal.
 */
public class DASWalker {

    /**
     * Download and Traverse the Specified XML File.
     */
    public void process(String systemID) {
        try {
        // Build Document via SAXBuilder
        SAXBuilder builder = new SAXBuilder();
        Document doc = builder.build(systemID);
        // Get Root Element
        Element root = doc.getRootElement();
        // Process Root Element
        processElement(root);
        } catch (JDOMException e) {
        system.out.println("JDOM Exception: " + e.getMessage());
        }
    }

    /**
     * Recursive Method to Process Elements.
     */
    private void processElement(Element element) {
        // Get Element Name and Normalized Text
        String elementName = element.getName();
        String text = element.getTextNormalize();
        System.out.println("Element: " + elementName);
        processAttributes(element);
        if (text != null && text.length() > 0)
            System.out.println("... Text: " + text);
        // Get all Element Children and pass to processElement
        List children = element.getChildren();
        for (int i = 0; i < children.size(); i++) {
            Element child = (Element) children.get(i);
            processElement(child);
        }
    }
```

Listing 8.3 *(cont.)*

```java
/**
 * Process Attributes.
 */
private void processAttributes(Element element) {
    List attributes = element.getAttributes();
    for (int i = 0; i < attributes.size(); i++) {
        Attribute attribute = (Attribute) attributes.get(i);
        String attributeName = attribute.getName();
        String attributeValue = attribute.getValue();
        System.out.println("... " + attributeName + " : " +
          attributeValue);
    }
}

/**
 * Prints Command Line Usage.
 */
private static void printUsage() {
    System.out.println("usage: DASWalker xml-file");
    System.exit(0);
}

/**
 * Main Method.
 */
public static void main(String[] args) throws Exception {
    if (args.length != 1) {
        printUsage();
    }
    DASWalker app = new DASWalker();
    app.process(args[0]);
}
}
```

and can be used to parse the contents of any DAS XML response document. For example, the following command line issues a *dsn* command, requesting a full list of data sources hosted by the Wormbase DAS server:

```
java org.xmlbio.jdom.DASWalker http://www.wormbase.org/db/das/dsn
```

In response to a *dsn* request, Wormbase returns the document shown in Listing 8.4. JDOM parses this document and outputs the following content:

```
Element:   DASDSN
Element:   DSN
Element:   SOURCE
... id:    elegans
... Text:  elegans
Element:   MAPMASTER
... Text:  http://www.wormbase.org/db/das/elegans
Element:   DESCRIPTION
... Text:  C. elegans annotations from WormBase
```

Listing 8.4 DAS DSN response from Wormbase.org

```
<?xml version="1.0" standalone="yes"?>
<!DOCTYPE DASDSN SYSTEM "http://www.biodas.org/dtd/dasdsn.dtd">
<DASDSN>
   <DSN>
      <SOURCE id="elegans">elegans</SOURCE>
      <MAPMASTER>http://www.wormbase.org/db/das/elegans</MAPMASTER>
      <DESCRIPTION>C. elegans annotations from WormBase
         </DESCRIPTION>
   </DSN>
</DASDSN>
```

Element Names and Namespaces

The `DASWalker` program works by traversing the complete contents of the JDOM tree. The core of the tree traversal occurs within the recursive `processElement()` method. This method first extracts the element name via the `getName()` method:

```
String elementName = element.getName();
```

If you want to determine the element's namespace, call the `getNamespace()` method. With the returned `Namespace` object, you can determine the Namespace prefix and the Namespace URI. For example:

```
Namespace namespace = element.getNamespace();
if (namespace != null) {
    System.out.println ("... Namespace --> " + namespace.getPrefix()
            + ": " + namespace.getURI());
}
```

Extracting Element Text

You have several options when extracting element text. First, the `getText()` method returns the text as it is stored in memory and may include multiple whitespace characters. The `getTextTrim()` method gets the same text string, but removes the leading and trailing whitespaces. Finally, the `getTextNormalize()` method replaces multiple whitespace characters with a single space character, and returns a more compact representation of the string.

Note that the `Element` API also includes a handy convenience method for extracting the text of a specific subelement. For example, this line of code extracts the text from the first element named "MAPMASTER":

```
String mapMaster = dsnElement.getChildText("MAPMASTER");
```

Traversing Element Children

To traverse through the children of an element, the `DASWalker` program calls the `getChildren()` method:

```
List children = element.getChildren();
for (int i = 0; i < children.size(); i++) {
```

```
        Element child = (Element) children.get(i);
        processElement(child);
}
```

The `getChildren()` method returns a list of *immediate* descendants, each of which is passed recursively to the `processElement()` method.

If you want to filter for specific elements, use the `getChildren (String name)` method. For example, as we will see in the next example, this line of code extracts all <DSN> elements from the root element:

```
List dsnElements = root.getChildren("DSN");
```

Working with Attributes

Working with XML attributes is very straightforward in JDOM. Each JDOM attribute is represented by an `Attribute` object, and you receive a complete list of attributes via the `getAttributes()` method:

```
List attributes = element.getAttributes();
for (int i = 0; i < attributes.size(); i++) {
    Attribute attribute = (Attribute) attributes.get(i);
    String attributeName = attribute.getName();
    String attributeValue = attribute.getValue();
    System.out.println("... " + attributeName + " : "
        + attributeValue);
}
```

As a convenience, the `Attribute` class has a number of get() methods for performing automatic type conversion. For example, there are methods for `getDoubleValue()`, `getFloatValue()`, and `getIntValue()`.

As an extra convenience, you can also extract a specific attribute value from the `Element` object by using the `getAttributeValue()` method. For example, the following line extracts the "id" attribute from the current `element` object:

```
String id = element.getAttributeValue("id")
```

Much like `Element` objects, you can determine an Attribute object's namespace information by calling the `getNamespace()` method. For example:

```
Namespace namespace = attribute.getNamespace();
String prefix = namespace.getPrefix();
String uri = namespace.getURI();
```

Traversing the Complete Contents of an XML Document

The `Element getChildren()` method only returns element children. However, it is sometimes useful to extract nonelement children, such as comments and processing instructions. To get the complete contents of an `Element` object, use the `getContent()` method. This method will return a list of JDOM objects, including objects of type `Comment`, `CDATA`, `ProcessingInstruction`, etc. To determine the type of each object in the list, you frequently need an if/else tree with multiple uses of the `instanceof` operator. For example:

```
List content = element.getContent();
for (int i = 0; i < content.size(); i++) {
   Objec object = content.get(i);
   if (object instanceof Element)
         ...
   else if (object instanceof Comment)
         ...
   else if (object instanceof CDATA)
         ...
   else if (object instanceof ProcessingInstruction)
         ...
   else if (object instanceof Text)
         ...
}
```

JDOM includes a `Filter` interface for filtering the direct descendants of an `Element` object. This can be useful for numerous scenarios. For example, you can filter for elements within a specific namespace, or filter for specific node types, such as processing instructions or comments.

To use a filter, you have the option of implementing the JDOM `Filter` interface directly or using one of two prebuilt JDOM filter classes. To build your own filter, you need to create a new class and implement the three required methods, as defined in the JDOM Filter `interface`. However, for most common tasks, you can use the prebuilt `ElementFilter` or `ContentFilter` classes. The `ElementFilter` is useful for filtering elements which match specific criteria, e.g., elements must exist within a specified namespace. The `ContentFilter` is useful for filtering specific types of nodes. For example, the following code extracts only comment nodes and hides all the rest:

```
// By passing false to constructor, all JDOM
// objects are filtered out.
ContentFilter filter = new ContentFilter(false);
// Turn Filtering on for Comments only
filter.setCommentVisible(true);
// Get the Comments
List content = element.getContent(filter);
```

As you can see from the code above, once you have a filter object, you pass it to the `getContent()` method. The method applies the filter to all direct descendants and returns only those objects which match the defined criteria.

8.2.3   Parsing DAS *dsn* Documents

The `DASWalker` program can traverse any arbitrary DAS XML response (or any arbitrary XML document, for that matter). Most of the time, however, you want to parse a specific XML document, which adheres to a predefined DTD or XML Schema. Given the XML document, you frequently want to extract a subset of the data and pass the data objects along to other components within your application.

We will see more examples of parsing specific DAS documents in our discussion of JDAS at the end of the chapter. However, for now, we consider a slightly simpler example. Our next example parses *dsn* responses from DAS servers and prints out a simple directory of available data sources. For example, the following command line:

```
java org.xmlbio.jdom.DAS_DSNProcessor http://www.wormbase.org/
   db/das/dsn
```

generates the following output:

```
Retrieving data sources from: http://www.wormbase.org/db/das/dsn
Data Source:
... ID: elegans --> elegans
... MapMaster: http://www.wormbase.org/db/das/elegans
... Description: C. elegans annotations from WormBase
```

The full code is shown in Listing 8.5.

Listing 8.5 DAS_DSNProcessor.java

```java
package org.xmlbio.jdom;

import org.jdom.Document;
import org.jdom.Element;
import org.jdom.JDOMException;
import org.jdom.input.SAXBuilder;

import java.util.Iterator;
import java.util.List;

/**
 * Parses DAS Data Source (DSN) Data.
 * Illustrates Basic use of JDOM Traversal Functionality.
 * Example usage: java DAS_DSNProcessor http://www.wormbase.org/
 *    db/das/dsn
 */
public class DAS_DSNProcessor {

    /**
     * Processes XML Document.
     */
    public void process(String systemID) {
        System.out.println("Retrieving data sources from:" + systemID);
        try {
            // Build Document
            SAXBuilder builder = new SAXBuilder();
            Document doc = builder.build(systemID);
            // Get Root Element
            Element root = doc.getRootElement();
            processRootElement(root);
```

Listing 8.5 *(cont.)*

```java
        } catch (JDOMException e) {
            System.out.println("JDOM Exception: " + e.getMessage());
        }
    }

    /**
     * Processes Root Element of DAS Response.
     */
    private void processRootElement(Element root) {
        List dsnElements = root.getChildren("DSN");
        Iterator iterator = dsnElements.iterator();
        while (iterator.hasNext()) {
            Element dsnElement = (Element) iterator.next();
            processDSNElement(dsnElement);
        }
    }

    /**
     * Processes DSN Element of DAS Response.
     */
    private void processDSNElement(Element dsnElement) {
        System.out.println("Data Source:");
        Element sourceElement = dsnElement.getChild("SOURCE");
        if (sourceElement != null) {
            String id = sourceElement.getAttributeValue("id");
            String version = sourceElement.getAttributeValue("version");
            String sourceText = sourceElement.getTextNormalize();
            System.out.print("... ID: " + id);
            if (version != null)
                System.out.print(", Version: " + version);
            System.out.println(" --> " + sourceText);
        }

        String mapMaster = dsnElement.getChildText("MAPMASTER");
        String description = dsnElement.getChildText("DESCRIPTION");

        if (mapMaster != null && mapMaster.length() > 0)
            System.out.println("... MapMaster: " + mapMaster);
        if (description != null && description.length() > 0)
            System.out.println("... Description: " + description);
    }

    /**
     * Print Command Line Usage.
     */
    private static void printUsage() {
        System.out.println("usage: DAS_DSNProcessor xml-file");
        System.exit(0);
    }
```

Listing 8.5 *(cont.)*

```
/**
 * Main Method.
 */
public static void main(String[] args) {
    if (args.length != 1) {
        printUsage();
    }
    DAS_DSNProcessor app = new DAS_DSNProcessor();
    app.process(args[0]);
}
}
```

The example in Listing 8.5 does a good job of recapping the main methods of the JDOM API. For example, the `processRootElement()` method takes the root element of the document and filters it for all elements named "DSN":

```
List dsnElements = root.getChildren("DSN");
```

Each matching element is then sent to the `processDSNElement()` method. Here, we extract information on three specific elements: <SOURCE>, <MAPMASTER>, and <DESCRIPTION>. For each of these elements, the code extracts the element text and any known attributes. For example, these lines extract the <SOURCE> element data:

```
Element sourceElement = dsnElement.getChild ("SOURCE");
if (sourceElement != null) {
    String id = sourceElement.getAttributeValue("id");
    String version = sourceElement.getAttributeValue("version");
    String sourceText = sourceElement.getTextNormalize();
    ...
}
```

JDOM and XPath

The very latest JDOM code now provides built-in support for XPath via the open source Jaxen Project (*http://jaxen.sourceforge.net*). As of this writing, XPath support is functional, but not fully complete, and is therefore not yet part of the official JDOM distribution. To get the very latest JDOM code with XPath functionality, you must download the source code from the JDOM CVS server directly. Instructions are available on the JDOM web site at: *http://www.jdom.org/downloads/source.html*.

Once you have downloaded the source code and generated the proper Jar files, using XPath is easy. To apply an XPath query to an existing document, you first need a JDOM `XPath` object. New instances of `XPath` are available via the static `XPath.newInstance()` method. For example, the following code instantiates an XPath object for locating DAS <DNA> sequence elements:

```
XPath xpath = XPath.newInstance ("/DASDNA/SEQUENCE/DNA");
```

The XPath `selectNodes()` method receives a `Document` object, applies the XPath query, and returns a matching list of objects. For example:

```
List list = xpath.selectNodes (doc);
```

You can then iterate through the matching nodes and extract any data you need. A more complete code excerpt is shown below:

```
SAXBuilder builder = new SAXBuilder ();
Document doc = builder.build (url);

// Create XPath Instance
XPath xpath = XPath.newInstance ("/DASDNA/SEQUENCE/DNA");

// Select Nodes which match XPath
List list = xpath.selectNodes (doc);

// Iterate through selected nodes
Iterator iterator = list.iterator ();
while (iterator.hasNext ()) {
    Element element = (Element) iterator.next ();
    System.out.println ("DNA: "
        + element.getTextNormalize ());
}
```

8.3   Creating DAS Documents with JDOM

So far, we have seen that JDOM provides a very simple, powerful API for parsing and traversing existing XML documents. We now turn to the use of JDOM in creating and modifying documents. To illustrate the concepts, we examine a simple program that generates DAS response documents using hard-coded data. Creating DAS documents like this may be quite useful to you, but the basic principles can be applied to any other bioinformatics application.

8.3.1   Creating New Documents

To create a new XML document via JDOM, you first need to instantiate a JDOM `Document` object:

```
Document document = new Document();
```

To specify a DTD for your document, create a `DocType` object and then call the `Document set-DocType()` method. For example, the following code creates a document with the dasdna.dtd file:

```
DocType docType = new DocType("DASDNA", "http://biodas.org/dtd/dasdna.dtd");
    document.setDocType(docType);
```

8.3.2   Creating New Elements

To create a new XML element, you instantiate a new `Element` object and specify the element name. For example, the following code instantiates a <DASDNA> element:

```
Element root = new Element("DASDNA");
```

If your element exists within an XML Namespace, you need to first create a JDOM `Namespace` object, and then pass this to the element constructor. For example, this code excerpt creates an <html> element within the XHTML namespace:

```
Namespace xhtmlNamespace = Namespace.getNamespace("xhtml",
    "http://www.w3.org/TR/REC-html40");
Element root = new Element("html", xhtmlNamespace);
```

Every document must have a root element, which is set via the `setRootElement()` method. For example, the following creates a <DASDNA> element and sets it as the root element:

```
Document document = new Document();
Element root = new Element("DASDNA");
document.setRootElement(root);
```

To create nested elements within your document, you call the `addContent()` method. For example, this code creates a <SEQUENCE> element and adds it to the root <DASDNA> element:

```
Element root = new Element("DASDNA");
Element sequenceElement = new Element("SEQUENCE");
root.addContent(sequenceElement);
```

To move an element within a document, you must first `detach()` it from its parent. It is then free to be attached anywhere within the document. For example, the following code swaps the first two elements within a document:

```
Element root = doc.getRootElement();
List children = root.getChildren();
Element element1 = (Element) children.get(0);
Element element2 = (Element) children.get(1);
element1.detach();
element2.detach();
root.addContent(element2);
root.addContent(element1);
```

Once you have an `Element` object, you can easily set its attributes via the `setAttribute()` method. For example, the following code adds two attributes to the <SEQUENCE> element:

```
Element sequenceElement = new Element("SEQUENCE");
sequenceElement.setAttribute("id", id);
sequenceElement.setAttribute("version", version);
```

To remove an attribute, use the `removeAttribute (String attributeName)` method.

Setting Element Text

To add text to an element, use the `setText()` method. For example:

```
Element dnaElement = new Element("DNA");
dnaElement.setText("taatttctcccattttgtaggttatca");
```

JDOM will automatically escape any special characters for you. For example, the following call adds sample HTML code to your document:

```
Element paragraph1 = new Element("p", xhtmlNamespace);
paragraph1.setText("Sample HTML Code: <B>This is Bold</B>");
```

If you subsequently call `paragraph1.getText()`, you get back the text string exactly like this: "Sample HTML Code: <B>This is Bold</B>." However, when you output the document via `XMLOutputter`, the angle brackets are automatically replaced with entity references. The outputted element therefore looks like this:

```
<xhtml:p>Sample HTML Code: &lt;B&gt;This is Bold&lt;/B&gt;</xhtml:p>
```

If you want to explicitly add a CDATA section to your document, instantiate a `CDATA` object and add it to the contents of your element. For example:

```
Element paragraph2 = new Element("p", xhtmlNamespace);
CDATA cdata = new CDATA("<B><I>More Sample Code</I></B>");
paragraph2.addContent(cdata);
```

If you subsequently call `paragraph2.getText()`, JDOM automatically detects the CDATA sections and returns the following string: "<B><I>More Sample Code</I></B>." However, when you output the document via `XMLOutputter`, the CDATA section is explicitly set like this:

```
<xhtml:p><![CDATA[<B><I>More Sample Code</I></B>]]></xhtml:p>
```

8.3.3   A Complete Example

To tie all these concepts together, let us look at a complete example. Listing 8.6 creates a valid XML document, which adheres to the dasdna.dtd file. For simplicity, the data is hard-coded. Note that the program uses `XMLOutputter` to pretty print the generated document. Sample output is provided below:

```
<?xml version="1.0" encoding="UTF-8"?>
<!DOCTYPE DASDNA SYSTEM "http://biodas.org/dtd/dasdna.dtd">
<DASDNA>
    <SEQUENCE id="1" version="8.30" start="1000" stop="1025">
        <DNA length="25">taatttctcccattttgtaggttat</DNA>
    </SEQUENCE>
    <SEQUENCE id="2" version="8.30" start="1000" stop="1025">
        <DNA length="25">taatgcaactaaatccaggcgaagc</DNA>
    </SEQUENCE>
</DASDNA>
```

Most of the code is self-explanatory. We first create a `Document` object, specify its DTD, and add a root <DASDNA> element. We then add two embedded <SEQUENCE> objects to the root <DASDNA> element. The `outputDocument()` method takes care of the `XMLOutputter` options and outputs the document to the screen.

Listing 8.6 DASCreator.java

```java
package org.xmlbio.jdom;

import org.jdom.DocType;
import org.jdom.Document;
import org.jdom.Element;
import org.jdom.output.XMLOutputter;


import java.io.IOException;

/**
 * JDOM Creator.
 * Illustrates how to create a new XML document from scratch via
   JDOM API.
 */
public class DASCreator {

    /**
     * Creates a Sample DAS DNA Document.
     */
    public void createXML() throws IOException {
        // Create New Document
        Document document = new Document();

        // Set DTD
        setDocType(document);

        // Create Root Element
        Element root = new Element("DASDNA");
        document.setRootElement(root);

        // Create Sequence Element #1
        Element sequenceElement = createSequenceElement
            ("1", "8.30", "1000", "1025", "taatttctcccattttgtaggttat");
        root.addContent(sequenceElement);
        // Create Sequence Element #2
        sequenceElement = createSequenceElement("2", "8.30", "1000",
            "1025", "taatgcaactaaatccaggcgaagc");
        root.addContent(sequenceElement);

        // Output Document in Pretty Print Format
        outputDocument(document);
    }

    /**
     * Sets the DTD.
     */
    private void setDocType(Document document) {
        DocType docType = new DocType("DASDNA",
```

Listing 8.6 (cont.)

```java
            "http://biodas.org/dtd/dasdna.dtd");
        document.setDocType(docType);
    }


    /**
     * Creates a DAS Sequence Element.
     */
    private Element createSequenceElement(String id, String version,
            String start, String stop, String dnaSequence) {
        Element sequenceElement = new Element("SEQUENCE");
        sequenceElement.setAttribute("id", id);
        sequenceElement.setAttribute("version", version);
        sequenceElement.setAttribute("start", start);
        sequenceElement.setAttribute("stop", stop);
        Element dnaElement = createDNAElement(dnaSequence);
        sequenceElement.addContent(dnaElement);
        return sequenceElement;
    }


    /**
     * Creates a DAS DNA Element.
     */
    private Element createDNAElement(String dnaSequence) {
        Element dn{Element = new Element("DNA");
        dnaElement.setAttribute("length",
                Integer.toString(dnaSequence.length()));
        dnaElement.setText(dnaSequence);
        return dnaElement;
    }


    /**
     * Outputs the New XML Document in "pretty-print" format.
     */
    private void outputDocument(Document document) throws
      IOException {
        XMLOutputter out = new XMLOutputter();
        out.setNewlines(true);
        out.setIndentSize(4);
        out.output(document, System.out);
    }


    /**
     * Main Method.
     */
    public static void main(String[] args) throws IOException {
        DASCreator creator = new DASCreator();
        creator.createXML();
    }
}
```

8.4   Building the JDAS Library

Now that we have covered the basics of JDOM, we are ready to move onto a larger case study. JDAS is an open source Java library for connecting to and querying remote Distributed Annotation System (DAS) servers. It was created by the author during the course of writing this book and can be downloaded from the web site that accompanies this book at: *http://www.xmlbio.org/jdas*.

JDAS aims to be fully compliant with the latest DAS specification and currently offers the following features:

- Full support for the following DAS Commands: dsn, dna, sequence, entry_points, types, and features.
- Easy access to the X-DAS Status code and detailed error messages.
- Easy access to the X-DAS-Capabilities header.
- Console logging feature for viewing all request/response data as it is transmitted over the wire.
- XML validation for all DAS responses and the ability to programmatically disable XML validation.

JDAS makes extensive use of several open source packages, including JDOM, the Apache Xerces XML parser, and the Jakarta Commons HttpClient library. JDOM is responsible for parsing DAS XML responses and packaging the data into intermediate data model objects. For example, the JDOM code parses DAS sequence documents and packages the data into `Sequence` objects. The HttpClient library is responsible for connecting to remote DAS servers and issuing DAS commands.

8.4.1   Using JDAS

The goal of JDAS is to provide flexible access to DAS data via a very simple API. To that end, JDAS consists of just four main packages:

- **org.xmlbio.jdas.request:** Request classes encapsulate requests sent to remote DAS servers. Request classes currently exist for six DAS commands, including: `DsnRequest`, `SequenceRequest`, and `DnaRequest`. All request classes extend the base `DASRequest` class, which provides methods for setting the DAS server URL, specifying the DAS data source, activating/deactivating XML validation, and setting timeouts for network connections. See Table 8.2 for details.
- **org.xmlbio.jdas.parameter:** Parameter classes encapsulate parameters sent to DAS servers. For example, the `Segment` parameter enables you to specify exact genomic coordinates, such as chromosome number and start/stop base pair values.
- **org.xmlbio.jdas.response:** Response classes encapsulate data returned by DAS servers. For each request class, there is a corresponding response class. For example, the package includes: `DsnResponse`, `SequenceResponse`, and `DnaResponse`. Each of the response classes provides accessor methods for extracting the underlying data. For example, the `SequenceResponse` class includes a `getSequenceList()` method, which returns an `ArrayList` of `Sequence` objects. All response classes extend the base `DasResponse` class, which provides access to the X-DAS status code, the X-DAS version number, and the X-DAS capabilities header. The base class also provides a convenience method for viewing the complete XML response document. See Table 8.3 for complete details.
- **org.xmlbio.jdas.datamodel:** Data model classes encapsulate the core data returned by DAS servers. For example, the package includes `Sequence` and `DataSource` classes. Each of the classes includes specific accessor methods for extracting individual pieces of data.

Table 8.2 The DasRequest class. Abstract base class for all DAS requests

	Method Summary
int	`getConnectionTimeOut ()`
	Gets the Connection Time Out
String	`getDataSource()`
	Gets the DAS Data Source
String	`getHttpGetURI()`
	Gets the complete HTTP URL used for Get Requests
NameValuePair[]	`getRequestParameters()`
	Gets the List of Request Parameters
java.net.URL	`getRequestURL()`
	Gets the DAS Request URL (not including request parameters)
String	`getURLBase()`
	Gets the URL Base of the DAS Server
boolean	`getValidateXMLResponse()`
	Gets the XML Validation Status
void	`setConnectionTimeOut(int timeOutInMilliseconds)`
	Sets the Connection Time Out
void	`setDataSource(java.lang.String dataSource)`
	Sets the DAS Data Source
void	`setURLBase(java.lang.String urlBase)`
	Sets the URL Base of the DAS Server
void	`setValidateXMLResponse(boolean validate)`
	Activates/Deactivates XML Validation of DAS Response

Table 8.3 The DasResponse class. Abstract base class for all DAS responses

	Field Summary
static int	`X_DAS_STATUS_BAD_COMMAND`
	X-DAS-Status: 400 Bad command (command not recognized)
static int	`X_DAS_STATUS_BAD_COMMAND_ARGS`
	X-DAS-Status: 402 Bad command arguments (arguments invalid)
static int	`X_DAS_STATUS_BAD_DATA_SOURCE`
	X-DAS-Status: 401 Bad data source (data source unknown)
static int	`X_DAS_STATUS_BAD_REFERENCE_OBJECT`
	X-DAS-Status: 403 Bad reference object (reference sequence unknown)
static int	`X_DAS_STATUS_BAD_STYLESHEET`
	X-DAS-Status: 404 Bad stylesheet (requested stylesheet unknown)
static int	`X_DAS_STATUS_COORDINATE_ERROR`
	X-DAS-Status: 405 Coordinate error (sequence coordinate is out of bounds/invalid)
static int	`X_DAS_STATUS_OK`
	X-DAS-Status: 200 OK, data follows
static int	`X_DAS_STATUS_SERVER_ERROR`
	X-DAS-Status: 500 Server error, not otherwise specified
static int	`X_DAS_STATUS_UNIMPLEMENTED_FEATURE`
	X-DAS-Status: 501 Unimplemented feature
	Method Summary
String[]	`getDasCapabilities()`
	Gets the X-DAS-Capabilities Header
int	`getDasStatusCode()`
	Gets the X-DAS-Status Code
String	`getDasStatusCodeDescription()`
	Gets the X-DAS-Status Code Description
String	`getDasVersion()`
	Gets the X-DAS-Version Number

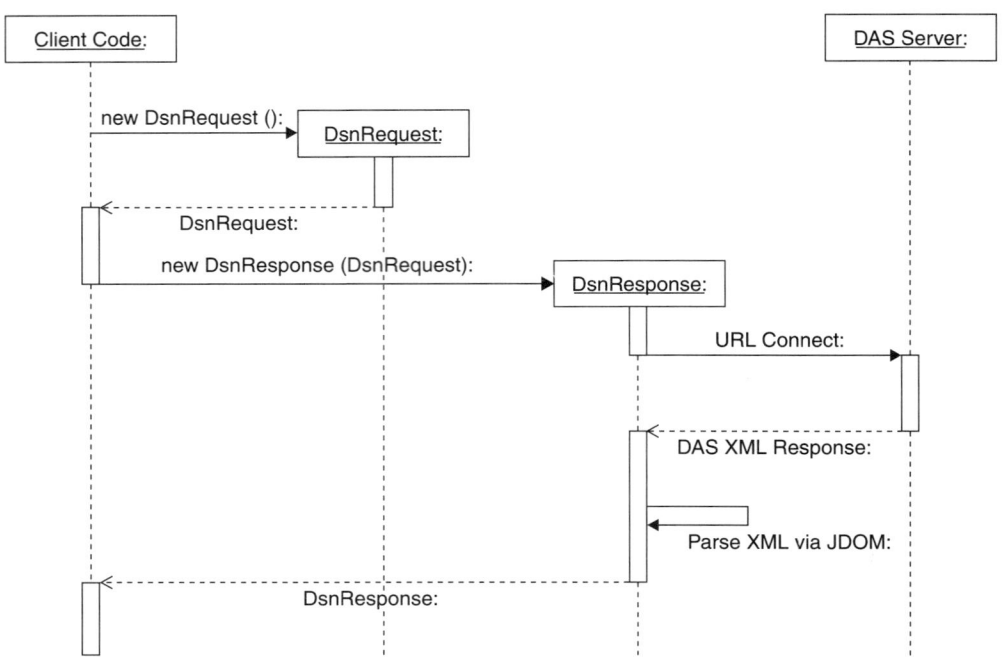

Figure 8.1 Sequence diagram, illustrating typical client interaction with the JDAS library.

To use JDAS, you generally follow four steps:

- First, instantiate a DAS request object. For example, to issue a DAS *sequence* command, instantiate a `SequenceRequest` object.
- Second, instantiate a response object. Each response constructor must be passed a corresponding request object. For example, the `SequenceResponse` class must be instantiated with a `SequenceRequest` parameter. Upon instantiation, JDAS connects to the remote server, sends the DAS request, and parses the XML response document. In the event of a network error or an XML validation error, JDAS will throw a `DasException`.
- Third, check the DAS response status code. If the server code is set to `X_DAS_STATUS_OK` (200), jump to the next step. Otherwise, check the status code and determine the source of the error.
- Process the DAS response data. All DAS response data is available via simple data model objects. For example, the DAS `Sequence` object contains data from a DAS Sequence command, and therefore contains accessor methods, such as `getLength()`, `getStart()`, and `getSequenceData()`.

A sequence diagram of the typical JDAS client interaction is provided in Figure 8.1.

To make these concepts more concrete, let us look at a sample JDAS client application. Listing 8.7 provides sample code for issuing a *dsn* request to the Ensembl DAS server. As a first step, we create a `DsnRequest` object:

```
DsnRequest dsnRequest = new DsnRequest
    (SampleConstants.ENSEMBL_BASE);
```

Listing 8.7 SampleDsnRequest.java

```java
package org.xmlbio.jdas.sample;

import org.xmlbio.jdas.datamodel.DataSource;
import org.xmlbio.jdas.request.DsnRequest;
import org.xmlbio.jdas.response.DasException;
import org.xmlbio.jdas.response.DasResponse;
import org.xmlbio.jdas.response.DsnResponse;

import java.util.ArrayList;

/**
 * Sample DSN Request to Ensembl.
 *
 * @author Ethan Cerami
 */
public class SampleDsnRequest {

    /**
     * Execute DSN Request.
     */
    public void execute() {
        try {
            DsnRequest dsnRequest = new DsnRequest
                (SampleConstants.ENSEMBL_BASE);
            dsnRequest.setValidateXMLResponse(false);
            DsnResponse dsnResponse = new DsnResponse(dsnRequest);
            if (dsnResponse.getDasStatusCode() == DasResponse.X_DAS_STATUS_OK){
                outputDataSources(dsnResponse);
            } else {
                System.out.println("DAS Status Code: "
                        + dsnResponse.getDasStatusCode());
                System.out.println("DAS Status Code Description: "
                        + dsnResponse.getDasStatusCodeDescription());
            }
        } catch (DasException e) {
            System.out.println("DASException: " + e.getMessage());
        }
    }

    /**
     * Outputs Data Source Objects.
     * @param dsnResponse DSNResponse Object.
     */
    private void outputDataSources(DsnResponse dsnResponse) {
        ArrayList dataSources = dsnResponse.getDataSources();
        System.out.println("Number of DataSources: " + dataSources.size());
        for (int i = 0; i < dataSources.size(); i++) {
            DataSource dataSource = (DataSource) dataSources.get(i);
            System.out.println("Data Source:");
```

Listing 8.7 *(cont.)*

```
        System.out.println("... ID: " + dataSource.getId());
        System.out.println("... Name: " + dataSource.getName());
        if (dataSource.getVersion() != null) {
            System.out.println("... Version: "
                    + dataSource.getVersion());
        }
        System.out.println("... Description: "
                + dataSource.getDescription());
        System.out.println("... MapMaster: "
                + dataSource.getMapMaster());
        }
    }

    /**
     * Main Method
     * @param args Command Line Arguments.
     */
    public static void main(String[] args) {
        SampleDsnRequest sample = new SampleDsnRequest();
        sample.execute();
    }
}
```

Second, we create a `DsnResponse` object:

```
DsnResponse dsnResponse = new DsnResponse(dsnRequest);
```

Upon instantiation, JDAS connects to the specified DAS server, issues a *dsn* request, and parses the XML response document. Third, we check the response code. If the status is set to X_DAS_STATUS_OK, we output the contents of the data source objects. Otherwise, we output the DAS status code number and description. Note that the code includes a try/catch block for catching DasException objects.

When you run the program, it will connect to the Ensembl DAS server and display a directory of registered data sources. Sample output is provided below:

```
Number of DataSources: 18
Data Source:
... ID: ens1431cds
... Name: Ensembl 14.31 CDS
... Version: 14
... Description: Ensembl CDS
... MapMaster: http://das.ensembl.org/das/ensembl1431/
Data Source:
... ID: ens1431snp
... Name: snps
... Version: 14.31
... Description: Human genome SNPs from Ensembl
... MapMaster: http://das.ensembl.org/das/ensembl1431/
[Output Continues... ]
```

8.4.2 The JDAS Source Code

A complete discussion of the JDAS source code is beyond the scope of this chapter. However, we can focus on the JDOM/XML parsing specific code. The core of XML parsing occurs within the JDAS response classes.

The base `DasResponse` code is responsible for connecting to the remote server, sending the request, and receiving the XML response document. The XML document is then sent to the JDOM `SAXBuilder` class, which validates the XML document, and returns a JDOM `Document` object. The `Document` object is then sent to the specific response subclass, where the relevant data is extracted and converted into data model objects.

For example, Listing 8.8 shows the complete source code for the `DsnResponse` class. The heart of the `DsnResponse` class occurs within the `parseDocument()` method. This method

Listing 8.8 DsnResponse.java

```java
package org.xmlbio.jdas.response;

import org.apache.log4j.Logger;
import org.jdom.Document;
import org.jdom.Element;
import org.jdom.JDOMException;
import org.xmlbio.jdas.datamodel.DataSource;
import org.xmlbio.jdas.request.DsnRequest;

import java.io.File;
import java.util.ArrayList;
import java.util.List;

/**
 * Encapsulates a DAS DSN Response.
 *
 * @author Ethan Cerami
 */
public class DsnResponse extends DasResponse {
    private static final String DSN_ELEMENT = "DSN";
    private static final String SOURCE_ELEMENT = "SOURCE";
    private static final String DESCRIPTION_ELEMENT = "DESCRIPTION";
    private static final String MAP_MASTER_ELEMENT = "MAPMASTER";
    private static final String ID_ATTRIBUTE = "id";
    private static final String VERSION_ATTRIBUTE = "version";
    /**
     * ArrayList of Data Source Objects.
     */
    private ArrayList dataSources;

    /**
     * Constructor.
     * @param dsnRequest DAS ENSEMBL_HUMAN Request Object.
     * @throws DasException Indicates Error Reading/Parsing XML Response.
     */
```

Listing 8.8 (*cont.*)

```java
public DsnResponse(DsnRequest dsnRequest)
        throws DasException {
    super(dsnRequest);
}

/**
 * Constructor.
 * @param file Local File containing XML Response.
 * @param validate Boolean flag to validate XML Response.
 * @throws DasException Indicates Error Reading/Parsing XML Response.
 */
public DsnResponse(File file, boolean validate)
        throws DasException {
    super(file, validate);
}

/**
 * Gets Complete List of Data Sources.
 * @return ArrayList of Data Source objects.
 */
public ArrayList getDataSources() {
    return this.dataSources;
}

/**
 * Parses DAS XML Response and extracts DataSource data.
 * @param document JDOM Document.
 * @throws JDOMException Indicates Error Parsing XML Document.
 */
protected void parseDocument(Document document) throws
  JDOMException {
    dataSources = new ArrayList();
    Element root = document.getRootElement();
    List children = root.getChildren(DsnResponse.DSN_ELEMENT);
    for (int i = 0; i < children.size(); i++) {
        Element dsnChild = (Element) children.get(i);
        DataSource dataSource = extractDataSource(dsnChild);
        dataSources.add(dataSource);
    }
}

/**
 * Extract Data Source Data.
 * @param dsnElement ENSEMBL_HUMAN Element
 * @return Populated DataSource object.
 */
private DataSource extractDataSource(Element dsnElement) {
    DataSource dataSource = new DataSource();
    extractSourceData(dsnElement, dataSource);
```

Listing 8.8 (*cont.*)

```java
        extractMapMasterData(dsnElement, dataSource);
        extractDescriptionData(dsnElement, dataSource);
        logDataSource(dataSource);
        return dataSource;
    }

    /**
     * Extracts MapMaster Data.
     * @param dsnElement ENSEMBL_HUMAN Element.
     * @param dataSource Data Source object.
     */
    private void extractMapMasterData(Element dsnElement,
            DataSource dataSource) {
        Element mapMasterElement = dsnElement.getChild(
                DsnResponse.MAP_MASTER_ELEMENT);
        if (mapMasterElement != null) {
            String text = mapMasterElement.getTextNormalize();
            dataSource.setMapMaster(text);
        }
    }

    /**
     * Extracts Description Data.
     * @param dsnElement ENSEMBL_HUMAN Element.
     * @param dataSource Data Source object.
     */
    private void extractDescriptionData(Element dsnElement,
            DataSource dataSource) {
        Element descElement = dsnElement.getChild(
                DsnResponse.DESCRIPTION_ELEMENT);
        if (descElement != null) {
            String text = descElement.getTextNormalize();
            dataSource.setDescription(text);
        }
    }

    /**
     * Extracts Source Data, including id, version, and name.
     * @param dsnElement ENSEMBL_HUMAN Element.
     * @param dataSource Data Source object.
     */
    private void extractSourceData(Element dsnElement,
            DataSource dataSource) {
        Element sourceElement =
                dsnElement.getChild(DsnResponse.SOURCE_ELEMENT);
        if (sourceElement != null) {
            String id = sourceElement.getAttributeValue(
                    DsnResponse.ID_ATTRIBUTE);
            String version = sourceElement.getAttributeValue(
                    DsnResponse.VERSION_ATTRIBUTE);
```

Listing 8.8 (*cont.*)

```
            String name = sourceElement.getTextNormalize();
            if (id != null) {
                dataSource.setId(id);
            }
            if (version != null) {
                dataSource.setVersion(version);
            }
            if (name != null) {
                dataSource.setName(name);
            }
        }
    }

    /**
     * Logs Data Source Data.
     * @param dataSource Data Source Object.
     */
    private void logDataSource(DataSource dataSource) {
        Logger log = Logger.getLogger(DsnResponse.class);
        log.info("Data Source Found:");
        log.info("...Source ID:      " + dataSource.getId());
        log.info("...Source Name:    " + dataSource.getName());
        log.info("...Source Version: " + dataSource.getVersion());
        log.info("...MapMaster:      " + dataSource.getMapMaster());
        log.info("...Description:    " + dataSource.getDescription());
    }
}
```

walks the document tree and extracts all the data source information. For example, we begin tree traversal by first locating all the <DSN> elements:

```
Element root = document.getRootElement();
List children = root.getChildren(DsnResponse.DSN_ELEMENT);
```

The code then extracts the <SOURCE>, <MAPMASTER>, and <DESCRIPTION> elements. For example, this code extracts the <MAPMASTER> element text:

```
Element mapMasterElement = dsnElement.getChild
    (DsnResponse.MAP_MASTER_ELEMENT);
if (mapMasterElement != null) {
    String text = mapMasterElement.getTextNormalize();
    dataSource.setMapMaster(text);
}
```

Each of the other response classes contains similar code for traversing the parsed response document. To take a look, you can download the complete source code from the web site that accompanies this book.

Hopefully you will agree that JDOM is an excellent toolkit for parsing XML documents in Java. If you have not already done so, download the JDOM distribution, try out the examples in this chapter, and see just how easy it is to get started.

Web Services for Bioinformatics 9

Web services represent a new paradigm for building distributed applications over the Internet. This chapter will introduce you to web services and explore the many ways in which web services are currently being used in bioinformatics. We examine several architectural alternatives for building web services, introduce the major web service specifications, and explore numerous coded examples. To make the concepts as concrete as possible, we explore a web service framework built by the National Cancer Institute (NCI), and explore several client applications that extract useful genomic data. Specific topics will include:

- Introduction to Web Services
- Case Study Overview: the National Cancer Institute caBIO Project
- Introduction to REST-Based Web Services
- Introduction to SOAP
- Introduction to the Apache Axis SOAP Library

9.1 Introduction to Web Services

We begin the chapter by first defining web services, examining a sample bioinformatics web service, and exploring the main architectural options for building web services.

9.1.1 Web Services Defined

A web service is any service that is available over the Internet, uses a standardized XML messaging system, and is not tied to any one operating system or programming language [4]. Ideally, a web service also has a public interface described in XML and is easily discoverable via a centralized or distributed web services registry. As officially defined by the Web Services Activity of the World Wide Web Consortium, a web service is therefore:

> a software system identified by a URI, whose public interfaces and bindings are defined and described using XML. Its definition can be discovered by other software systems. These systems may then interact with the Web service in a manner prescribed by its definition, using XML based messages conveyed by Internet protocols. [95]

The goal of web services is to marry the success of the web with the strengths of XML. The World Wide Web provides a global infrastructure for distributing and linking documents throughout the world. XML offers a platform-independent framework for describing a diverse set of data, and

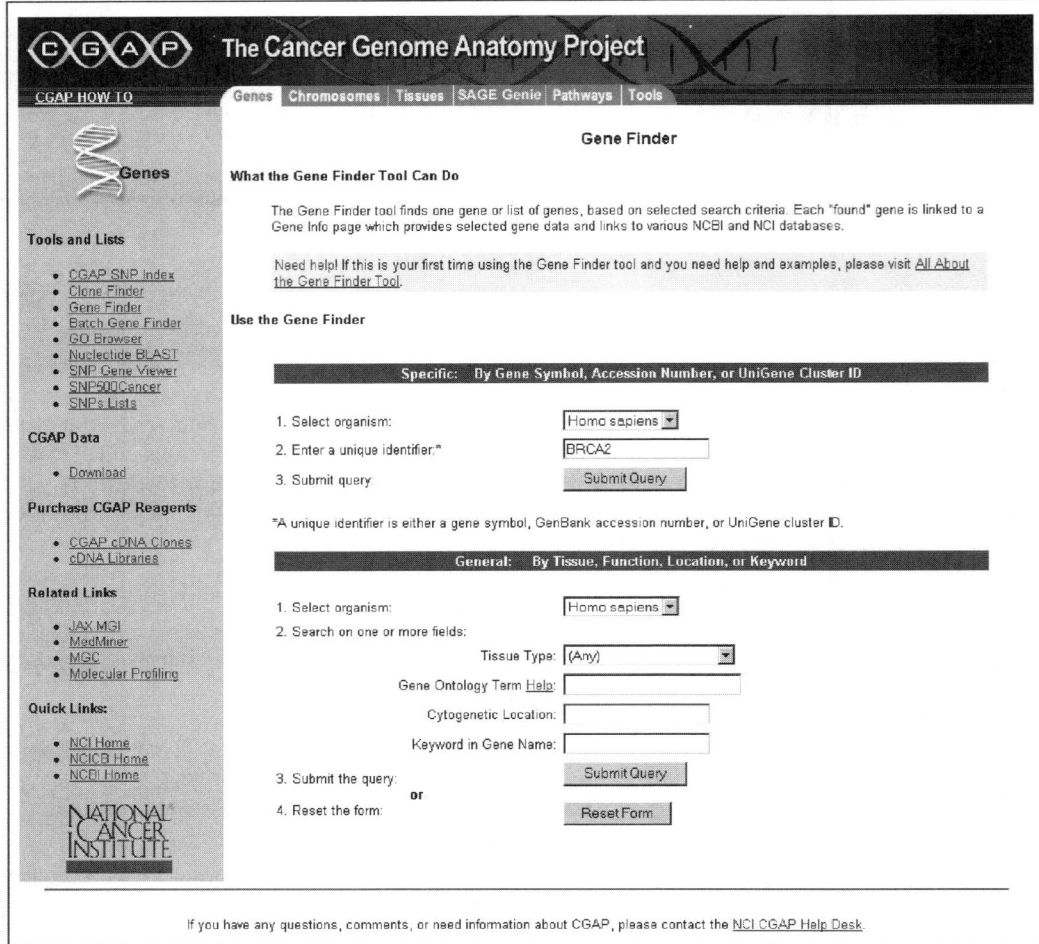

Figure 9.1 The Cancer Genome Anatomy Project (CGAP) Gene Finder web page.

provides a growing set of tools, parsers, and other related technologies. When used in combination, XML delivered over the web provides a robust solution for delivering structured content and enabling interapplication communication.

In the predominant model of the World Wide Web, we have what is commonly referred to as a *human-centric* web. Documents are distributed around the world, web clients connect to web servers via the HTTP protocol, and most documents are created for easy human consumption. For example, web servers serve up HTML documents, images, and multimedia clips. Web clients receive these resources, render them within a browser, and present the results to human readers.

Human readers have little difficulty processing and extracting data from rendered HTML pages. However, HTML is not ideal for application consumption, and XML is much better suited to this kind of task. By delivering XML via web protocols, we therefore move from a purely human-centric web to an *application-centric* web. Data that was formerly only available in HTML format is now made available in XML format, and applications can now use this data for their own purposes.

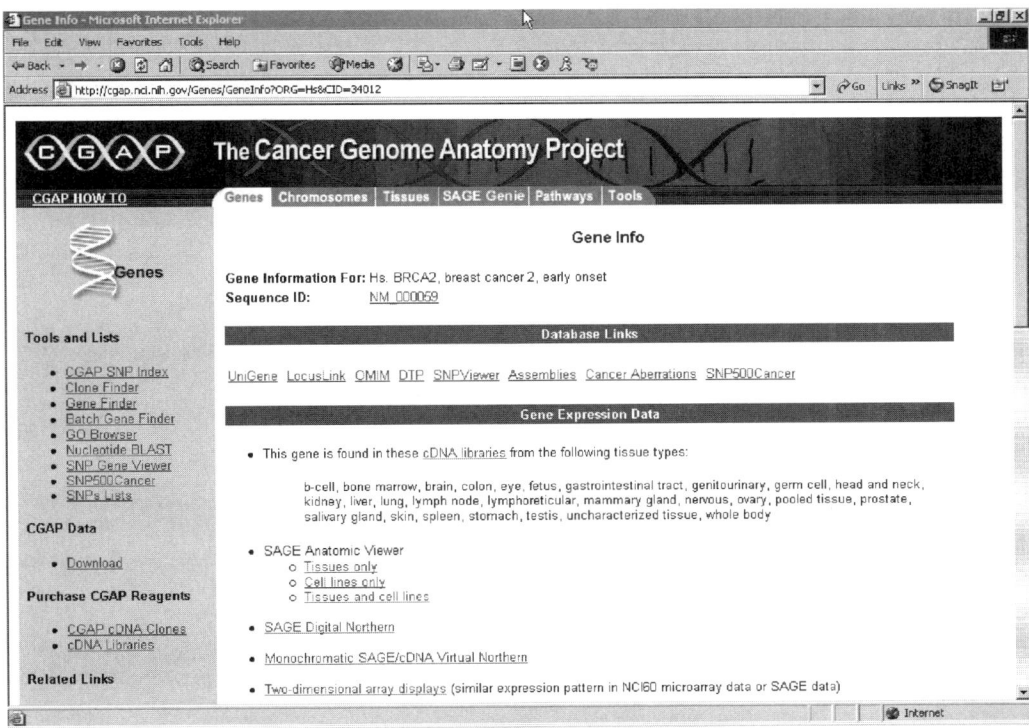

Figure 9.2 The Cancer Genome Anatomy Project (CGAP) Gene Info web page. Information regarding the BRCA2 gene is shown.

As a concrete example, consider the Cancer Genome Anatomy Project (CGAP), hosted by the U.S. National Cancer Institute. We will explore CGAP in more detail in our case study below. For now, let us focus on the sample screenshot provided in Figure 9.1. The screenshot shows a copy of the CGAP Gene Finder page. From this page you can search for genes based on multiple criteria, such as organism, unique identifier, or Gene Ontology term. As an example, we are searching for the BRCA2 gene in human. (BRCA2 has been characterized as a tumor suppressor gene, and mutations in the gene have been linked to an increased risk of breast cancer.)

Clicking the "Submit Query" button, we get a list of matching results (in this case, we get only one result). Selecting this one result provides us with the "Gene Info" page for BRCA2 (shown in Figure 9.2). This page includes several important data elements, including: the gene name and short description, links to external databases, a list of tissues where the gene is expressed, a list of orthologous genes in mouse, and a list of Gene Ontology terms associated with the gene. For example, we can see that BRCA2 is involved in apoptosis or programmed cell death.

From an application perspective, it would be much more convenient if all the data in Figure 9.2 were also available in an easily digestible XML format. In fact, it is. Through the National Cancer Institute caBIO interface, applications can programmatically interface with CGAP and extract result sets directly. For example, consider the following URL:

http://cabio.nci.nih.gov/servlet/GetXML?query=Gene&crit_name=BRCA2

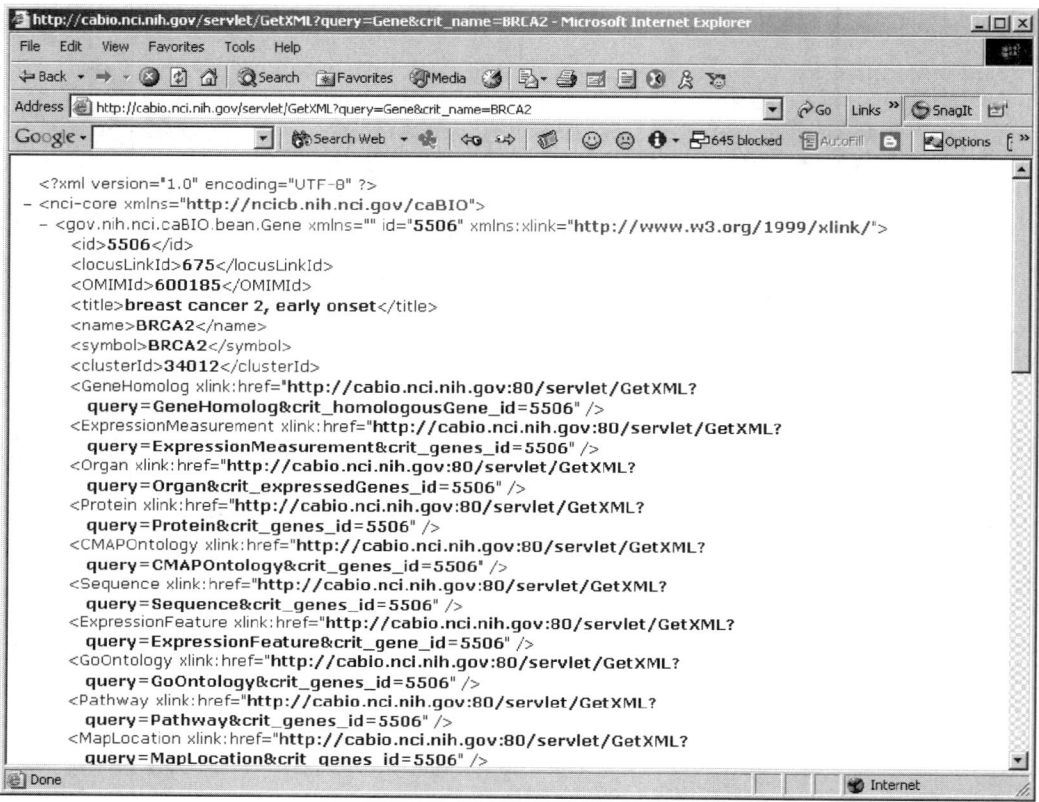

Figure 9.3 The Cancer Genome Anatomy Project (CGAP) getXML service. Information regarding the BRCA2 gene is shown.

This URL requests the BRCA2 gene for all cataloged organisms and receives back two hits: one for mouse and one for human. The results of the query are shown in Figure 9.3. As you can see, the browser has issued a regular URL request, but has received an XML document back from the caBIO server. This XML document contains the same data as the HTML document, including the gene title, name, and symbol, as well as links to additional sets of data, such as associated Gene Ontology terms. By making data available in XML format, the caBIO framework makes it much easier for other institutions, labs, and researchers to extract relevant data and use this data for their own purposes. Furthermore, since XML is not tied to any one platform, the caBIO framework is open to a diverse set of operating systems and programming languages.

9.1.2   Architectural Options

There are several different architectural options for building web services. These options are separated into three general categories:

- **XML via HTTP, or REST**: In this option, XML documents are identified with Uniform Resource Identifiers (URIs), just like regular HTML documents. However, instead of returning

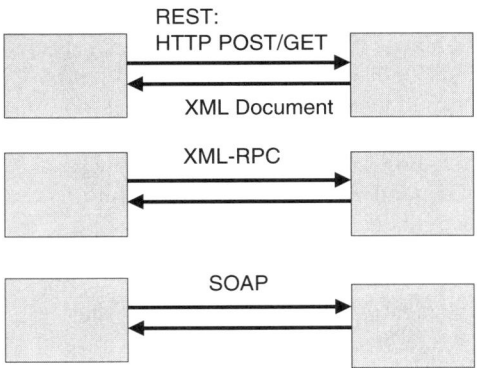

Figure 9.4 Three options for building web services: XML via HTTP (REST), XML-RPC, and SOAP.

regular HTML documents, the web server returns easily digestible XML documents. To build a complete, robust web service, all you need is an understanding of URIs, HTTP, and XML. This approach is most commonly referred to as REST, or Representational State Transfer. We will define REST and the properties of REST-based web services in the first half of this chapter.

- **XML-RPC**: Created by Dave Winer, XML-RPC [110] is an XML-based protocol that enables remote procedure calls (RPC). In XML-RPC, a client application can invoke a remote procedure by sending a standard XML document via HTTP. This standard XML document contains the procedure or method name to invoke and an optional list of parameters. The server receives the XML payload, invokes the right service, and returns the results in another standard XML document.
- **SOAP**: SOAP enables applications to communicate via standard XML messaging, and is capable of working on multiple transport protocols, including HTTP and SMTP (Simple Mail Transfer Protocol). Unlike XML-RPC, SOAP is an official recommendation of the World Wide Web Consortium (W3C). We explore SOAP in the second half of this chapter.

Each of these web service options is summarized in Figure 9.4. To manage the scope of coverage, this chapter will cover REST and SOAP, but will not cover XML-RPC. For an overview of XML-RPC, see Chapter 2 of *Web Services Essentials* [4] or *Programming Web Services with XML-RPC* [109].

Fortunately, our bioinformatics case study provides multiple interfaces, including a REST-based interface and a SOAP-based interface. You can therefore try out both interfaces and compare and contrast the two architectural options. For background on our case study, we now turn to an overview of the caBIO project.

9.2   Case Study: Introduction to the NCI caBIO Project

For our case study, we will be examining the Cancer Bioinformatics Infrastructure Objects (caBIO) [100] project, created by the U.S. National Cancer Institute (NCI). caBIO provides a programmatic interface to caCORE, a comprehensive set of interlinked databases hosted at NCI. In its initial incarnation, caBIO was built to provide programmatic access to several NCI resources, including the Cancer Genome Anatomy Project (CGAP), the Cancer Molecular Analysis Project (CMAP), and the Genetic Annotation Initiative (GAI) [100]. Since its early stages, however, caCORE has been expanded to include a number of other data resources, including those hosted at NCI, the National Center for Biotechnology Information (NCBI), and other public repositories. For example, caCORE now provides access to cancer clinical trial data, the NCI thesaurus of cancer-specific terminology, and biological pathways from BioCarta [97]. caCORE also aggregates several NCBI databases, including Unigene, LocusLink, and RefSeq [97]. All software related to the caCORE project is available under an open source license.

The caBIO framework is organized around objects referred to as "domain objects." Domain objects encapsulate biological, laboratory, or clinical entities [100], such as genes, chromosomes, sequences, diseases, and biological pathways. Behind the scenes, each of these domain objects is defined as a Java class with numerous accessor methods. For example, the caBIO `Gene` class has a `getName()` method, which returns a string indicating the name of the gene. Many of these accessor methods, however, return other caBIO domain objects. For example, the `getChromosome()` method returns a caBIO `Chromosome` object; likewise, the `getProteins()` method returns an array of caBIO `Protein` objects.

Starting with a `Gene` object, you can navigate through a web of interconnected data. Accordingly, "domain objects are related to each other, and examining these relationships can bring to the surface biomedical knowledge that was previously buried in the various primary data sources" [98].

The complete set of caBIO domain objects and the relationships between these objects are formally defined within a series of Unified Modeling Language (UML) diagrams. These UML diagrams are used to automatically generate the Java classes, and therefore form the basis of all the caBIO APIs described below. A PDF file of the complete caBIO UML model is included in the caBIO distribution.

To access and search domain objects, caBIO provides three different interfaces (see Figure 9.5):

- Java RMI: Clients written in Java can download the caBIO distribution and access domain objects directly. Behind the scenes, client applications use Java Remove Method Invocation (RMI) to connect to caBIO and retrieve the data.
- XML via HTTP: Clients issue specific URL requests to caBIO, and receive XML documents in response. This corresponds to the REST-based approach defined briefly in the section above.
- SOAP: Clients connect to caBIO via SOAP, and receive XML SOAP messages in response.

As we proceed through the chapter, we will examine each of these interfaces. But, first, let us examine the Java RMI interface. This is not a web service interface per se, but by examining the Java RMI interface first, you can gain a much greater insight into the two other interfaces that follow.

The National Cancer Institute maintains a mailing list for caBIO users. Mailing list archives and subscription information are available online at: *http://list.nih.gov/archives/cabio_users.html.*

The examples in this chapter are based on caBIO version 2.0. As this book goes to press, however, caBio is finalizing caBIO version 2.1. Check the caBIO web site for complete, up-to-date information.

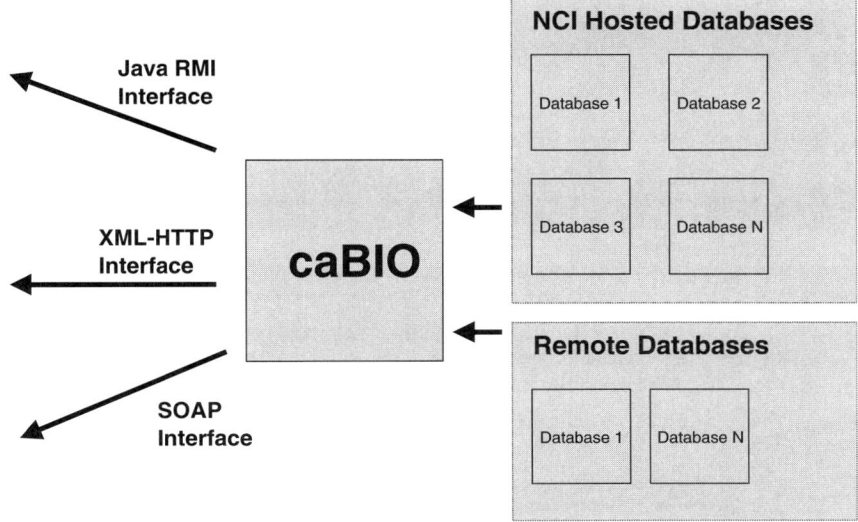

Figure 9.5 Schematic overview of the three caBIO programmatic interfaces. Each interface provides access to the same core set of databases.

9.2.1 Background: Connecting to caBIO via the Java RMI Interface

We begin by connecting to caBIO via the Java RMI interface, and requesting data regarding the BRCA2 gene in human. Step-by-step instructions are provided below:

1. Connect to the caBIO web site: *http://ncicb.nci.nih.gov/core/caBIO*, and download the caBIO distribution package.
2. Unpack the caBIO distribution to a directory of your choosing.
3. Set up your classpath. Check the caBIO `jars` directory and make sure to include the following four jar files in your classpath:
   - caBIO.jar: contains the core caBIO library, including copies of all caBIO domain objects.
   - xercesImpl.jar: contains implementation code for the Apache Xerces Java XML parser.
   - xml-apis.jar: contains standard XML parsing interfaces, including the Simple API for XML (SAX) and the Document Object Model (DOM).
   - jaxb-rt-1.0.ea.jar: contains code for the Java Architecture for XML Binding (JAXB), used to automatically transform XML documents to Java classes and vice versa.
4. Type in the following code (or download it from xmlbio.org):

```
package org.xmlbio.cabio;

import gov.nih.nci.caBIO.bean.*;
import gov.nih.nci.caBIO.util.ManagerException;
import gov.nih.nci.common.exception.OperationException;

/**
 * Sample caBio Application.
 * Illustates use of the caBIO Java RMI Interface.
```

```
 *
 * @author Ethan Cerami
 */
public class GeneRmi {

    /**
     * Executes a caBIO Search for the BRCA2 Gene in Human.
     */
    public void execute() throws ManagerException,
      OperationException {
        // Instantiate a new Gene Object.
        Gene gene = new Gene();

        // Instantiate a new Gene Search Criteria Object
        // and search for the BRCA2 Gene in Human
        GeneSearchCriteria criteria = new GeneSearchCriteria();
        criteria.setSymbol("BRCA2");
        criteria.setOrganism("Homo Sapiens");

        // Execute Search
        SearchResult result = gene.search(criteria);
        if (result != null) {
            // Get Result Set
            Gene[] genes = (Gene[]) result.getResultSet();
            for (int i = 0; i < genes.length; i++) {
                Gene geneCandidate = genes[i];
                outputGene (geneCandidate);
                System.out.println("---------------");
            }
        }
    }

    /**
     * Outputs Gene Information.
     */
    private void outputGene (Gene gene) throws OperationException {
        String locusLinkId = gene.getLocusLinkId();
        Long id = gene.getId();
        Taxon taxon = gene.getTaxon();
        Chromosome chr = gene.getChromosome();
        String title = gene.getTitle();
        String scientificName = taxon.getScientificName();

        System.out.println("Internal ID: " + id);
        System.out.println("LocusLink ID: " + locusLinkId);
        System.out.println("Title: " + title);
        System.out.println("Organism: " + scientificName);
        System.out.println("Chromosome: " + chr.getName());

        outputHomologousGenes (gene);
    }
```

```
    /**
     * Outputs Homologous Genes.
     */
    private void outputHomologousGenes(Gene gene) throws
        OperationException {
          GeneHomolog geneHomologs[] = gene.getGeneHomologs();
          for (int i=0; i < geneHomologs.length; i++) {
              GeneHomolog geneHomolog = geneHomologs[i];
              String name = geneHomolog.getName();
              Long id = geneHomolog.getId();
              String organism = geneHomolog.getTaxon().
                getScientificName();
              Float similarity = geneHomolog.
                getSimilarityPercentage();
              System.out.println("Homologous Gene: " + name);
              System.out.println("... Internal ID: " + id);
              System.out.println("... Organism: " + organism);
              System.out.println("... Similarity Percentage: " +
                similarity);
          }
    }


    /**
     * Main Method.
     */
    public static void main(String args[]) throws Exception {
        GeneRmi geneRmi = new GeneRmi();
        geneRmi.execute();
    }
}
```

This example application just touches the surface of what is possible with caBIO. However, it does a good job of illustrating basic caBIO usage. The most important item to note is the caBIO Gene object. The first line of the `execute()` method instantiates an empty Gene object:

```
Gene gene = new Gene();
```

Within caBIO, each domain object has an associated "Search Criteria" object. For example, the gene domain object has a `GeneSearchCriteria` object; likewise, the chromosome domain object has a `ChromosomeSearchCriteria` object. Each of these search criteria objects enables you to specify one or more search parameters. For example, the `GeneSearchCriteria` object enables you to search by gene symbol, chromosome number, and organism. In our case, we search by gene symbol and organism:

```
criteria.setSymbol("BRCA2");
criteria.setOrganism("Homo Sapiens");
```

Note that if you specify two or more criteria, caBIO will connect each search criterion with a logical AND operator.

To execute a search for matching genes, we execute the `search()` method, and pass in the `GeneSearchCriteria` object:

```
SearchResult result = gene.search(criteria);
```

In response, we receive a caBIO `SearchResult` object, containing any matching domain objects. The actual domain objects are available via the `getResultSet()` method:

```
Gene[] genes = (Gene[]) result.getResultSet();
```

Note that you must cast to the correct caBIO domain object, in this case, we cast the result set to an array of caBIO `Gene` objects. We then iterate through all the matching results and display several `Gene` properties, including the gene title, Locus Link ID (as specified within the NCBI Locus Link database), and the organism. We also display information on all homologous genes. There are many other additional properties available, but we have not included these to keep the overall example more concise.

5. To run the sample program, you must grant the application explicit security permissions to connect to caBIO. This is a security requirement of Java RMI and is not specific to caBIO. To do so, you will need to copy or write a java security file. Fortunately, the caBIO distribution includes a sample java.policy file, which you can use for this purpose. To execute the sample program, you must specify the caBIO java.policy file via the –D system property argument. For example:

```
java -Djava.security.policy=java.policy org.xmlbio.cabio.GeneRmi
```

Upon execution, the sample program will display the following:

```
Internal ID: 5506
LocusLink ID: 675
Title: breast cancer 2, early onset
Organism: Homo sapiens
Chromosome: 14
Homologous Gene: Brca2
... Internal ID: 108157
... Organism: Mus musculus
... Similarity Percentage: 79.71
--------------
```

As you can see, caBIO has found one gene match in human. This gene has a 79.71% similarity with a homologous mouse gene.

If you fail to specify a security policy file, you will see a stack trace containing the following error:

```
java.lang.RuntimeException: Error getting object manager:
Couldn't instantiate remote object manager.
```

> For an in-depth discussion of the complete Java API, you can consult the caCORE Technical Guide [99], available in PDF format at: *http://ncicb.nci.nih.gov/core/caBIO*. You can also reference the complete Java API, also available online.

Behind the scenes, the caBIO REST and SOAP interfaces programmatically interact with the Java domain objects. More specifically, each of the caBIO domain objects implements an `XMLInterface`, which provides a number of useful XML-related methods. For example, each domain object must implement a `toXML()` method, which provides an XML representation of the object. Both the REST and SOAP interfaces invoke the `toXML()` method to obtain XML representations of domain objects. Likewise, both interfaces also use `SearchCriteria` objects to search for matching domain objects. Therefore, the better you understand the Java API, the better you can understand the XML-based interfaces. And, now that we understand the broad outlines of the Java API, we are ready to tackle the REST-based interface.

9.3 Introduction to REST-Based Web Services

As stated at the beginning of the chapter, there are a number of different architectural options for building web services. The first option is to build web services using just Uniform Resource Identifiers (URIs), Hypertext Transfer Protocol (HTTP), and XML. This approach is usually referred to as REST. This section explores the underlying concepts of REST, and explains how to connect to the caBIO REST interface.

9.3.1 Introduction to REST

REST stands for Representational State Transfer, and was originally coined by Roy T. Fielding in his doctoral dissertation from the University of California, Irvine [102]. Fielding developed REST over a period of several years, while he was busy building the Apache web server, serving as Chairman of the Apache Software Foundation, and editing or co-editing several major Internet specifications, including HTTP/1.1 and Uniform Resource Identifiers (URIs) [102].

REST is an attempt to formalize the architecture of the World Wide Web, and to formally describe those architectural elements, which are most responsible for the success of the web. As Fielding writes in his dissertation:

> The name "Representational State Transfer" is intended to evoke an image of how a well-designed Web application behaves: a network of web pages (a virtual state-machine), where the user progresses through the application by selecting links (state transitions), resulting in the next page (representing the next state of the application) being transferred to the user and rendered for their use. [102]

Within his work, Fielding identifies a number of successful properties of the web, including: support for caching and scalability to large audiences, independent evolution of web clients and web servers, a low entry barrier for new developers, and a single addressing scheme for all resources [102]. Also critical to REST is the highly decentralized nature of the web itself, and the ability to evolve protocols, even when there is no centralizing force which can mandate global upgrades [102].

According to REST, all of these successful properties are based on the three most important elements of web architecture: URIs, HTTP, and HTML. URIs provide a global method for identifying resources anywhere on the globe. HTTP provides an application protocol for web clients to retrieve resources from web servers. And, HTML provides a simple common language for creating web content.

Since Fielding's original conceptualization of REST, a number of developers have applied REST principles to the world of web services. Henceforth, we therefore refer to these services as REST-based web services. These developers argue that XML messaging systems, such as XML-RPC or SOAP, add a level of unnecessary overhead, and that you can build robust web services with just URIs, HTTP, and XML. According to one proponent of REST architecture:

> The best part about REST is that it frees you from waiting for standards like SOAP and WSDL to mature. You do not need them. You can do REST today, using W3C and IETF standards that range in age from 10 years (URIs) to 3 years (HTTP 1.1). [108]

The Distributed Annotation System (DAS) [6] is a prime example of a REST-based web service. As described in Chapter 6, all DAS data is addressable via a URL. For example, the following URL requests annotation features from UCSC for a portion of human chromosome 3:

http://genome.cse.ucsc.edu/cgi-bin/das/hg12/features?segment=3:50000,100000

DAS clients issue requests via HTTP and DAS servers return XML documents encapsulating the requested data. DAS is built entirely on URL requests, HTTP, and a small set of Document Type Definitions (DTDs). With just these three elements, DAS is able to create a highly decentralized system of annotation servers, and is able to do so without the need of a more formalized XML messaging system, such as SOAP.

9.3.2   Connecting to the caBIO REST Interface

To explore REST-based services in more detail, let us now return to the caBIO framework. As described above, caBIO provides three main interfaces for obtaining data. The second of these interfaces uses only HTTP and XML, and is therefore considered a REST-based web service.

To access the caBIO REST interface, you need to issue a request to the caBIO GetXML servlet. As of this writing, the GetXML servlet is available at the following URL:

http://cabio.nci.nih.gov/servlet/GetXML

The GetXML servlet expects a series of URL parameters. Generally, requests consist of a domain object, e.g., gene, chromosome, or sequence, and a list of one or more search criteria. For example, in the following URL:

http://cabio.nci.nih.gov/servlet/GetXML?query=Gene&crit_symbol=BRCA2

the *query* parameter refers to a caBIO domain object and we are requesting gene domain objects only. The *crit_symbol* parameter further narrows the search to those genes with the symbol "BRCA2." The URL parameters and their values are case sensitive.

Here is another sample query:

http://cabio.nci.nih.gov/servlet/GetXML?query=Gene&crit_symbol=
BRCA2&crit_taxon_scientificName=homo+sapiens

This time, we have provided two search criteria, and have requested the BRCA2 gene for humans only. Note that you must properly encode the values of all URL parameters. By URL encoding conventions, you are required to replace all spaces with + signs, and provide hexadecimal representations for all nonalphanumeric characters. In response, the caBIO server will return the following document. For brevity, only an excerpt of the full response document is shown:

```
<nci-core xmlns="http://ncicb.nih.nci.gov/caBIO">
    <gov.nih.nci.caBIO.bean.Gene xmlns="" id="5506"
        xmlns:xlink="http://www.w3.org/1999/xlink/">
        <id>5506</id>
        <locusLinkId>675</locusLinkId>
        <OMIMId>600185</OMIMId>
        <title>breast cancer 2, early onset</title>
        <name>BRCA2</name>
        <symbol>BRCA2</symbol>
        <clusterId>34012</clusterId>
        <GeneHomolog
            xlink:href="http://cabio.nci.nih.gov:80/servlet/GetXML?
            query=GeneHomolog&amp; crit_homologousGene_id=5506"/>
```

```
        <ExpressionMeasurement
            xlink:href="http://cabio.nci.nih.gov:80/servlet/GetXML?
            query=ExpressionMeasurement&amp;crit_genes_id=5506"/>
        <Organ
            xlink:href="http://cabio.nci.nih.gov:80/servlet/GetXML?
            query=Organ&amp;crit_expressedGenes_id=5506"/>
        <Protein
            xlink:href="http://cabio.nci.nih.gov:80/servlet/GetXML?
            query=Protein&amp;crit_genes_id=5506"/>
        [For brevity, not all elements are shown .]
    </gov.nih.nci.caBIO.bean.Gene>
    <searchResult xmlns="">
        <hasMore>false</hasMore>
        <startsAt>1</startsAt>
        <endsAt>2</endsAt>
    </searchResult>
</nci-core>
```

As you can see, caBIO has returned an XML document with one gene element. The specified gene matches the one discovered earlier in the chapter when we used the caBIO Java RMI interface.

The returned XML document contains two types of elements. The first type encapsulates simple character data, such as `id`, `title`, `name`, and `symbol`. The second type provides links to related objects. For example, in the caBIO object model, `Gene` objects can reference 0 or more homologous genes. Rather than including information about all homologous genes within a single XML document, caBIO provides pointers to these related objects. These pointers are specified as XLinks (see the sidebar for a quick overview of the XLink specification), which specify additional search queries against caBIO. For example, our sample document includes the following link to homologous genes:

```
<GeneHomolog
    xlink:href="http://cabio.nci.nih.gov:80/servlet/GetXML?
    query=GeneHomolog&amp;crit_homologousGene_id=5506"/>
```

If you reconnect to caBIO and specify the new XLink URL, caBIO will return the following XML document:

```
<?xml version="1.0" encoding="UTF-8"?>
<nci-core xmlns="http://ncicb.nih.nci.gov/caBIO">
    <gov.nih.nci.caBIO.bean.GeneHomolog xmlns="" id="108157"
    xmlns:xlink="http://www.w3.org/1999/xlink/">
        <id>108157</id>
        <similarityPercentage>79.71</similarityPercentage>
        <Gene xlink:href="http://cabio.nci.nih.gov:80/servlet/GetXML?
        query=Gene&amp;crit_geneHomologs_id=108157"/>
    </gov.nih.nci.caBIO.bean.GeneHomolog>
    <searchResult xmlns="">
        <hasMore>false</hasMore>
        <startsAt>1</startsAt>
        <endsAt>2</endsAt>
    </searchResult>
</nci-core>
```

This document indicates that the human BRCA2 gene has one homolog, which has a 79.91 similarity match. The homologous gene is identified with an ID of 108157. You can then retrieve information about this gene by issuing yet another request:

http://cabio.nci.nih.gov/servlet/GetXML?query=Gene&crit_id=108157

Although caBIO makes extensive use of XLinks, it also provides the option of returning "heavy" XML. Using this option, caBIO will expand the first level of XLinks and return XML representations for these objects. For example, instead of including an XLink to homologous genes, caBIO will include information about the homologous genes directly. To activate heavy XML, set the *returnHeavyXML* parameter to 1. For example:

http://cabio.nci.nih.gov/servlet/GetXML?query=Gene&crit_symbol= BRCA2&returnHeavyXML=1

When using this option, the returned XML documents are significantly larger and the response time from caBIO is slower. If you only want to expand specific XLinks, you can optionally specify a *fillInObjects* parameter. With this parameter, you can specify a comma-separated list of related objects. For example, this URL:

http://cabio.nci.nih.gov/servlet/GetXML?query=Gene&crit_symbol= BRCA2&fillInObjects=GeneHomolog

will return gene homology data directly, but will return XLinks for other related objects.

Error Handling

A major element of web services is planning for when things go wrong, and propagating error messages back to client applications. As we will soon see, SOAP has very specific rules for encoding and propagating error messages. By contrast, REST-based web services do not have a well-defined convention for returning error messages, and a number of alternative options are currently in use. For example, we have already seen that the DAS protocol uses its own set of HTTP headers (e.g., the X-DAS-Status header), to inform clients of errors (see Chapter 6 for details). By contrast, caBIO uses a mini XML document to inform users of error messages. For example, try issuing the following query:

http://cabio.nci.nih.gov/servlet/GetXML?query=Gene&crit_sumbol=BRCA2

We have misspelled the word "symbol" and caBIO responds by returning a mini XML document with a single `error` element:

```
<error>Your search has failed for the following reason:
Couldn't find getter for sumbol. Please refer to
http://ncicb.nci.nih.gov/content/coreftp/caBIO_JavaDocs/index.html
for more information.
</error>
```

Applications connecting to caBIO can therefore check for the `error` element and take appropriate action, as needed.

Sidebar: XLinks

XML Linking Language, or XLink [101], is a W3C Recommendation that enables you to embed links within XML documents. *Simple* XLinks provide the same functionality as HTML hypertext links, but *extended* XLinks enable much more complex linking capability. For example, an XML document with extended XLinks can link one XML element to multiple resources. XLinks even enable you to create separate linker documents that describe bidirectional connections between resource. Unlike HTML links, which are supported by web browsers and have very well-defined user semantics, support for XLinks is not yet widely supported, and the user experience of navigating XLinks varies from application to application.

To create XLinks within an XML document, you must first declare an XML namespace for the XLink specification. You can specify whatever namespace prefix you like, but by convention, most people and applications set the namespace prefix to: xlink . The value of the namespace must be set to *http://www.w3.org/1999/xlink.*

Having defined an XML namespace, you can then attach an XLink to any XML element. At a minimum, each XLink must include an *xlink:type* attribute, which can be set to one of several values; the two most commonly specified values are "simple" and "extended." Depending on the type, you may set other XLink attributes or subelements. For example, if you are specifying a simple XLink, you may also want to specify an *xlink:href* attribute. The *href* attribute specifies the location of the linked resource and is usually specified with an absolute or relative URL.

For example, the following document includes two simple XLinks pointing to NCBI resources:

```
<?xml version="1.0" encoding="UTF-8"?>
<?xml-stylesheet href="xlink.css" type="text/css" ?>
<resources xmlns:xlink="http://www.w3.org/1999/xlink">
    <resource xlink:type="simple"
        xlink:href="http://www.ncbi.nlm.nih.gov/omim/"
        xlink:title="Online Mendelian Inheritance in Man"
        xlink:show="replace"
        xlink:actuate="onRequest">OMIM</resource>
    <resource xlink:type="simple"
        xlink:href="http://www.ncbi.nlm.nih.gov/RefSeq/"
        xlink:title="NCBI Reference Sequences"
        xlink:show="new"
        xlink:actuate="onRequest">RefSeq</resource>
</resources>
```

Both of these links include an *xlink:type* attribute set to "simple" and an *xlink:href* attribute set to an absolute URL value. The *title* attribute can be used by browsers to display a mouse over description or a tool-tip. The *show* attribute provides general semantics for how the link should be shown within the application browser. For example, a value of "replace" indicates that the content of the link should replace the content of the current window. A value of "new" indicates that the content of the link should be displayed in a new, separate application window. A value of "embed" indicates that the content of the link should be embedded directly within the page, much like an image embedded inside of HTML.

Lastly, the *actuate* attribute indicates the conditions under which the link will be followed. For example, if this value is set to "onRequest," the application will wait for the user to click the link before following it. If the value is set to "onLoad," the application will immediately follow the link as soon as it loads the container XML document.

Extended XLinks require a few additional attributes and additional work. For complete details, the full W3C specification is online at: *http://www.w3.org/TR/xlink.* For an example application that illustrates one potential use of extended XLinks, see Simon St. Laurent's image map example, available online at: *http://www.simonstl.com/buildxml/extended2/links.htm.*

9.3.3   Example Application: Command Line caBIO Browser

As a next step toward exploring the caBIO REST interface, we now explore a simple command line browser application. The goal of the application is to present a user with caBIO data, and to allow the user to navigate all the XLinks contained within the response document. By simply selecting which links to follow, the user can navigate through a wealth of caBIO data.

To build the command line application, we combine our previous knowledge of JDOM from Chapter 8, and our new understanding of REST and XLinks. Before looking at any code, however, let us first look at a sample application run.

The application begins by prompting the user for a gene symbol. In the sample below, the user has typed in brca2 (user input is denoted in bold):

```
Enter a Gene Symbol: brca2
```

Based on the gene symbol, the application makes its first request to caBIO, parses the XML response document, and presents a text version of the caBIO data:

```
id: 5506
locusLinkId: 675
OMIMId: 600185
title: breast cancer 2, early onset
name: BRCA2
symbol: BRCA2
clusterId: 34012
1. Link: GeneHomolog
2. Link: ExpressionMeasurement
3. Link: Organ
4. Link: Protein
5. Link: CMAPOntology
6. Link: Sequence
7. Link: ExpressionFeature
8. Link: GoOntology
9. Link: Pathway
10. Link: MapLocation
11. Link: GeneAlias
12. Link: Taxon
13. Link: Chromosome
14. Link: SNP
15. Link: Library
16. Link: Target
hasMore: false
startsAt: 0
endsAt: 1
Select a link #:        12
```

As you can see, the user is presented with top-level information about the gene, including the gene title, name, and symbol. All XLinks are also presented to the user, and are numbered in the order they appear in the document. The user is then prompted to select a link number. In the case above, the user has selected Link #12 to retrieve taxonomy information. Based on this selection, the application follows the specified XLink, parses the response document, and again presents the contained data. For example:

```
id: 5
scientificName: Homo sapiens
abbreviation: Hs
isPreferred: true
1. Link: Pathway
2. Link: Protein
3. Link: Tissue
4. Link: Gene
5. Link: Chromosome
hasMore: false
startsAt: 0
endsAt: 1
0. Go Back
Select a link #:
```

As you can see, information about Homo sapiens is presented as are additional XLinks. The user can then choose to follow these new XLinks or select 0 to go back to the previous document.

The complete code for the command line browser is presented in Listing 9.1. Take a moment now to skim over the code, and we will explore the most important parts in detail below.

Listing 9.1 Command line caBIO browser

```java
package org.xmlbio.cabio;

import java.io.IOException;
import java.io.InputStreamReader;
import java.io.BufferedReader;
import java.util.List;
import java.util.ArrayList;
import java.util.Stack;

import org.jdom.*;
import org.jdom.input.SAXBuilder;

/**
 * A Bare Bones caBIO Command Line Browser.
 *
 * @author Ethan Cerami
 */
public class caBioBrowser {
    private static final String baseUrl =
            "http://cabio.nci.nih.gov/servlet/GetXML?";
    private ArrayList links;
    private Stack history = new Stack();
    private int linkCount;

    /**
     * Executes a caBio Search with the specified URL.
     */
    private void executeQuery(String url) throws JDOMException,
      IOException {
```

Listing 9.1 *(cont.)*

```
        // Reset Links
        linkCount = 1;
        links = new ArrayList();

        // Connect to caBio and read into a JDOM Document.
        SAXBuilder builder = new SAXBuilder();
        Document doc = builder.build(url);

        // Process all Elements
        Element rootElement = doc.getRootElement();
        processElement(rootElement);

        // Prompt for Next User Choice
        navigate(url);
    }

/**
 * Navigates to Next Link
 */
private void navigate(String url) throws IOException,
  JDOMException {
    if (history.size() > 0) {
        System.out.println("0. Go Back");
    }
    String choice = promptUser("Select a link #");
    int option = Integer.parseInt(choice);
    String nextUrl;
    if (option == 0) {
        nextUrl = (String) history.pop();
    } else {
        history.push(url);
        nextUrl = (String) links.get(option - 1);
    }
    executeQuery(nextUrl);
}

/**
 * Prompts for User Input.
 */
private String promptUser(String prompt) throws IOException {
    System.out.print(prompt+": ");
    InputStreamReader isr = new InputStreamReader(System.in);
    BufferedReader stdin = new BufferedReader(isr);
    return stdin.readLine();
}

/**
 * Recursively Processes Elements.
 */
private void processElement(Element element) {
    String text = element.getTextNormalize();
```

Listing 9.1 (cont.)

```java
        String elementName = element.getName();
        Namespace xlink = Namespace.getNamespace
                ("http://www.w3.org/1999/xlink/");
        String href = element.getAttributeValue("href", xlink);
        if (text != null && text.length() > 0) {
            System.out.println(elementName + ": " + text);
        }
        if (href != null) {
            links.add(href);
            System.out.println(linkCount + ". Link: " + elementName);
            linkCount++;
        }
        List children = element.getChildren();
        for (int i = 0; i < children.size(); i++) {
            Element child = (Element) children.get(i);
            processElement(child);
        }
    }

    /**
     * Initializes Browser with First caBIO Query.
     */
    public void init() throws JDOMException, IOException {
        String symbol = promptUser("Enter a Gene Symbol");
        StringBuffer url = new StringBuffer(baseUrl);

        // Specify the Domain Object to retrieve
        url.append("query=Gene&");

        // Specify the Initial Search Criteria
        url.append("crit_symbol="+symbol);
        url.append("&crit_taxon_scientificName=homo+sapiens");
        executeQuery(url.toString());
    }

    /**
     * Main Method.
     */
    public static void main(String args[]) throws Exception {
        caBioBrowser browser = new caBioBrowser();
        browser.init();
    }
}
```

The heart of the application occurs within two methods: executeQuery() and processElement(). The executeQuery() method is responsible for issuing a URL request to caBIO, parsing the response XML document, and prompting for user input regarding XLink navigation. We are using the JDOM API to connect to caBIO and parse the XML response document:

```
// Connect to caBio and read into a JDOM Document.
SAXBuilder builder = new SAXBuilder();
Document doc = builder.build(url);
// Process all Elements
Element rootElement = doc.getRootElement();
processElement(rootElement);
```

As a quick refresher, the `SAXBuilder` object reads in an XML document and constructs an internal JDOM representation of the document. After construction, the JDOM `Document` object contains the complete contents of the caBIO XML response document, making it available for easy traversal. Document traversal begins by passing the root document element to the recursive `processElement()` method.

The `processElement()` method extracts the element name, any associated text, and any embedded XLinks. XLinks are all defined within the namespace of *http://www.w3.org/1999/xlink/*, and we use this namespace to extract all *href* attributes:

```
Namespace xlink = Namespace.getNamespace
    ("http://www.w3.org/1999/xlink/");
String href = element.getAttributeValue("href", xlink);
```

All *href* links are then added to an `ArrayList` object for subsequent traversal. Finally, the `processElement()` method extracts all child elements, and passes each child recursively back to the same method. This enables us to traverse this entire document tree, display all embedded data, and extract all XLinks. The remainder of the code in Listing 9.1 provides for user input and XLink navigation. Note also that a stack of XLink URLs is maintained; this enables the application to record a history of documents browsed, and thereby enables users to navigate back to previously viewed documents.

Side Bar: Bio Browser

Our command line caBIO browser does a good job of illustrating the caBIO REST interface, but it is certainly rudimentary in design and functionality. For a visual browser with much more robust functionality, check out the Bio Browser [129] application, written by Jonny Wray.

Bio Browser is a visual Java client application that enables users to search caBIO, and easily navigate through embedded XLinks with a simple click of the mouse. It even includes support for Scalable Vector Graphics (SVG), which enables users to view the interactive pathway diagrams that are served up by caBIO. Bio Browser is currently made available as a Java Web Start application, making it extremely easy to install. To get started, go to *http://www.jonnywray.com/java*. A sample screenshot of the Bio Browser application is shown in Figure 9.6.

As explored in Chapter 5, the National Center for Biotechnology Information (NCBI) maintains a REST-based web service, called EFetch. EFetch represents a core element in NCBI's effort to provide programmatic access to a wide set of interconnected databases, and currently provides access to sequence, literature, and taxonomy databases. EFetch is also currently capable of returning documents in multiple file formats. For example, you can retrieve sequence records in FASTA, GenBank, ASN.1, GenBank XML, or TinySeq XML. For additional details, refer back to Chapter 5 or go to the EFetch Help page at: *http://eutils.ncbi.nlm.nih.gov/entrez/query/static/efetchseq_help.html*.

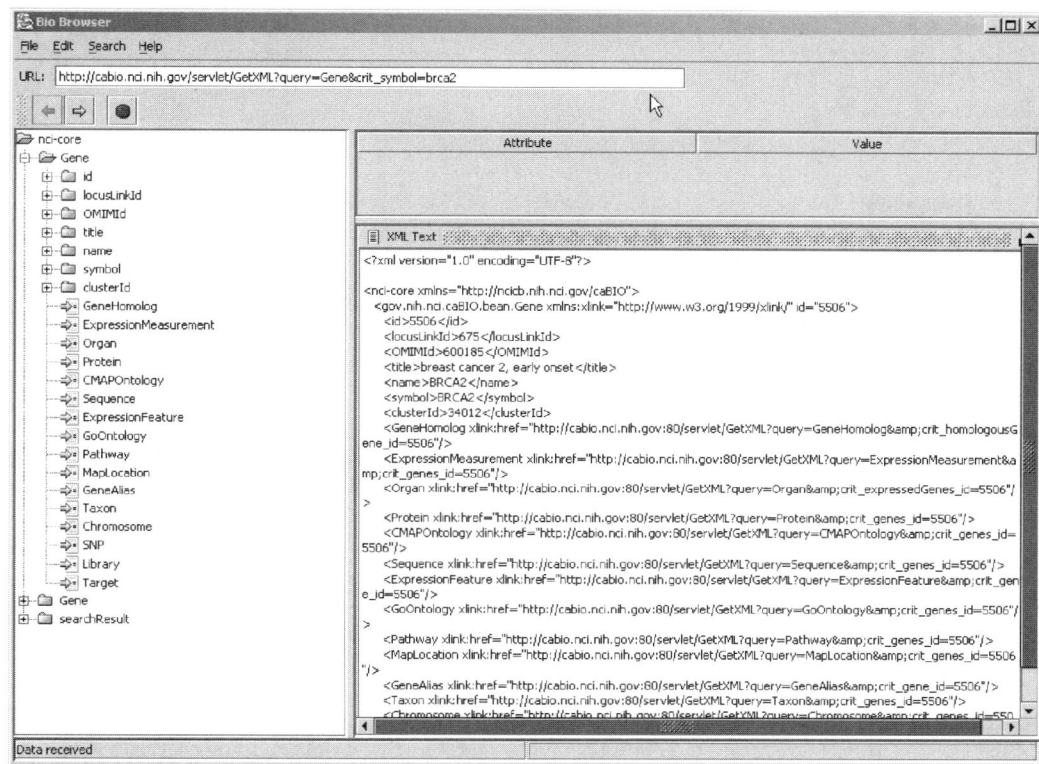

Figure 9.6 Sample screenshot of the Bio Browser application. Reprinted with permission of Jonny Wray.

9.4   Introduction to SOAP

SOAP [103; 104; 106] is an XML-based protocol which enables applications to communicate and share data. SOAP is currently available as version 1.2, and is formally endorsed by the World Wide Web Consortium (W3C).* Since its introduction, a number of major computer vendors, including Microsoft, IBM, and Sun Microsystems, have announced broad support for SOAP and have provided a number of SOAP implementations and toolkits. A number of bioinformatics resources are also currently available via SOAP. For example, the European Bioinformatics Institute (EBI) provides a SOAP-based interface for retrieving nucleotide sequence records [26], and the DNA Database of Japan (DDBJ) provides a SOAP-based interface for retrieving sequence records and running BLAST queries [107]. As discussed above, the National Cancer Institute caBIO project also provides a SOAP-based interface.

This section provides an introduction to the SOAP protocol and includes a number of complete, working examples. We begin with a description of the SOAP message format and examine a sample

---

* In previous versions, SOAP stood for Simple Object Access Protocol. However, the W3C SOAP 1.2 working group decided that the acronym was misleading, and decided to drop the acronym altogether [105].

SOAP request and response. We also discuss the encoding of error messages and the use of SOAP via HTTP. We then provide an introduction to Apache Axis, an open-source Java toolkit for creating SOAP services and clients. The section concludes with a complete example for connecting to the caBIO SOAP interface.

Information about the EBI XEMBL SOAP interface is available at: *http://www.ebi.ac.uk/xembl*. Information about the DNA Database of Japan (DDBJ) SOAP project is available at: *http://xml.ddbj.nig.ac.jp/soapp.html*.

9.4.1   SOAP Overview

SOAP formally defines a framework for computers to communicate via XML. The SOAP specification is designed to accommodate a number of usage scenarios, and is therefore open-ended on a number of implementation issues. For example, SOAP can be used to transmit any arbitrary payload of XML between any two computers. SOAP is also not tied to a specific network protocol, and one can theoretically deliver SOAP messages via HTTP, Simple Mail Transfer Protocol (SMTP), or Blocks Extensible Exchange Protocol (BEEP). The SOAP framework also supports the use of SOAP intermediaries. For example, an initial SOAP sender can send a message to a SOAP intermediary, which can forward it to a second SOAP intermediary, and so on, until the message reaches the ultimate receiver.

While the SOAP specification can accommodate a diverse set of usage scenarios, most "real-world" SOAP applications are much more focused in application. For example, most SOAP applications are specifically designed to enable Remote Procedure Calls (RPC). In this scenario, a client invokes a remote method by sending a SOAP message with the method name and any number of method arguments. In response, the receiver invokes the correct method and encodes the method response in a second SOAP message. While SOAP can also be transmitted via several protocols, nearly all real-world SOAP messages are transmitted via the very familiar HTTP protocol. And, finally, while it is possible to use SOAP intermediaries, most current services stick to one SOAP sender and one SOAP receiver, and avoid the use of SOAP intermediaries altogether.

Most SOAP applications are therefore focused on three elements: using XML to perform Remote Procedure Calls, transmitting SOAP messages via HTTP, and using just two computers—one sender and one receiver. SOAP is capable of much more, but we focus on these three properties for our introductory discussion. Furthermore, it is important to note that while it certainly helps to understand the intricacies of the SOAP protocol in detail, most developers will use toolkits, which shield them from SOAP-specific details. For example, you will not need to construct a SOAP message from scratch—in fact, most SOAP toolkits will pack, transmit, and unpack SOAP messages for you automatically.

A Sample SOAP Session

The best way to learn SOAP is to examine a sample SOAP session. Later in this chapter, we will build a simple bioinformatics web service that calculates the GC content of a specified sequence. Our remote service provides a single method: `getGCContent()`, which expects a single string argument and returns a double value. Assuming we have already deployed our service, a client

Listing 9.2 A SOAP request

```
<?xml version="1.0" encoding="UTF-8"?>
<soapenv:Envelope
    xmlns:soapenv="http://schemas.xmlsoap.org/soap/envelope/"
    xmlns:xsd="http://www.w3.org/2001/XMLSchema"
    xmlns:xsi="http://www.w3.org/2001/XMLSchema-instance">
 <soapenv:Body>
    <getGCContent
      soapenv:encodingStyle="http://schemas.xmlsoap.org/soap/
      encoding/">
     <arg0 xsi:type="xsd:string">ATGTACCCCG</arg0>
    </getGCContent>
 </soapenv:Body>
</soapenv:Envelope>
```

Listing 9.3 A SOAP response

```
<?xml version="1.0" encoding="UTF-8"?>
<soapenv:Envelope
    xmlns:soapenv="http://schemas.xmlsoap.org/soap/envelope/"
    xmlns:xsd="http://www.w3.org/2001/XMLSchema"
    xmlns:xsi="http://www.w3.org/2001/XMLSchema-instance">
 <soapenv:Body>
  <getGCContentResponse
      soapenv:encodingStyle="http://schemas.xmlsoap.org/soap/
      encoding/">
      <getGCContentReturn xsi:type="xsd:double">0.6</getGCContentReturn>
  </getGCContentResponse>
 </soapenv:Body>
</soapenv:Envelope>
```

application can invoke the GC content method by sending a properly formatted SOAP message. A sample SOAP request is shown in Listing 9.2.

A number of important elements are shown in Listing 9.2. First, all SOAP messages must be specified with a root `Envelope` element. The `Envelope` element contains the SOAP message, much as a regular envelope might contain a letter.

Second, within the Envelope, we find a `Body` element, where we specify the main "payload" of the SOAP message. In our case, the `Body` element contains a single `getGCContent` element, indicating the name of the remote method we want to invoke. This element also contains a single method argument, where we have specified a short nucleotide sequence.

Third, SOAP messages make extensive use of namespaces (for a quick review of XML Namespaces, refer to Chapter 2). For example, the SOAP `Envelope` , `Body` , and `Header` elements must be defined in the SOAP Envelope namespace. This enables SOAP processors to easily identify SOAP-specific elements and process them accordingly. SOAP messages may also include additional namespaces. For example, in Listing 9.2, we declare namespaces for XML Schema and XML Schema instance documents. This enables us to reference data types specified in the XML Schema specification. For example, we can define `arg0` as an XML Schema `string` data type.

In response to our SOAP request, our web service generates a corresponding SOAP response. See Listing 9.3.

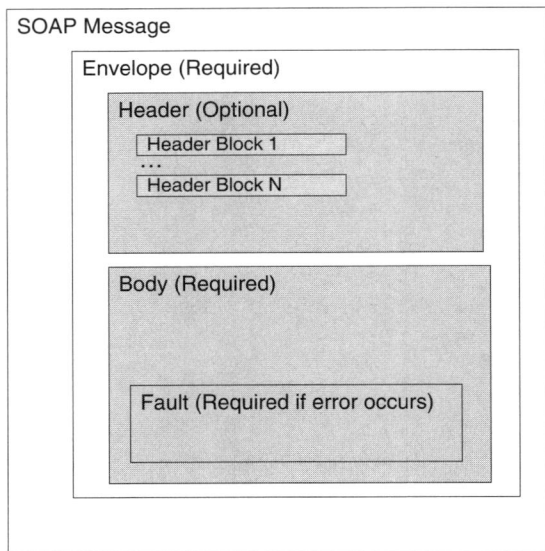

Figure 9.7 The SOAP Envelope.

Again, the SOAP message contains a SOAP `Envelope,` and the same namespace declarations. However, the `Body` element now contains the response for our remote method invocation. The `getGCContentReturn` element specifies an *xsi:type* attribute of `xsd:double`, indicating that our method has returned a double value of 0.6.

9.4.2   Constructing SOAP Messages

The W3C SOAP specification provides very explicit rules for constructing SOAP messages. This includes specific rules for constructing SOAP envelopes, headers, bodies, and faults.

SOAP Envelope

Every SOAP message must have a root SOAP `Envelope` element, and this element must be defined within the SOAP Envelope namespace. The SOAP Envelope namespace is used to specify the SOAP version. For SOAP 1.1, the namespace must be set to: *http://schemas.xmlsoap.org/soap/envelope.* For SOAP 1.2, the namespace must be set to: *http://www.w3.org/2003/05/soap-envelope.*

The SOAP Envelope can contain one optional `Header` element, and one mandatory `Body` element. The `Body` element can in turn contain the message payload or an optional `Fault` element. A schematic diagram of the SOAP Envelope is provided in Figure 9.7.

SOAP Header

The SOAP `Header` element is used to convey additional information or meta-data about the SOAP message. For example, the header element may be used to convey session information, account information, or transaction identification. The `Header` element must be specified within the SOAP Envelope namespace and can contain any number of subelements, referred to as *header blocks*. Header blocks can be specified with zero or more SOAP-specific attributes. For example, header blocks can specify a *mustUnderstand* attribute. If set to "true,"* the SOAP receiver must understand and process the SOAP header block. Otherwise, the SOAP receiver must return a fault.

Header blocks can also include other SOAP-specific attributes, such as *role* and *relay*. However, these attributes are specific to using SOAP intermediaries. For complete details, refer to the SOAP specification, Part I, available online at: *http://www.w3.org/TR/SOAP.*

Here is an example header:

```
<env:Header xmlns:env="http://www.w3.org/2003/05/soap-envelope" >
    <t:Transaction xmlns:t="http://www.xmlbio.org"
            env:mustUnderstand="true">705
    </t:Transaction>
</env:Header>
```

This header specifies one header block, where we have specified a transaction identifier. The transaction header block must be understood and processed by the receiving SOAP server. Otherwise, the SOAP receiver must return a fault.

SOAP headers are not yet widely used. For example, none of the bioinformatics services referenced in this chapter use SOAP headers. As SOAP services mature, this may change, and header elements may become more prevalent.

SOAP Body

The `Body` element is used to encapsulate the "payload" of the SOAP message. For example, in a client request, the `Body` may contain the method to invoke, along with a list of method arguments. In a SOAP response, the `Body` may contain the results of the method invocation or a fault element.

You can include any well-formed XML within a SOAP body element. However, an important aspect of SOAP processing is the automatic transformation of XML data into language-specific data structures.

In XML applications, *marshaling* is the process of transforming language-specific data structures into a cross-platform XML representation. For example, the Axis toolkit will automatically marshal a Java array of `double` values into an XML representation. This XML representation can then be *unmarshaled* at the other end and converted back into a different programming language, such as C#. This is the magic that enables cross-platform communication.

In order to effectively marshal and unmarshal XML data, SOAP senders and receivers must agree on a convention for encoding XML data. For example, both ends must agree on a convention for encoding primitive data types, such as integers and doubles. Both ends may also need to agree

---

* SOAP 1.1 uses the integer values of 1/0 for boolean types; SOAP 1.2 uses the boolean values of true/1/false/0.

on a convention for encoding compound data types, such as arrays and hash maps. While the SOAP specification does not mandate the use of one encoding convention, the specification does define a built-in set of encoding rules, referred to as the *SOAP encoding style*.

The SOAP encoding style defines rules for marshaling and unmarshaling XML data and is not tied to any programming language. For primitive data types, the SOAP encoding style leverages XML Schema data types and therefore includes support for primitive types, such as integers, floats, and doubles. The SOAP encoding style also includes detailed support for compound data types, such as structs and arrays.

By leveraging the SOAP encoding style, SOAP toolkits can automatically marshal and unmarshal XML encoded data. Best of all, the transformation details are usually completely hidden from the developer. For example, the Apache Axis toolkit can automatically marshal a Java array of `double` values and transform it into a SOAP array of XML Schema `xsd:double` values. At the other end, a Microsoft .NET application can automatically unmarshal the SOAP message into a C# array of `double` values.

For SOAP 1.1, the SOAP Encoding style must be set to: *http://schemas.xmlsoap.org/soap/encoding*. For SOAP 1.2, the encoding style must be set to: *http://www.w3.org/2003/05/soap-encoding*.

SOAP Faults

One of the primary advantages of using SOAP is that it has a set of well-defined rules for encoding error messages and propagating them back to clients. To indicate an error, the SOAP receiver must return a SOAP message with a SOAP Envelope and Body. However, the `Body` element must contain a single SOAP `Fault` element.

In SOAP 1.1, the fault element must include a mandatory `faultCode` and a mandatory `faultString`. The `faultCode` must contain a SOAP-specific value and must be chosen from one of the values specified in Table 9.2. For example, a value of `Client` indicates an error in the client request, whereas a value of `Server` indicates that the server encountered an internal error. The `faultString` element is used to convey a human readable explanation of the error. For example, here is a sample SOAP 1.1 fault message:

```
<?xml version="1.0" encoding="UTF-8"?>
<soapenv:Envelope
    xmlns:soapenv="http://schemas.xmlsoap.org/soap/envelope/"
    xmlns:xsd="http://www.w3.org/2001/XMLSchema"
    xmlns:xsi="http://www.w3.org/2001/XMLSchema-instance">
  <soapenv:Body>
   <soapenv:Fault>
    <faultcode>soapenv:Server</faultcode>
    <faultstring>Database down for required maintenance.</faultstring>
   </soapenv:Fault>
  </soapenv:Body>
</soapenv:Envelope>
```

This SOAP message indicates that a Server error occurred, and that the database is currently down for required maintenance.

Table 9.1 SOAP 1.1 and 1.2 fault elements

SOAP 1.1 Element	Corresponding SOAP 1.2 Element
`FaultCode` Indicates an error code, suitable for algorithmic processing. The value of the `faultCode` must match one of the values in Table 9.2	`env:Code` Indicates an error code, suitable for algorithmic processing. The `env:Code` element must contain a mandatory `env:Value` element and an optional `env:Subcode` element. The `env:Value` element must match one of the values in Table 9.2
`FaultString` Indicates a human-readable explanation of the error	`env:Reason` Indicates a human-readable explanation of the error. The `env:Reason` element must contain one or more `Text` elements. Each `Text` element must contain an XML *lang* attribute, indicating the language of the error message. For example, a value of "en" indicates an English error message. By returning multiple `Text` elements, you can return error messages in multiple languages
`FaultActor` Indicates which SOAP node in the message path caused the fault. Used for SOAP intermediaries	`env:Node` and `env:Role` The `env:Node` element identifies which SOAP node in the message path caused the fault. The `env:Role` element identifies the role of the SOAP node when the fault was generated. Used for SOAP intermediaries
`Detail` Carries application specific error messages, such as a program stack trace	`env:Detail` Carries application-specific error messages, such as a program stack trace

SOAP fault processing is significantly different in SOAP 1.2 [105], and represents one of the biggest changes between the two versions. Below is a sample SOAP 1.2 fault:

```
<?xml version="1.0" encoding="UTF-8"?>
<soapenv:Envelope xmlns:soapenv="http://www.w3.org/2003/05/soap-
   envelope">
    <soapenv:Body>
        <soapenv:Fault>
            <soapenv:Code>
                <soapenv:Value>soapenv:Receiver</soapenv:Value>
            </soapenv:Code>
            <soapenv:Reason>
                <soapenv:Text xml:lang="en"
                   xmlns:xml="http://www.w3.org/XML/1998/namespace">
                Database down for required maintenance.
                </soapenv:Text>
            </soapenv:Reason>
        </soapenv:Fault>
    </soapenv:Body>
</soapenv:Envelope>
```

This document contains the same error as the previous example. However, the `faultCode` and `faultReason` elements have been replaced with `Code` and `Reason` elements, respectively. Additional differences are summarized in Tables 9.1 and 9.2.

9.4.3   Transporting SOAP via HTTP

SOAP messages can be delivered via a variety of network protocols. However, transporting SOAP via HTTP remains the most popular option. When using HTTP, the sender will usually send the

Table 9.2 SOAP 1.1 and 1.2 fault codes

SOAP 1.1 Fault Code	Corresponding SOAP 1.2 Fault Code
`VersionMismatch` Indicates an invalid SOAP Envelope Namespace value or a SOAP version mismatch between SOAP client and server	`VersionMismatch` No Change
`MustUnderstand` Indicates that the SOAP receiver is unable to process a header block element with a *mustUnderstand* attribute set to "1/true." This ensures that *mustUnderstand* elements are not silently ignored	`MustUnderstand` No Change
`Client` Indicates that the client request contained an error. For example, the client has specified a nonexistent method name, or has supplied the incorrect parameters to the method	`Sender` In SOAP 1.2, `Client` is changed to `Sender`
`Server` Indicates that the server is unable to process the client request. For example, a central database may be down for routine maintenance	`Receiver` In SOAP 1.2, `Server` is changed to `Receiver`
	`DataEncodingUnknown` New to SOAP 1.2. Indicates that the incoming SOAP message uses an encoding style, which is either unknown or not supported at the receiving node

SOAP message via HTTP POST. The receiver will respond by transmitting a SOAP message in the body of the HTTP response.

For example, here is a complete SOAP request, as transmitted via HTTP:

```
POST /axis/services/BioService1 HTTP/1.0
Content-Type: text/xml; charset=utf-8
Accept: application/soap+xml, application/dime, multipart/related, text/*
User-Agent: Axis/1.1
Host: localhost
Cache-Control: no-cache
Pragma: no-cache
SOAPAction: ""
Content-Length: 417
<?xml version="1.0" encoding="UTF-8"?>
<soapenv:Envelope
    xmlns:soapenv="http://schemas.xmlsoap.org/soap/envelope/"
    xmlns:xsd="http://www.w3.org/2001/XMLSchema"
    xmlns:xsi="http://www.w3.org/2001/XMLSchema-instance">
  <soapenv:Body>
   <getGCContent
    soapenv:encodingStyle="http://schemas.xmlsoap.org/soap/encoding/">
    <arg0 xsi:type="xsd:string">ATGTACCCCG</arg0>
   </getGCContent>
  </soapenv:Body>
</soapenv:Envelope>
```

In SOAP 1.1, applications are required to specify a `Content-Type` of `text/xml` and clients are required to specify a `SOAPAction` HTTP Header. The `SOAPAction` header is used to indicate the "intent" of the SOAP message [96]. For example, SOAP servers can inspect the `SOAPAction` header, and automatically route the message to the correct web service. Firewalls can also check for the existence of the header and automatically filter out all SOAP messages. The SOAP specification does not define any rules for valid SOAPAction header values, and the interpretation of the header is entirely server dependent. However, even if the server does not require a specific SOAPAction header value, the specification requires that clients specify an empty string (" "), or a null value. For example:

```
SOAPAction: ""
```

or

```
SOAPAction:
```

Here is a sample HTTP response:

```
HTTP/1.1 200 OK
Content-Type: text/xml; charset=utf-8
Date: Wed, 14 Jan 2004 15:55:58 GMT
Server: Apache Coyote/1.0
Connection: close
<?xml version="1.0" encoding="UTF-8"?>
<soapenv:Envelope
    xmlns:soapenv="http://schemas.xmlsoap.org/soap/envelope/"
    xmlns:xsd="http://www.w3.org/2001/XMLSchema"
    xmlns:xsi="http://www.w3.org/2001/XMLSchema-instance">
  <soapenv:Body>
   <getGCContentResponse
    soapenv:encodingStyle="http://schemas.xmlsoap.org/soap/encoding/">
    <getGCContentReturn xsi:type="xsd:double">0.6</getGCContentReturn>
   </getGCContentResponse>
  </soapenv:Body>
</soapenv:Envelope>
```

Note that the server response also specifies a Content-Type of text/xml . If the request is successful, the server must return an HTTP status code of 200 OK. Otherwise, the server must return SOAP fault and an HTTP status code of 500 Internal Server Error.

In SOAP 1.2, the Content-Type has been changed from text/xml to application/soap+xml [105]. The SOAPAction Header has also been deprecated and is no longer required. In its place, clients can specify an optional *action* parameter to the Content-Type [105]. Furthermore, SOAP 1.2 requires a finer grained mapping between SOAP fault codes and HTTP status codes [105]. More specifically, a SOAP fault code of env:Sender must trigger an HTTP 400 Bad Request status code; all other faults must trigger an HTTP 500 Internal Server Error status code [104].

9.5 Introduction to Apache Axis

We are now ready to apply our knowledge of the SOAP protocol and start building our own web services. This section provides an introduction to the Apache Axis [93] web services toolkit. Axis is an open source Java toolkit, hosted by the Apache Software Foundation and currently maintained by several dozen dedicated volunteers. Using Axis, you can build and deploy SOAP services, create SOAP clients, and even create SOAP intermediaries.

9.5.1 Building a Web Service with Axis

To explore Axis in detail, we explore the complete lifecycle of building a new web service. We begin by creating a web service and deploying it locally, and then move onto creating a typical SOAP client. We also discuss options for debugging SOAP services and capturing SOAP messages as they are transmitted over the wire.

To get started, you will need to first download a copy of the Axis distribution from the Axis web site at: *http://ws.apache.org/axis*. Complete installation instructions are included in the distribution and are also available on the Axis web site.

Building the Service

Building a new web service is remarkably simple in Axis. Listing 9.4 provides the complete source code for our new bioinformatics service. Our service provides a single public method, getGCContent(). This method receives a sequence string argument, counts up the total number of Gs and Cs, and calculates the total GC content for the sequence. The calculated result is returned as a double value.

Most noticeably, the code in Listing 9.4 does not import any Axis-specific libraries or utilities. Each web service class is simply a public class, with any number of public methods. When you deploy the class, Axis takes care of wrapping your class and SOAP-enabling it for you [94].

Listing 9.4 A bioinformatics web service

```
package org.xmlbio.axis;

/**
 * Sample Web Service: Determines GC Content.
 *
 * @author Ethan Cerami
 */
public class BioService1 {

    /**
     * Determines GC Content for the specified sequence string.
     */
    public double getGCContent (String sequence) {
        sequence = sequence.trim().toUpperCase();
        int counter = 0;
        for (int i=0; i<sequence.length(); i++) {
            char c = sequence.charAt(i);
            if (c == 'G' || c == 'C') {
                counter++;
            }
        }
        return counter / (double) sequence.length();
    }
}
```

Deploying the Service

Axis provides a number of options for hosting and deploying web services. The first and simplest option is to host your services within the standalone Axis server. The second option is to host your services within a servlet engine, such as Apache Tomcat. In the interests of getting you up and running quickly, we explore the first option here. However, the stand-alone Axis server is designed for small-scale development use only [94]. For production-level services, you will need to follow the second option. For complete details, refer to the Axis installation notes.

To run the stand-alone Axis server, make sure your CLASSPATH is set correctly, open a new terminal or DOS window, and then type:

```
java org.apache.axis.transport.http.SimpleAxisServer -p 8080
```

This will start the stand-alone Axis server on port 8080, and you will see the following message:

```
- SimpleAxisServer starting up on port 8080.
```

To keep the server running, keep the terminal or DOS window open as you proceed with the rest of the section.

Axis provides an Admin Client, which is capable of deploying, undeploying, and listing web services. To deploy a new web service, you must invoke the Admin Client and specify a Web Services Deployment Descriptor (WSDD) file. The WSDD file contains information about your web service and directs the server to immediately deploy it. A sample WSDD file for our bioinformatics web service is presented in Listing 9.5.

Listing 9.5 deploy1.wsdd. Web Services Deployment Descriptor (WSDD) file

```
<?xml version="1.0" encoding="UTF-8"?>
<!-- Deploys the BioService to the Axis Engine -->
<deployment name="BioService1"
   xmlns="http://xml.apache.org/axis/wsdd/"
    xmlns:java="http://xml.apache.org/axis/wsdd/providers/java">
    <service name="BioService1" provider="java:RPC">
       <parameter name="className"
         value="org.xmlbio.axis.BioService1"/>
       <parameter name="allowedMethods" value="*"/>
    </service>
</deployment>
```

Our sample WSDD file has a root `deployment` element and one `service` element. The `service` element specifies the name of the web service, identifies the Java class that implements the service, and specifies that all methods in the class are to be made public. It also specifies that our service is to be deployed as a SOAP Remote Procedure Call (RPC).

To deploy the service, invoke the AdminClient program and pass the name of the WSDD file. Again, make sure your CLASSPATH is set correctly, and then type:

```
java org.apache.axis.client.AdminClient deploy1.wsdd
```

Assuming your stand-alone server is still running, you will see the following message:

```
Processing file deploy1.wsdd
<Admin>Done processing</Admin>
```

Listing 9.6 undeploy1.wsdd: Undeploys the BioService

```
<?xml version="1.0" encoding="UTF-8"?>
<!-- Undeploys the BioService1 -->
<undeployment xmlns="http://xml.apache.org/axis/wsdd/">
    <service name="BioService1"/>
</undeployment>
```

To verify that your service is indeed available, you can request a full list of currently deployed services:

```
java org.apache.axis.client.AdminClient list
```

You will receive a large XML document with several `service` elements. Verify that the newly deployed service is listed here.

To undeploy a web service, you invoke the AdminClient again. This time, however, you must specify a WSDD file with a root `undeployment` element. For example, Listing 9.6 specifies a WSDD file for undeploying the BioService.

To undeploy the BioService, type:

```
java org.apache.axis.client.AdminClient undeploy1.wsdd
```

You will see the following message:

```
Processing file undeploy1.wsdd
<Admin>Done processing</Admin>
```

> Axis also supports Java Web Services (JWS) files for instant deployment [94]. To use this option, you must first install a servlet engine, such as Apache Tomcat, and then make a few minor changes to your service class. First, remove any package declarations from your service class, rename it from .java to .jws, and place it in the `webapps/axis` directory. Upon restart, Axis will automatically discover all JWS files and instantly deploy them [94]. Instant deployment only works for nonpackaged classes and is intended for simple development purposes only [94]. For production-level web services, you are advised to use WSDD files, as described in this section.

Building the Client

We now have an Axis server running, and our sample web service is fully deployed. The next step is to create a SOAP client and test out the connection. Listing 9.7 provides the full source code for a typical SOAP client.

The core of the Axis client code is creating and configuring an Axis `Call` object. You can instantiate a `Call` object by passing in the absolute URL of the web service. In this case, we specify a localhost URL running on port 8080. Once you have a `Call` object, you need to specify the remote method name and any method arguments. Method arguments are specified as an array of Objects and you can therefore include arbitrary Java objects, Strings, or wrapper classes such as `Integer` and `Double`. In our case, we specify a single String parameter:

```
Object params[] = new Object[1];
params[0] = SEQUENCE;
```

For simplicity, the sequence value is hard-coded.

Listing 9.7 BioClient1.java

```java
package org.xmlbio.axis;

import org.apache.axis.client.Call;

import java.net.MalformedURLException;
import java.net.URL;
import java.rmi.RemoteException;

/**
 * Sample Client to the Bio Web Service.
 *
 * @author Ethan Cerami
 */
public class BioClient1 {
    private static final String SEQUENCE = "ATGTACCCCG";

    public static void main(String[] args) {
        try {
            // Specify URL to Localhost BioService1
            URL url = new URL
                    ("http://localhost:8080/axis/services/BioService1");
            // Create a New Call Object
            Call call = new Call (url);

            // Set Method Name and Parameters
            call.setOperationName("getGCContent");
            Object params[] = new Object[1];
            params[0] = SEQUENCE;

            // Invoke Remote Service
            Double gc = (Double) call.invoke(params);
            System.out.println("Sequence: " + SEQUENCE
                    + " has GC Content: " + gc);
        } catch (MalformedURLException e) {
            System.out.println(e.getMessage());
        } catch (RemoteException e) {
            System.out.println(e.getMessage());
        }   }
    }
}
```

With the method name and argument set, we can execute the remote method by calling the `invoke()` method. This method will automatically create the correct SOAP request message, send it to the remote server, and parse the corresponding SOAP response message. The `invoke()` method returns a Java `Object` , which you can cast at runtime. For example, we know that the `getGCContent()` method returns a double value, and we therefore cast to the corresponding wrapper class:

```java
Double gc = (Double) call.invoke(params);
```

When you run the client, you should therefore see the following output:

```
Sequence: ATGTACCCCG has GC Content: 0.6
```

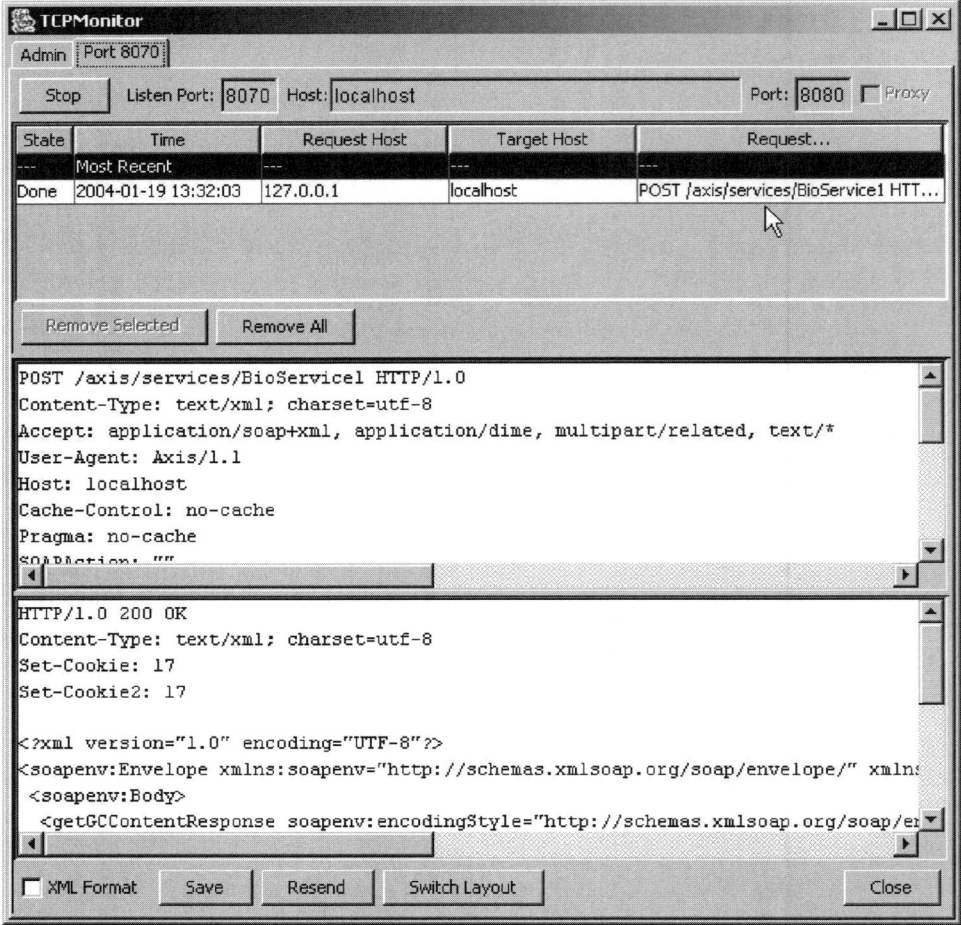

Figure 9.8 The Axis TCPMonitor tool in action.

In the event of a SOAP fault, the `invoke()` method will throw a `RemoteException`. You can then inspect the `RemoteException` for a specific error message.

Axis also includes built-in support for the Web Services Description Language (WSDL). WSDL is beyond the scope of this chapter, but you can find complete details in *Web Services: Essentials*, by Ethan Cerami [4].

Viewing SOAP Messages with TCPMonitor

As you build web services, it is very useful to monitor and debug SOAP messages as they are transmitted over the wire. Axis includes a handy tool called TCPMonitor, which enables you to do just that. To use it, type:

```
java org.apache.axis.utils.tcpmon 8070 localhost 8080
```

This will direct the TCPMonitor tool to listen for requests on port 8070 and to forward those requests to the localhost running on port 8080. Once the monitoring tool is running, update the URL of you web client to point to port 8070. For example, modify Listing 9.7 as follows:

```
URL url = new URL ("http://localhost:8070/axis/services/BioService1");
```

Then, rerun the client application, and TCPMonitor will capture the entire SOAP conversation. A sample screenshot is shown in Figure 9.8.

9.5.2 Connecting to caBIO with Axis

As a final topic, we now come full circle to our caBIO case study. caBIO maintains SOAP interfaces for all of the domain objects described at the beginning of the chapter. For each domain object, there is an associated SOAP method name. For example, to retrieve `Gene` domain objects, you invoke the `getGenes` SOAP method. Each of the domain-specific methods takes in a hash map of name/value pairs, and returns a single XML string. The valid name/value arguments match public methods specified in the corresponding `SearchCriteria` object. For example, the `GeneSearchCriteria` object has a `setSymbol()` method. The hash map for invoking the SOAP service can therefore include a member named "symbol."

Sample client code for connecting to the caBIO SOAP interface via Axis is provided in Listing 9.8. The client code in Listing 9.8 is nearly identical to our earlier example. As before, we instantiate a `Call` object, set the method name, and method parameters. This time, however, the method parameters consist of a `HashMap` object, containing a single name/value pair. Note also that caBIO requires that the method name be namespace qualified, and that we use the caBIO namespace of "urn:nci-gene-service."

Listing 9.8 GeneSoap.java

```java
package org.xmlbio.cabio;

import org.apache.axis.client.Call;
import javax.xml.namespace.QName;
import java.net.MalformedURLException;
import java.net.URL;
import java.rmi.RemoteException;
import java.util.HashMap;

/**
 * Sample caBio Application.
 * Illustates use of the caBio SOAP Interface.
 *
 * @author Ethan Cerami
 */
public class GeneSoap {
    String caBioUrl = "http://cabio.nci.nih.gov:80/soap/servlet/
        rpcrouter";
```

Listing 9.8 (*cont.*)

```
/**
 * Executes a caBio Search for the BRCA2 Gene.
 */
public void execute() throws MalformedURLException,
  RemoteException {
     // Create the Axis Call Object
     URL url = new URL(caBioUrl);
     Call call = new Call(url);

     // Set Method Name
     QName operationName = new QName("urn:nci-gene-service",
        "getGenes");
     call.setOperationName(operationName);

     // Set Method Parameters
     HashMap params [] = new HashMap[1];
     params[0] = new HashMap();
     params[0].put("symbol", "brca2");

     // Invoke Remote Method and print XML Response
     String response = (String) call.invoke(params);
     System.out.println(response);
}

/**
 * Main Method.
 */
public static void main(String args[]) throws Exception {
     GeneSoap geneSoap = new GeneSoap();
     geneSoap.execute();
}
}
```

In Listing 9.8, we also cast the invoke() response object to a String object. This String object contains a complete XML representation of all matching Gene objects and is identical to the XML one would receive via the REST-based caBIO interface.

Appendix: IUPAC Code Tables

1 Nucleotide Base Codes

Authority: Nomenclature Committee of the International Union of Biochemistry [112].

Symbol	Meaning
A	Adenine (A)
C	Cytosine (C)
G	Guanine (G)
T	Thymine in DNA (T); Uracil in RNA
M	A or C
R	A or G
W	A or T
S	C or G
Y	C or T
K	G or T
V	A, C, or G; not T
H	A, C, or T; not G
D	A, G, or T; not C
B	C, G, or T; not A
N	A, C, G, or T

2 Amino Acid Codes

Authority: IUPAC-IUB Joint Commission on Biochemical Nomenclature [113].

1-Letter Code	3-Letter Code	Description
A	Ala	Alanine
R	Arg	Arginine
N	Asn	Asparagine
D	Asp	Aspartic acid (Aspartate)
C	Cys	Cysteine
Q	Gln	Glutamine
E	Glu	Glutamic acid (Glutamate)

(cont.)

(cont.)

1-Letter Code	3-Letter Code	Description
G	Gly	Glycine
H	His	Histidine
I	Ile	Isoleucine
L	Leu	Leucine
K	Lys	Lysine
M	Met	Methionine
F	Phe	Phenylalanine
P	Pro	Proline
S	Ser	Serine
T	Thr	Threonine
W	Trp	Tryptophan
Y	Tyr	Tyrosine
V	Val	Valine
B	Asx	Aspartic acid or Asparagine
Z	Glx	Glutamine or Glutamic acid
X	Xaa	Any amino acid

Bibliography

[1] Achard, F., G. Vaysseix, and E. Barillot, "XML, bioinformatics and data integration," *Bioinformatics* 2001; 17 (2):115–125.

[2] Barillot, E. and F. Achard, "XML: a lingua franca for science?," *Trends in Biotechnology* 2000; 18 (8):331–333.

[3] Bray, Tim, xml.com. *Annotated XML Specification. http://www.xml.com/axml/testaxml.htm*

[4] Cerami, Ethan, *Web Services: Essentials*. 1st edn. Beijing; Sebastopol, CA: O'Reilly, 2002.

[5] Chicurel, M., "Bioinformatics: bringing it all together," *Nature* 2002; 419 (6908):751, 753, 755 passim.

[6] Dowell, R. D., R. M. Jokerst, A. Day, S. R. Eddy, and L. Stein, "The distributed annotation system," *BMC Bioinformatics* 2001; 2 (1):7.

[7] Dumbill, Edd, Whither Web Services? *xml.com*, 2002.

[8] I3C FAQs. Interoperable Informatics Infrastructure Consortium. *http://i3c.org/about/faq.asp*

[9] Martin, A. C., "Can we integrate bioinformatics data on the Internet?," *Trends Biotechnol* 2001; 19 (9):327–328.

[10] Stein, L., "Creating a bioinformatics nation," *Nature* 2002; 417 (6885):119–120.

[11] Stein, L. D., "Integrating biological databases," *Nature Rev Genet* 2003; 4 (5):337–345.

[12] Bioinformatic Sequence Markup Language—BSML 3.1 Reference Manual. LabBook, Inc. *http://www.bsml.org/i3c/docs/BSML3_1_Reference_Manual.pdf*

[13] Bioinformatic Sequence Markup Language—BSML 3.1 Tutorials. LabBook, Inc. *http://www.bsml.org/i3c/docs/BSML3_1_Tutorials.pdf*

[14] Bray, Tim, Dave Hollander, Andrew Layman, and Richard Tobin, World Wide Web Consortium (W3C). *Namespaces in XML 1.1 (W3C Recommendation). http://www.w3.org/TR/2004/REC-xml-names11-20040204*

[15] Bray, Tim, Dave Hollander, and Andrew Layman, World Wide Web Consortium (W3C). *Namespaces in XML (W3C Recommendation). http://www.w3.org/TR/REC-xml-names*

[16] Bray, Tim, Jean Paoli, C.M. Sperberg-McQueen, and Eve Maler, World Wide Web Consortium (W3C). *Extensible Markup Language (XML) 1.0 (W3C Recommendation). http://www.w3.org/TR/2000/REC-xml-20001006*

[17] Cibulskis, Kristian, "An introduction to BSML," *XML Journal* 4 (3).

[18] Cowan, John, World Wide Web Consortium (W3C). *Extensible Markup Language (XML) 1.1 (W3C Recommendation). http://www.w3.org/TR/2004/REC-xml11-20040204*

[19] Dublin Core Metadata Element Set, Version 1.1: Reference Description. Dublin Core Metadata Initiative. *http://dublincore.org/documents/dces*

[20] Flynn, Peter, *The XML FAQ. http://www.ucc.ie/xml*

[21] Gilbert, Howard, *Character Encoding and the Web. http://www.yale.edu/pclt/encoding*

[22] Hillmann, Diane, Dublin Core Metadata Initiative, *Using Dublin Core. http://dublincore.org/documents/2003/08/26/usageguide*

[23] Korpela, Jukka, *A Tutorial on Character Code Issues. http://www.cs.tut.fi/~jkorpela/chars.html*

[24] Powell, Andy, Dublin Core Metadata Initiative. *Guidelines for Implementing Dublin Core in XML.* *http://dublincore.org/documents/dc-xml-guidelines*

[25] Spitzner, Joseph, LabBook, Inc. Bioinformatic Sequence Markup Language—BSML Document Type Definition DTD Version 3.1. *http://www.labbook.com/dtd/bsml3_1.dtd*

[26] Wang, L., J. J. Riethoven, and A. Robinson, "XEMBL: distributing EMBL data in XML format," *Bioinformatics* 2002; 18 (8):1147–1148.

[27] AGAVE—Architecture for Genomic Annotation, Visualization and Exchange 2.3 DTD. DoubleTwist, Inc. *http://www.lifecde.com/products/agave/schema/v2_3/agave.dtd*

[28] AGAVE—Architecture for Genomic Annotation, Visualization and Exchange 3.0 Beta XML Schema. DoubleTwist, Inc. *http://www.lifecde.com/products/agave/schema/v3_0/agave.xsd*

[29] Baxevanis, Andreas D., and B. F. Francis Ouellette, "Bioinformatics: a practical guide to the analysis of genes and proteins," 2nd edn., *Methods of Biochemical Analysis; vol. 43*, New York: Wiley-Interscience, 2001.

[30] Bray, Tim, Jean Paoli, C.M. Sperberg-McQueen, and Eve Maler, World Wide Web Consortium (W3C). *Extensible Markup Language (XML) 1.0 (2nd edn.).* *http://www.w3.org/TR/2000/REC-xml-20001006*

[31] *CellML 1.0 Specification.* CellML.org. *http://www.cellml.org/public/specification/index.html*

[32] Fenyo, D. 1999. "The biopolymer markup language," *Bioinformatics* 1999; 15 (4):339–340.

[33] *Gene Ontology File Format Guide.* Gene Ontology Consortium. *http://www.geneontology.org/GO.format.html*

[34] Harold, Elliotte Rusty and W. Scott Means, *XML in a Nutshell.* 2nd edn. Sebastopol, CA: O'Reilly, 2002.

[35] *Introduction to ASN.1.* ASN.1 Information Site. *http://asn1.elibel.tm.fr/en/introduction/index.htm*

[36] *NCBI Data in XML.* National Center for Biotechnology Information (NCBI), National Institutes of Health, U.S. Department of Health and Human Services. *http://www.ncbi.nih.gov/IEB/ToolBox/XML/ncbixml.txt*

[37] Ostell, James, National Center for Biotechnology Information (NCBI), National Institutes of Health, U.S. Department of Health and Human Services. *NCBI TinySeq DTD.* *http://www.ncbi.nih.gov/dtd/NCBI_TSeq.dtd*

[38] Ostell, James, National Center for Biotechnology Information (NCBI), National Institutes of Health, U.S. Department of Health and Human Services. *NCBI GBSeq DTD.* *http://www.ncbi.nih.gov/dtd/NCBI_GBSeq.dtd*

[39] Spellman, P. T., M. Miller, J. Stewart, C. Troup, U. Sarkans, S. Chervitz, D. Bernhart, G. Sherlock, C. Ball, M. Lepage, M. Swiatek, W. L. Marks, J. Goncalves, S. Markel, D. Iordan, M. Shojatalab, A. Pizarro, J. White, R. Hubley, E. Deutsch, M. Senger, B. J. Aronow, A. Robinson, D. Bassett, C. J. Stoeckert, Jr., and A. Brazma, "Design and implementation of microarray gene expression markup language (MAGE-ML)," *Genome Biol* 2002; 3 (9):RESEARCH0046.

[40] St. Laurent, Simon, *XML: A Primer.* New York: MIS:Press, 1998.

[41] Ashburner, M., C. A. Ball, J. A. Blake, D. Botstein, H. Butler, J. M. Cherry, A. P. Davis, K. Dolinski, S. S. Dwight, J. T. Eppig, M. A. Harris, D. P. Hill, L. Issel-Tarver, A. Kasarskis, S. Lewis, J. C. Matese, J. E. Richardson, M. Ringwald, G. M. Rubin, and G. Sherlock, "Gene ontology: tool for the unification of biology," The Gene Ontology Consortium. *Nature Genet* 2000; 25 (1):25–29.

[42] Biron, Paul V. and Ashok Malhotra, World Wide Web Consortium. *XML Schema Part 2: Datatypes (W3C Recommendation).* *http://www.w3.org/TR/xmlschema-2*

[43] Fallside, David C., World Wide Web Consortium. *XML Schema Part 0: Primer (W3C Recommendation).* *http://www.w3.org/TR/xmlschema-0*

[44] Finney, A. and M. Hucka, "Systems biology markup language: Level 2 and beyond," *Biochem Soc Trans* 2003; 31 (6):1472–1473.

[45] Harris, M. A., J. Clark, A. Ireland, J. Lomax, M. Ashburner, R. Foulger, K. Eilbeck, S. Lewis, B. Marshall, C. Mungall, J. Richter, G. M. Rubin, J. A. Blake, C. Bult, M. Dolan, H. Drabkin, J. T. Eppig, D. P. Hill, L. Ni, M. Ringwald, R. Balakrishnan, J. M. Cherry, K. R. Christie, M. C. Costanzo, S. S.

Dwight, S. Engel, D. G. Fisk, J. E. Hirschman, E. L. Hong, R. S. Nash, A. Sethuraman, C. L. Theesfeld, D. Botstein, K. Dolinski, B. Feierbach, T. Berardini, S. Mundodi, S. Y. Rhee, R. Apweiler, D. Barrell, E. Camon, E. Dimmer, V. Lee, R. Chisholm, P. Gaudet, W. Kibbe, R. Kishore, E. M. Schwarz, P. Sternberg, M. Gwinn, L. Hannick, J. Wortman, M. Berriman, V. Wood, N. de la Cruz, P. Tonellato, P. Jaiswal, T. Seigfried, and R. White, "The Gene Ontology (GO) database and informatics resource," *Nucleic Acids Res* 2004; 32 Database issue:D258–261.

[46] Hermjakob, H., L. Montecchi-Palazzi, G. Bader, J. Wojcik, L. Salwinski, A. Ceol, S. Moore, S. Orchard, U. Sarkans, C. von Mering, B. Roechert, S. Poux, E. Jung, H. Mersch, P. Kersey, M. Lappe, Y. Li, R. Zeng, D. Rana, M. Nikolski, H. Husi, C. Brun, K. Shanker, S. G. Grant, C. Sander, P. Bork, W. Zhu, A. Pandey, A. Brazma, B. Jacq, M. Vidal, D. Sherman, P. Legrain, G. Cesareni, I. Xenarios, D. Eisenberg, B. Steipe, C. Hogue, and R. Apweiler, "The HUPO PSI's molecular interaction format—a community standard for the representation of protein interaction data," *Nat Biotechnol* 2004; 22 (2):177–183.

[47] Hucka, M., A. Finney, H. M. Sauro, H. Bolouri, J. C. Doyle, H. Kitano, A. P. Arkin, B. J. Bornstein, D. Bray, A. Cornish-Bowden, A. A. Cuellar, S. Dronov, E. D. Gilles, M. Ginkel, V. Gor, Goryanin, II, W. J. Hedley, T. C. Hodgman, J. H. Hofmeyr, P. J. Hunter, N. S. Juty, J. L. Kasberger, A. Kremling, U. Kummer, N. Le Novere, L. M. Loew, D. Lucio, P. Mendes, E. Minch, E. D. Mjolsness, Y. Nakayama, M. R. Nelson, P. F. Nielsen, T. Sakurada, J. C. Schaff, B. E. Shapiro, T. S. Shimizu, H. D. Spence, J. Stelling, K. Takahashi, M. Tomita, J. Wagner, and J. Wang, "The systems biology markup language (SBML): a medium for representation and exchange of biochemical network models," *Bioinformatics* 2003; 19 (4):524–531.

[48] Malik, Ayesha, "Create flexible and extensible XML schemas: building XML schemas in an object-oriented framework," *IBM DeveloperWorks*, October 2002.

[49] Murray-Rust, P. and H. S. Rzepa, "Chemical markup, XML, and the World Wide Web. 4. CML schema," *J Chem Inf Comput Sci* 2003; 43 (3):757–772.

[50] Orchard, S., H. Hermjakob, and R. Apweiler, "The proteomics standards initiative," *Proteomics* 2003; 3 (7):1374–1376.

[51] Smith, Donald, "Understanding W3C schema complex types," *xml.com* August 2001.

[52] Thompson, Henry S., David Beech, Murray Maloney, and Noah Mendelsohn, World Wide Web Consortium. *XML Schema Part 1: Structures (W3C Recommendation). http://www.w3.org/TR/xmlschema-1*

[53] Van der Vlist, Eric, *XML Schema*. 1st edn. Sebastopol, CA: O'Reilly, 2002.

[54] Walmsley, Priscilla, *Definitive XML Schema*. Upper Saddle River, NJ: Prentice-Hall PTR, 2002.

[55] Apweiler, R., A. Bairoch, C. H. Wu, W. C. Barker, B. Boeckmann, S. Ferro, E. Gasteiger, H. Huang, R. Lopez, M. Magrane, M. J. Martin, D. A. Natale, C. O'Donovan, N. Redaschi, and L. S. Yeh, "UniProt: the universal protein knowledgebase," *Nucleic Acids Res* 2004; 32 Database issue:D115–119.

[56] Christiansen, Tom and Nathan Torkington, *Perl Cookbook*. 2nd edn. Beijing; Sebastopol, CA: O'Reilly, 2003.

[57] Harold, Elliotte Rusty, *Processing XML with Java: A Guide to SAX, DOM, JDOM, JAXP, and TrAX*. Boston: Addison-Wesley, 2003.

[58] McLean, Grant, *Perl-XML frequently asked questions. http://perl-xml.sourceforge.net/faq*

[59] McLean, Grant, XML::Simple—Easy API to maintain XML. *http://www.cpan.org*

[60] Newby, Adam, et al. *LWP—The World-Wide Web library for Perl. http://cpan.org*

[61] *Perl SAX 2.0 Binding*. Perl XML Project. *http://perl-xml.sourceforge.net/sax*

[62] Rodriguez, Michel, *Processing XML efficiently with perl and XML::Twig. http://xmltwig.com/xmltwig/tutorial*

[63] Rodriguez, Michel, *XML::Twig—A perl module for processing huge XML documents in tree mode. http://www.xmltwig.com*

[64] *SAX 2.0 Changes*. SAX Project *http://www.saxproject.org/?selected=sax2*

[65] *SAX Genesis*. SAX Project. *http://www.saxproject.org/?selected=history1*

[66] Schwartz, Randal L., Perlboot—Beginner's Object-Oriented Tutorial.

[67] Schwartz, Randal L. and Tom Phoenix, *Learning Perl*. 3rd edn. Sebastopol, CA: O'Reilly, 2001.

[68] Sergeant, Matt, *XML::SAX::PurePerl*. *http://www.cpan.org*

[69] Sergeant, Matt and Christian Glahn. *XML::LibXML—Perl binding for libxml2*. *http://www.cpan.org*

[70] Sergeant, Matt, Kip Hampton, and Robin Berjon, *XML::SAX—Simple API for XML*. *http://www.cpan.org*

[71] *Simple API for XML (SAX)*. SAX Project. *http://www.saxproject.org*

[72] Stajich, J. E., D. Block, K. Boulez, S. E. Brenner, S. A. Chervitz, C. Dagdigian, G. Fuellen, J. G. Gilbert, I. Korf, H. Lapp, H. Lehvaslaiho, C. Matsalla, C. J. Mungall, B. I. Osborne, M. R. Pocock, P. Schattner, M. Senger, L. D. Stein, E. Stupka, M. D. Wilkinson, and E. Birney, The Bioperl toolkit: Perl modules for the life sciences. *Genome Res* 2002; 12 (10):1611–1618.

[73] Tisdall, James D., *Beginning Perl for Bioinformatics*. 1st edn. Beijing; Sebastopol, CA: O'Reilly, 2001.

[74] Wall, Larry, Tom Christiansen, and Jon Orwant, *Programming Perl*. 3rd edn. Beijing; Cambridge, MA: O'Reilly, 2000.

[75] Wall, Larry, Clark Cooper, and Matt Sergeant, *XML::Parser— A perl module for parsing XML documents*. *http://www.cpan.org*

[76] Wood, Lauren, et al. The World Wide Web Consortium (W3C). *Document Object Model (DOM) Level 1 Specification* (2nd edn.). *http://www.w3.org/TR/2000/WD-DOM-Level-1-20000929*

[77] GFF (General Feature Format) Specifications Document. Wellcome Trust Sanger Institute. *http://www.sanger.ac.uk/Software/formats/GFF/GFF_Spec.shtml*

[78] Gibas, Cynthia and Per Jambeck, *Developing Bioinformatics Computer Skills*. 1st edn. Beijing; Cambridge, MA: O'Reilly, 2001.

[79] Hubbard, T., D. Barker, E. Birney, G. Cameron, Y. Chen, L. Clark, T. Cox, J. Cuff, V. Curwen, T. Down, R. Durbin, E. Eyras, J. Gilbert, M. Hammond, L. Huminiecki, A. Kasprzyk, H. Lehvaslaiho, P. Lijnzaad, C. Melsopp, E. Mongin, R. Pettett, M. Pocock, S. Potter, A. Rust, E. Schmidt, S. Searle, G. Slater, J. Smith, W. Spooner, A. Stabenau, J. Stalker, E. Stupka, A. Ureta-Vidal, I. Vastrik, and M. Clamp, "The Ensembl genome database project," *Nucleic Acids Res* 2002; 30 (1):38–41.

[80] Kent, W. J., C. W. Sugnet, T. S. Furey, K. M. Roskin, T. H. Pringle, A. M. Zahler, and D. Haussler, "The human genome browser at UCSC," *Genome Res* 2002; 12 (6):996–1006.

[81] Stein, L., "Genome annotation: from sequence to biology," *Nature Rev Genet* 2001; 2 (7):493–503.

[82] Stein, L., S. R. Eddy, and R. D. Dowell, *Distributed Sequence Annotation System (DAS), Version 1.53*. *http://biodas.org/documents/spec.html*

[83] Stein, L., P. Sternberg, R. Durbin, J. Thierry-Mieg, and J. Spieth, "WormBase: network access to the genome and biology of Caenorhabditis elegans," *Nucleic Acids Res* 2001; 29 (1):82–86.

[84] *BioJava*. Open Bioinformatics Foundation. *http://biojava.org*

[85] Mordani, Rajiv and Scott Boag, Sun Microsystems. *Java API for XML Processing (Version 1.2 Final Release)*. *http://java.sun.com/xml/jaxp/index.jsp*

[86] Oren, Yuval, *Piccolo XML Parser for Java*. *http://piccolo.sourceforge.net*

[87] *Xerces2 Java Parser*, Apache Software Foundation. *http://xml.apache.org/xerces-2-j/index.html*

[88] Hunter, Jason, Java Community Process. *JSR 102: JDOM 1.0*. *http://www.jcp.org/en/jsr/detail?id=102*

[89] Hunter, Jason, "JDOM and XML Parsing," Parts I, II, and III. *Oracle Technology Network*, 2002.

[90] Hunter, Jason and Brett McLaughlin, "Easy Java/XML integration with JDOM," Parts 1 and 2. *JavaWorld* May 2000.

[91] Hunter, Jason and Brett McLaughlin. *JDOM*. *http://jdom.org*

[92] McLaughlin, Brett, *Java & XML*. 2nd edn. Sebastopol, CA; Cambridge, MA: O'Reilly, 2001.

[93] *Apache Axis*. The Apache Software Foundation. *http://ws.apache.org/axis*

[94] *Apache Axis User's Guide*. The Apache Software Foundation. *http://ws.apache.org/axis/java/user-guide.html*

[95] Austin, Daniel, Abbie Barbir, Christopher Ferris, and Sharad Garg, Web Services Architecture Requirements (W3C Working Group Note), 2004.

[96] Box, Don, David Ehnebuske, Gopal Kakivaya, Andrew Layman, Noah Mendelsohn, Henrik Frystyk Nielsen, Satish Thatte, and Dave Winer, World Wide Web Consortium (W3C). *Simple Object Access Protocol (SOAP) 1.1 (W3C Note)*. *http://www.w3.org/TR/2000/NOTE-SOAP-20000508*

[97] *caBIO Data Sources,* National Cancer Institute, Center for Bioinformatics, National Institutes of Health. *http://ncicb.nci.nih.gov/core/caBIO/core/caBIO/technical_resources/system_architecture/caBIO/ data_sources*

[98] *caBIO Overview.* National Cancer Institute. *http://ncicb.nci.nih.gov/initiatives/core/caBIO*

[99] *caCore 2.0 Technical Guide.* National Cancer Institute, Center for Bioinformatics, National Institutes of Health. *ftp://ftp1.nci.nih.gov/pub/cacore/caCORE2.0_Tech_Guide.pdf*

[100] Covitz, P. A., F. Hartel, C. Schaefer, S. De Coronado, G. Fragoso, H. Sahni, S. Gustafson, and K. H. Buetow, "caCORE: a common infrastructure for cancer informatics," *Bioinformatics* 2003; 19 (18):2404–2412.

[101] DeRose, Steve, Eve Maler, and David Orchard, World Wide Web Consortium (W3C). *XML Linking Language (XLink) Version 1.0 (W3C Recommendation)*. *http://www.w3.org/TR/xlink*

[102] Fielding, Roy Thomas, *Architectural Styles and the Design of Network-based Software Architectures*, University of California, Irvine, 2001.

[103] Gudgin, Martin, Marc Hadley, Noah Mendelsohn, Jean-Jacques Moreau, and Henrik Frystyk Nielsen, World Wide Web Consortium (W3C). *SOAP Version 1.2 Part 1: Messaging Framework (W3C Recommendation)*. *http://www.w3.org/TR/2003/REC-soap12-part1-20030624*

[104] Gudgin, Martin, Marc Hadley, Noah Mendelsohn, Jean-Jacques Moreau, and Henrik Frystyk Nielsen, World Wide Web Consortium. *SOAP Version 1.2 Part 2: Adjuncts (W3C Recommendation)*. *http://www.w3.org/TR/2003/REC-soap12-part2-20030624*

[105] Hadley, Marc, *What's New in SOAP 1.2*. *http://www.hadleynet.org/marc/whatsnew.html*

[106] Mitra, Nilo, World Wide Web Consortium (W3C). *SOAP Version 1.2 Part 0: Primer (W3C Recommendation)*. *http://www.w3.org/TR/2003/REC-soap12-part0-20030624*

[107] Miyazaki, S., H. Sugawara, T. Gojobori, and Y. Tateno, "DNA Data Bank of Japan (DDBJ) in XML," *Nucleic Acids Res* 2003; 31 (1):13–16.

[108] Prescod, Paul, REST and the Real World. *xml.com* February 2002.

[109] St. Laurent, Simon, Joe Johnston, and Edd Dumbill, *Programming Web Services with XML-RPC*. 1st edn. Beijing; Sebastopol, CA: O'Reilly, 2001.

[110] Winer, Dave, *XML-RPC Specification*. *http://www.xmlrpc.com/spec*

[111] Wray, Jonny, *Bio Browser*. *http://www.jonnywray.com/java*

[112] Cornish-Bowden, A., "Nomenclature for incompletely specified bases in nucleic acid sequences: recommendations 1984," *Nucleic Acids Res* 1985; 13 (9):3021–3030.

[113] IUPAC-IUB Joint Commission on Biochemical Nomenclature (JCBN), "Nomenclature and symbolism for amino acids and peptides. Recommendations 1983," *Biochem J* 1984; 219 (2):345–373.

Index